Health and Sustenance in Extreme Environment

Swati Srivastava • Manish Sharma
Rajeev Varshney
Editors

Health and Sustenance in Extreme Environment

A Military Perspective

Springer

Editors
Swati Srivastava
Defence Institute of Physiology and Allied Sciences (DIPAS)
Defence Research and Development Organization (DRDO)
Delhi, Delhi, India

Manish Sharma
Defence Institute of Physiology and Allied Sciences (DIPAS)
Defence Research and Development Organization (DRDO)
Delhi, Delhi, India

Rajeev Varshney
Defence Institute of Physiology and Allied Sciences (DIPAS)
Defence Research and Development Organization (DRDO)
Delhi, Delhi, India

ISBN 978-981-96-8641-4 ISBN 978-981-96-8642-1 (eBook)
https://doi.org/10.1007/978-981-96-8642-1

This Springer imprint is published by the registered company Springer Nature Singapore Pte Ltd.
The registered company address is: 152 Beach Road, #21-01/04 Gateway East, Singapore 189721, Singapore

This Book is dedicated to the soldiers standing in extreme environments to safeguard their country borders

Preface

Extreme and challenging environments are nurseries of natural selection—involving short-term and long-term adaptive mechanisms. Environmental extremes manifest by way of changes in temperature, oxygen (hypoxia/hyperoxia), availability of food, loss of habitat, pollutants, and predators, among others. Even today, intriguingly, situations such as sudden induction of low-lander natives to high/extreme altitude appear to ping such dormant adaptive pathways—deeply engrained in our genomes by evolutionary processes. To a basic scientists' delight, the modern-day requirements (owing to increase in exotic tourism, strategic military operations) have necessitated organized, fundamental studies in such terrains—a window to peep into an exciting natural laboratory. In this regard, Indian military operations—which involve requirements of rapid and/or extended deployment of troops to high altitude—have proven to be situations with distinctive scientific wealth. The decades of accumulating, overwhelming information in this field has led to recognition of key adaptive pathways and steadfast physiological potential dormant in all living beings. It is intriguing that the exposure to high altitude engages multiple organ systems—including lungs, heart, gut, brain, and muscles and thus, lack of acclimatization precipitate illnesses concerning the same. It is, therefore, not surprising that besides direct manifestation for health at high altitude, this information would be of great value to critical care medicine as well. The motivation to compile the present book was thus imminent—intent to bring up-to-date information and expert views at a common place, accessible to other researchers across the globe. We have reached out basic and applied scientists to share their understanding in other challenging environments as well—both natural and man-made. Besides high altitude, we could bring here relevant insight from environment posed by desert, sea, and jungles. For the sake of continuity of description, we have segregated the contents of this book into three sections—addressing issues related to high altitude, other extreme environments (including desert, marine, jungle), and problems common to all environments.

The first chapter by *Dr. Rajesh Arora* and team summarizes the impact of high altitude on military operations. They begin by describing geographical and historical perspective of northern/northeastern border of India followed by various illnesses reported in such situations. The authors also describe specific acclimatization protocol to circumvent illnesses encountered at high altitude. In subsequent chapters, the authors detail physiological effects of high altitude exposure on major organ systems in requisite details. *Dr. Rajender Gupta* and team describe physiological basis of cardiopulmonary responses at high altitude with special emphasis on HAPE. They also describe an interesting protein marker related to high altitude pulmonary edema. The subsequent chapter by *Dr. Ramesh Meena* and team describe other unresolved conditions affecting heart and lungs under hypoxic environment. They illustrate various inhibitors (either in use or under consideration) for intervention in these conditions. Besides lungs and heart, our vascular system also bears the brunt of pathological effects of hypoxia. Under these conditions, venous thrombosis manifests during extended stay but is a condition of greatest concern—principally because it can culminate in rapid mortality. *Dr. Iti Garg* and team put this information in perspective—adding a primer to this condition with complex pathoetiology.

Both heart and brain share the features of unique bioenergetic requirements—consuming the greatest amount of glucose and oxygen in our body. It is, therefore, expected that both organ systems would be vulnerable during hypoxia. The issues related to hypoxic brain are enumerated by *Dr. Kaushik Ray* and team. They describe specific neurophysiological processes and molecular mechanisms underlying cognitive issues encountered during high altitude. This chapter also describes specific intervention strategies to mitigate these neuropathological illnesses. *Dr. Pawan Dhull* and team add an interesting psychological perspective on common neurological ailments across various groups of workers (ranks) and ways to maintain optimal mental health in stressful environment. It is interesting to note that gut is increasingly being recognized as second brain of our body. An interesting contribution in this book by *Dr. Ramesh Meena* and team describes the role of gut and oral microbiome in high altitude acclimatization and illnesses. They also touch upon strategies for prevention of dysbiosis and microbiome as potential interventional route to preventing high altitude illnesses.

A relatively lesser-known aspect about the effects of environmental hypoxia on skeletal muscles is described by *Dr. Richa Rathore* and team. While there exist volumes of information on effects of hypoxia on cardiac muscles, the information pertaining to skeletal muscles is limited and hence, making an interesting contribution to this book. It enumerates interesting interjections about oxidative stress, ER stress, and proteostasis in context to skeletal muscle physiology. The authors also touch upon the role of specific miRNAs as regulators of this condition. This aspect, related to role of miRNAs in hypoxia, is however addressed with significant details by *Dr. Swati Srivastava* in a subsequent chapter. Besides basics of miRNA biology, this chapter has wealth of information about newly identified, hypoxia-specific miRNAs, their biomarker value and therapeutic potential in high altitude settings.

As mentioned above, heat (both dry and humid) poses a great challenge to military operations. One chapter each by *Dr. Abhishek Bharadwaj* and *Dr. Abhinav Dixit and Dr. Kalpana Baharwal* describes these aspects in significant details. These chapters contain information pertaining to epidemiology, physiological responses, threats, and mitigation strategies in environment with high temperatures (heat) with or without humidity. The chapter by *Dr. Vraj Kamboj* provides some unique insights into challenges for soldiers operating in jungle environment. Besides specific issues, they describe strategies and remedies to cope up in such environmental conditions.

Dr. P. Gokulakrishnan and team summarize issues related to marine operations—to cover effect of motion (sea sickness) on crew performance. This chapter describes unique information—the effect of motion on the vestibular system of mariners, serious safety concern and its possible amelioration by therapeutics or design based preventive measures.

It is less recognized that adverse environment could be man-made—emanating from poor anthropometric design of work stations and military equipment. These pitfalls could lead to short-term and long-term operational issues and stressors. Various critical strategies for human-machine environment interactions are highlighted by *Dr. Robert Varte* and team. They describe specific anthropometric and ergonomic approaches for optimal soldier performance in field while handling various military platforms.

Finally, two very interesting chapters by *Dr. Niroj Sethy* and *Dr. Gaurav Kumar* describe the spotlight, modern-day research tools—the multi-OMIC strategies and unbiased data analytics (including AI, ML) to address complex questions—such as those posed by multifaceted effects of hypobaric hypoxia (high altitude) on various organ systems and progress toward precision, personalized medicine.

We believe that the blend of aforementioned information—entailing classical physiology, molecular biology, medical issues, and modern research tools—in this compilation would prove to be a valuable resource for all high altitude researchers. We hope that various aspects have been covered in requisite details to incite interest of both young and seasoned researchers in this field. Due to volumes of information in this field, it is possible that contributions and other relevant details might have been missed in various accounts described by our contributing authors in this book. We tender an unconditional apology for any inadvertent omission of information or exclusion of any original references in view of space constraint. We solicit feedbacks from our readers and contributors and also, seek their support to the cause of continual improvement—both in content and spectrum of environmental challenges described in this book. We wish our audiences a very happy reading!

Delhi, Delhi, India

Swati Srivastava
Manish Sharma
Rajeev Varshney

Acknowledgments

At the outset, we sincerely thank Springer Nature for providing us with this platform to communicate to vivid audiences in the field of high altitude biology. The relentless, prompt support from the handling editors and their team at all stages of publication of this book is deeply appreciated and acknowledged. Specially, the services related to proof reading—to include plagiarism checks, guidelines pertaining digital object production, and final approval of manuscript from the authors—is sincerely acknowledged. We are very grateful to Defence Institute of Physiology and Allied Sciences (DIPAS) and Defence Research and Development Organization (DRDO), India, for providing us requisite permission, time, and opportunity to take up this scientific endeavor. The editors of this book are deeply grateful to all our authors for heartily agreeing to contribute by way of specific chapters, enumerating their own research and information available in the field. We are truly grateful for their time, effort, and commitment to this endeavor. Such contributions cannot be thanked enough in words as it benefits numerous researchers in the field, across the globe. We also gratefully acknowledge their respective departments, competent authorities, and funding bodies to have allowed them to contribute to this book. We are deeply grateful to all researchers, clinicians (specifically from Indian Armed Forces), human volunteers in various studies who have helped amass invaluable information compiled in this book titled *Health and Sustenance in Extreme Environment: A Military Perspective*. This book is dedicated to all the soldiers across the world who serve in difficult terrains above all personal issue in the service of nation. The book is also intended to benefit sojourners on expedition to high altitude and seeks to bring them to safety at such locations through access to right knowledge and inculcating appropriate scientific approach. Any inadvertent omissions in the acknowledgment is truly regretted and we remain committed to continued learning and improvement of content of this book in future volumes.

About the Editors

Swati Srivastava completed her PhD in environmental studies from the University of Delhi, New Delhi, India, and her master's in environmental biology with a specialization in biotechnology from the department of environmental studies, University of Delhi, India. She is currently a senior scientist at the Defence Institute of Physiology and Allied Sciences (DIPAS), Defence Research and Development Organization (DRDO), Delhi, India. Her research interests include the identification of biomarkers for susceptibility and genetic resistance to various high-altitude maladies, such as venous thrombosis (VT) and high-altitude pulmonary edema (HAPE), under extreme environmental conditions. She has published several book chapters and papers in peer-reviewed national and international journals. Her research also includes multi-omics-based studies on high-altitude maladies, during which she has identified various potential miRNAs, genomic and exomic variants, and methylation patterns that could serve as early predictive and diagnostic biomarkers for high-altitude disease management. She has also worked on identifying biomarkers for COVID-19 susceptibility and severity. She has been a university topper and gold medalist and has received several awards and honors during her research career, including the Best Research Publication Award and the Technology Group Award, among others.

Manish Sharma completed his PhD in molecular biology and immunology from the International Center for Genetic Engineering and Biotechnology (ICGEB), Jawaharlal Nehru University (JNU), New Delhi, India, and post-doctoral studies from Mass General Cancer Center, Boston, USA. Currently, he is serving as a senior scientist, heading a research group engaged in understanding the role of gas messengers in high-altitude illnesses at the Defense Institute of Physiology and Allied Sciences (DIPAS), Delhi, India. He has published several articles in peer-reviewed national and international journals. He has received awards including DRDO scientist of the year, Laboratory scientist of the year, Best publication, National Science Day, and Technology group award.

Rajeev Varshney completed his postgraduate in biochemistry from Aligarh Muslim University, Aligarh, India, and doctoral studies from Jawaharlal Nehru University, New Delhi, India. He served as the director of the Defence Institute of

Physiology and Allied Sciences (DIPAS), Delhi, India. He has contributed to the development of mechanism-based drugs for countering radiation injuries and understanding human responses to high altitude adaptation and maladies using "omics" approaches. He started a program for the inclusion of human factors engineering for indigenously developed military platforms. He has been awarded the DRDO Spin-off award, DRDO Lab-level Technology Group award, and the Young Scientist award (twice) for presenting his work in Japan and Italy. He has contributed toward the transfer of over 16 technologies developed at DIPAS through LA ToT agreements to approximately 132 industries for bulk production and the induction of products with users. He used his experience and physiological understanding of high altitudes in establishing the COVID-19 research center at DIPAS. He has published several papers in peer-reviewed national and international journals.

Contents

Part I

High Altitude (Hypobaric Hypoxia, Cold)

Impact of High Altitude on Military Operations: An Overview

1

Renu Bala Yadav, Arun Kumar Yadav, Virat Kuntalam, Vinay Kumar Yadav, Anuj Singhal, Palak Kapoor, and Rajesh Arora

Abstract

High altitude poses problems for mountaineers, local inhabitants, and sojourners and can also substantially affect military expeditions, as it presents an assortment of difficulties that can be exacerbated during medical evacuation, particularly in exploration, humanitarian aid during disasters and combat situations. These impacts fall under three categories: physiological, environmental, and tactical. The impact of high altitude on physiological aspects includes: (i) Altitude sickness: troops at high altitude encounter acute mountain sickness (AMS), common sicknesses such as headache, nausea, vomiting, dizziness, and fatigue. However, prophylaxis with acetazolamide and dexamethasone, followed by proper acclimatization of troops, could alleviate the symptoms of altitude sicknesses. (ii) Severe cases of altitude sickness can progress into high-altitude maladies, which include: high-altitude pulmonary edema (HAPE), high-altitude cerebral edema (HACE), thromboembolic disorders, and cold injuries like frostbite, chilblain, etc. (iii) Decreased physical performance: reduced oxygen levels impair endurance and physical stamina. Individuals may experience early fatigue and a progressive

R. B. Yadav · P. Kapoor · R. Arora (✉)
Defence Institute of Physiology and Allied Sciences (DIPAS), Defence Research and Development Organization (DRDO), Timarpur, Delhi, India
e-mail: rajesharora.dipas@gov.in

A. K. Yadav · V. Kuntalam · A. Singhal
Department of Internal Medicine, Army Hospital Research & Referral, New Delhi, Delhi, India

V. K. Yadav
Research & Development Establishment (Engineers), Defence Research and Development Organization, Pune, Maharashtra, India

S. Srivastava et al. (eds.), *Health and Sustenance in Extreme Environment*,
https://doi.org/10.1007/978-981-96-8642-1_1

decline in physical strength. (iv) Impact on cognitive functions: decision-making, memory, and focus are three cognitive processes susceptible to compromise by hypoxia and are vital in effective military operations. On the other hand, the impact of environmental conditions on high altitudes includes extremes of weather, terrain, etc. Military expeditions can be affected due to tactical and decision-making abilities, e.g, logistics supply, high-altitude communication systems, aviation, training and acclimatization. The present chapter gives insight into overcoming the challenges of high altitude on the human system by considering the above-discussed factors that can affect military expeditions at high altitudes.

Keywords

Military operations · High-altitude physiology · HAPE · HACE · Cold injuries · HAPS

1.1 Introduction

Since times immemorial, wars have been fought by military forces in the high mountainous regions to gain strategic advantage over the adversaries. In the modern day as well, the military continues to operate in some of the world's extreme high-altitude regions, enduring and adapting to the harsh terrains, to secure the strategic borders. High altitude is defined as an altitude above which one's ability to perform physically and mentally is limited due to a relative shortage of oxygen. High altitude is defined as heights between 2500 and 5500 m, or around 8000 and 18,000 ft. This is the range of heights where hypoxia can limit physical activity and cause disease, but where people can effectively acclimate if they have enough time [1]. Operations at high altitudes need specific knowledge, tools, and techniques. It is crucial to comprehend the climatic and physiological effects of high altitude while living, organizing, and carrying out even day-to-day activities in these demanding conditions. Developing new strategies and technologies to overcome the problems inherent due to high-altitude terrains is essential [2]. Strategic considerations are also important for dealing with operations in high-altitude regions, viz-à-viz. (i) Hypoxia and acclimatization: high-altitude oxygen deprivation impairs both mental and physical function. Understanding the physiological impact of hypoxia is crucial for effective planning and risk mitigation. Allowing sufficient time for acclimatization before high-altitude deployment is essential to minimize health risks and maintain performance. Before beginning high-altitude missions, troops should have enough time to acclimatize [3]. Various acclimatization strategies have been developed to support military operations in high-altitude environments, aiming to enhance physiological adaptation and operational effectiveness under hypoxic conditions. By following these acclimatization protocols, troops are less likely to suffer from altitude-related disorders such as acute mountain sickness (AMS), high-altitude pulmonary edema (HAPE), high-altitude cerebral edema (HACE), and cold injuries such as frostbite. (ii) Geopolitical importance: high-altitude areas are quite

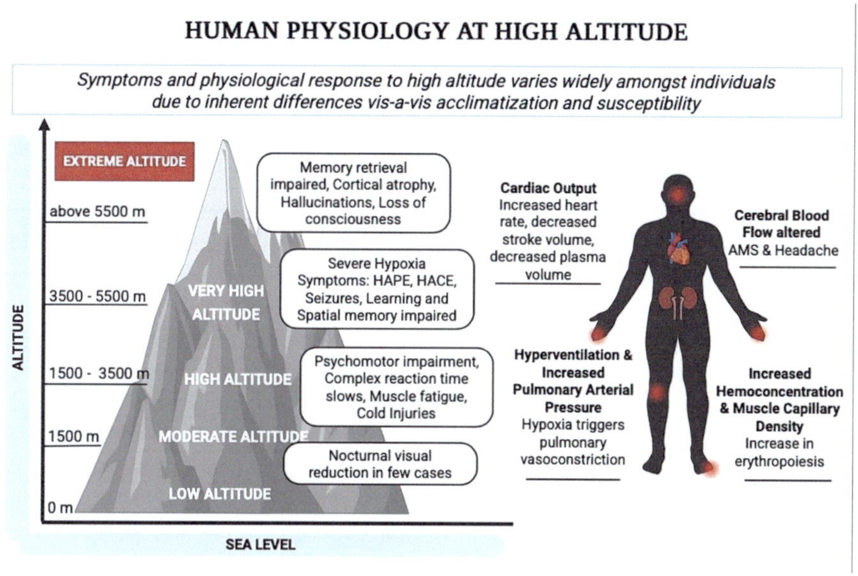

Fig. 1.1 Representative image of different stages of altitude and their common hypoxic effects

important from a geopolitical standpoint. Benefits, such as control over important mountain routes, passes and borders, can be gained by gaining control over strategic locations. (iii) Technological adaptations: some of the technical and logistical difficulties associated with high-altitude operations can be counteracted by technological advancements like drones that can operate at high altitude, portable oxygen supplies, and modern operational and protective gear. (iv) Strategic planning and implementation: to properly plan, strategists need to have precise knowledge of geography, natural and artificial resources, meteorological and environmental conditions, adversary movement, and conditioning. From a medical standpoint, prior to deployment in high-altitude regions, soldiers are recommended to increase their fluid intake to prevent the chances of dehydration. Additionally, travel to cold or high-altitude areas should be avoided until the lean body mass has been adequately restored [4]. The present chapter explores the impact of high-altitude environment on military expeditions, with a focus on physiological and pathophysiological responses, the effects of environmental stressors, and influence of these factors on overall performance, tactical decision making, and operational effectiveness. The Indian Armed Forces have been engaged in safeguarding the borders in extremely high altitude regions, which incidentally happens to be one of the world's most environmentally hostile terrain. The physiology of the soldier at extreme high altitude undergoes immense changes and the acclimatization/adaptation is dependent on a multitude of factors. Different stages of altitude and their related sickness are summarized in Fig. 1.1.

1.2 Challenges and Considerations for Soldiers at High-Altitude

Since time immemorial there have been many difficulties and losses in the past when high altitude has hindered military expeditions. The following are a few noteworthy instances.

Military expeditions at high-altitude regions have been shown to exert influence on overall human performance. Reports indicate that inadequate acclimatization prior to deployment at elevations ranging from 3000–5000 m, result in significant physiological impairment, adversely affecting physical performance, resulting in cognitive decline, and overall mission effectiveness. Rapid deployment to high-altitude regions, in some instances, has resulted in significant proportion of casualties due to acute mountain sickness (AMS). In response to these risks, focused research initiatives have been undertaken to better understand the altitude-related stressors, in order to develop mitigation strategies for safeguarding personnel health and sustaining performance in such extreme environments. These efforts have been of great importance, specially in regions where exposure to cold, low oxygen levels, temperature fluctuations, and humidity presents challenges to human endurance and operational efficiency [5]. In addition to the mentioned stressors, other factors such as carbon monoxide (CO) poisoning and nutritional challenges further compromise human health and operational performance in such environments. The use of combustion-based heating devices in shelters that are poorly ventilated, often results in build up of CO- a colourless, and odorless gas, significantly posing a severe risk of poisoning, which can be fatal, if undetected and untreated. Nutritional stress, on the other hand is a common concern, as the extreme cold and low oxygen conditions, often elevate the caloric requirement of personnels deployed, often exceeding 4000-5000 kCal per day. Despite the provision of specially formulated and energy dense rations, reduced appetite and limited access to fresh food items results in significant weight loss and nutrient deficiencies among personnel [6]. Maintaining hygiene and managing human waste is yet another persistent challenge faced by soldiers at high-altitude regions. The sub-zero temperature of these regions, lack the effective microorganisms, thereby hindering the process of natural decomposition. Traditional methods are at large ineffective in such areas, thereby resulting in accumulation of waste products, posing significant risk to both human health and environment. Addressing these sanitation challenges requires innovative approaches that are both feasible and eco-friendly under such extreme climatic conditions [6]. These challenges underscore the need for comprehensive physiological support systems in high-altitude operations. The Defence Research and Development Organization (DRDO) in this regard, has played a pivotal role in advancing interventions and technologies, aiming to mitigate and enhance the overall resilience, performance, sustenance and survivability of troops deployed at high-altitude regions. Recognizing the operational challenges associated with rapid deployment to high-altitude zones, the Life Sciences group of laboratories of DRDO has proposed a pre-conditioning strategy to help mitigate the altitude-related risks. One such contribution is the development of Normobaric Hypoxia Chamber, which delivers Intermittent Hypoxic Exposure (IHE), simulating the high-altitude environment, in order to induce adaptive physiological response. This method has been scientifically validated for training and acclimatization, improving physical endurance, and minimizing the risk of AMS. Another remarkable contribution is the development of portable One-Man High Altitude Pulmonary Oedema (HAPO) Chamber, a telemedicine system to provide emergency treatment to HAPO victims by enhancing the partial pressure of the breathing air inside the chamber. To bridge the gap in medical accessibility, DRDO has implemented a robust Telemedicine system, capable of supporting tele-consultation and diagnosis across pan-naval deployments. Apart from this, to address the nutritional challenges faced by the troops at high-altitude conditions, trench-based vegetable cultivation has been introduced as an innovative solution to ensure year-round access to fresh and balanced food. Additionally, Meals-Ready-To-Eat (MREs) ration, which are shelf-stable, safe, palatable, and nutritionally balanced and dense has been developed to meet the operational dietary needs of soldiers. Apart from this, a range of herbal and nutraceutical interventions tailored for high-altitude deployments have been developed by DRDO, to enhance the overall health and operational readiness of soldiers. The Composite Indian Herbal Preparation I (CIHP-I), a polyherbal formulation and DIP-G-Fit adaptogenic formulation developed by Defence Institute of Physiology and Allied Sciences (DIPAS) aimed at boosting physical performance under extreme environmental conditions. Another notable development with antistress properties, DIP-91, acts by promoting endurance, reduce stress and anxiety, without altering baseline metabolic activity. A sea buckthorn seed-oil derived gel based capsules rich in dietary fatty acids and vitamins has been formulated to provide antioxidant protection to support cardiovascular functioning, blood circulation, and cognitive health. A combination of spray-dried sea buckthorn and apricot pulp with *Rhodiola* extract has been developed as a neutraceutical herbal appetizer for its adaptogenic and nutritional properties. To address and effectively mitigate the cold-weather induced injuries such as frostbite, trench foot, and frostnip, an *Aloe vera* -based topical formulation (ALOCAL) in combination with pentoxifylline, vitamin C, and low dose asprin, has been developed, which has demonstrated effectiveness in managing both acute and chronic non-healing wounds owing to its high antioxidant, polyphenol and flavonoid content. Another remarkable breakthrough in the field of high-altitude research by DRDO, is the rapid advancement of military exoskeletons, which signifies a transformative leap in the soldier augmentation, seamlessly integrating biomechanics, robotics, cognitive technologies, and human-machine interface in order to boost strength, endurance and cognitive function. These powered exo-suits, particularly designed to increase the load carrying capacity for extended durations, are crucial for high-altitude operations, as it enables soldiers to perform strenuous tasks with greater efficiency, thereby minimising the chances of early fatigue and injury risk. This advancement will not only improve the operational effectiveness, but will also increase the soldiers endurance, enabling them to remain active for longer durations without frequent rest breaks in high-altitude terrains [7, 8]. Collectively, these innovations represents DRDO's integrative approach of combining traditional knowledge with the modern science to enhance overall heAloe veraalth, resilience, and performance of soldiers at high-altitude and other extreme environmental conditions. While significant progress has been made in this area, continuous research, field validation, and user feedback remain essential to optimize these solutions to be used under various occupational scenarios.

1.3 Understanding Human Limits in High-Altitude Conditions

The study of physiological processes that occur at high altitudes is gaining increasing attention as mountaineering, a recreational and adventurous activity, is on the rise. In addition, the aviation and space flight industries also suffer from these physiological changes. Persons who reside in high-altitude areas, either as permanent residents of high lands or as lowlanders posted temporarily in the military for their operations, undergo physiological changes as an adaptive process.

The effect of the partial pressure of oxygen, responsible for various illnesses occurring at high altitudes [9]. According to Dalton's law, total pressure of air is the sum of all gases present. At high altitudes, pH_2O and pCO_2 do not change, as they are determined by the metabolic process of the body itself, however, pO_2 and PN_2 do change. With increasing altitude, barometric pressure decreases, so the partial pressure of gases decreases. At sea level, the barometric pressure is 760 mmHg (1 atm) and the corresponding pO_2 is 159 mm Hg. As altitude increases, pO_2 decreases, at 10,000 ft., which is considered safe for long-term inhabitation, pO_2 drops further down to 110 mmHg. At extreme altitudes (>26,000 ft), i.e. at the summit of Mount Everest, atmospheric pressure drops to about one-third of the sea level. At this level, pO_2 is 56 mmHg, which is dangerously low and can cause unconsciousness in less than 45 s and death within 4–6 min in an unacclimatized individual. The primary cause of physiological changes is the reduced pO_2 in inspired air. These physiological changes can be in the form of an acute response of the body termed as adaptation, which happens in tourists and pilgrims visiting high-altitude places with rapid ascent and descent after staying for some time. The body immediately makes reflex adjustments in the respiratory and cardiovascular systems in response to a hypoxic environment. In addition to accommodation, due to long-term exposure to hypoxia, several changes take place in the body, which is termed acclimatization, these terms can be defined as follows:

Accommodation: This refers to the immediate changes happening acutely in the body due to sudden hypoxic exposure. The body senses decreased pO_2 and causes an increase in respiratory rate and depth of respiration. Furthermore, heart rate increases to increase cardiac output and meet the peripheral demands of tissues. Within hours of hypoxic exposure, 2,3-diphosphoglycerate (2,3 DPG) level increases and causes a decrease in the O_2 affinity of hemoglobin so that oxygen can be made available to tissues [10].

Acclimatization: This refers to physiological changes and compensatory mechanisms that occur in response to prolonged exposure to a hypoxic environment. The classic examples are individuals living at high altitude and military personnel posted at high altitude. Individuals permanently residing in these conditions have these changes since childhood, but for military persons who are posted at high altitudes, these physiological and compensatory mechanisms help them to reduce the ill effects of hypoxia. This is accomplished by adaptations in various systems [11]. The optimal rate of ascent (sleep altitude) should not be more than 500 m per day at levels greater than 2500 m. In addition to this, allowing at least one day of acclimatization at around 2500 m before further ascent and then again for every additional 1000 m of ascent, can reduce the risk of altitude sickness. Thorough acclimatization is crucial, and no undue physical exertion should be undertaken during the first 2 days after arrival. Individuals suffering from nose and throat infections should avoid traveling to high altitude by air.

The different stages of acclimatization at high altitudes are described in Table 1.1.

Table 1.1 Different stages of acclimatization protocols

Total Days	Days	Brief descriptions
Stage I acclimatization (applicable for individuals posted above 2700–3600 m height)		
Six days	Days 1 and 2	Rest is recommended, except for short walks within the unit line
	Days 3 and 4	Individuals should walk at a slow pace for 1.5–3 km. Avoid steep climbs.
	Days 5 and 6	Individuals can walk up to 5 km and can climb up to 300 m at a slow pace
Stage II acclimatization (above 3600 m and up to 4500 m)		
Four days	Days 1 and 2	Individuals should engage in slow walking for 1.5–3 km, avoiding steep climbings.
	Day 3	Should perform a slow walk and climb up to 300 m.
	Day 4	Should climb up to 300 m without pieces of equipment.
Stage III acclimatization (above 4500 m)		
Four days	Same as stage II acclimatization	

1.3.1 Impact of High Altitude on the Respiratory System

As the partial pressure of oxygen in the ambient air goes down with rising altitude, the immediate response to resulting hypoxia is an increase in ventilation known as hypoxic ventilatory response (HVR). Tidal volume and minute ventilation also increase following the decrease in SaO_2 [12].

It is mediated by peripheral chemoreceptors in the carotid bodies that sense low oxygen levels. Resultant hyperventilation reduces arterial CO_2, leading to hypocapnia, which in turn leads to respiratory alkalosis. The initial rise in pH partially offsets the stimulatory effect of hypoxia on the chemoreceptors by reducing the stimulatory effect on central chemoreceptors located in the medulla, which are primarily sensitive to changes in partial pressure of CO_2 (pCO_2) and pH in CSF [13]. With rising pH and central inhibitory effect, kidneys compensate for alkalosis by excreting bicarbonate and bringing back pH within the normal range, which in turn activates hyperventilation. In addition to the rate, the depth of respiration is also increased as a response to hypoxia.

In contrast to systemic vasculature, pulmonary vessel exposure to hypoxia leads to constriction of pulmonary arteries, thus directing blood flow more toward better-ventilated areas in the lung [14]. This improves ventilation–perfusion (V/Q) mismatching, better oxygen uptake by circulating blood, and ultimately delivery to the tissues. Hypoxia-inducible factors (HIFs) induce the expression of endothelin-1 (ET-1), which is a potent vasoconstrictor. HIFs also suppress NO production. HIFs also cause depolarization of smooth muscle cells of pulmonary arterioles by affecting ion channels (K^+ and Ca^{++}). While these vasoconstriction mechanisms help reduce V/Q mismatch and improve efficient oxygen uptake and delivery to tissue level, chronic vasoconstriction leads to remodeling of vessel walls and resultant pulmonary hypertension.

1.3.2 Impact of High Altitude on the Cardiovascular System

Hypoxia leads to increased sympathetic discharge, leading to tachycardia to maintain cardiac output and adequacy of oxygen delivery to tissue. Thus, the deficiency of oxygen in inspired air is compensated by increased blood volume delivery to the tissue, maintaining the net oxygen content at the tissue level unperturbed. Hypoxia-induced vasoconstriction in the pulmonary arteries increases pulmonary artery pressure, potentially leading to high-altitude pulmonary edema (HAPE) in susceptible individuals [15].

Cardiac output is maintained by an increased heart rate, but stroke volume may decrease due to a decrease in intravascular volume caused by hyperventilation and renal bicarbonate diuresis. There is increased angiogenesis due to persistently low oxygen tension. This leads to an increase in capillary density in tissues, thus improving oxygen delivery by efficient oxygen diffusion to cells [16].

1.3.3 Impact of High Altitude on Neurological Parameters

The human brain is extremely sensitive to hypoxia and with a gain in altitude, partial pressure of oxygen decreases causing hypoxia, however, compensatory mechanism increases cerebral blood flow thus maintaining sufficient oxygen delivery to the brain. Neurological consequences occur either when the ascent is rapid or the compensatory mechanism has failed. The symptoms vary accordingly with the rate of ascent and altitude gained. With rapid ascent at extreme altitudes loss of consciousness can occur within minutes. In addition to high-altitude illnesses like AMS, and HACE, individuals suffer from impairment of reaction time, short-term memory impairment, and learning deficit [17]. Neuropsychiatric symptoms like déjà vu, emotional lability, cognitive impairment, and hallucinations are also noticed. Transient ischemic attacks, cerebral infarction, and venous thrombosis can result due to the prothrombotic effects of hypoxia-induced polycythemia and dehydration. Migraine and seizure disorders are precipitated due to the consequence of hypoxia.

Ophthalmological disturbances include transient vision loss due to impairment of blood flow to the retina due to thrombosis. On the other hand, retinal hemorrhages can occur due to increased blood flow and disruption of the blood-retinal barrier [18].

Sleep disturbances include recurrent awakening at night due to periodic breathing. Periodic breathing occurs due to increased ventilatory drive due to hypoxia. Hyperventilation, in turn, leads to hypocapnia, which leads to central inhibition of respiration, resulting in episodes of apnea. Apnea for a brief moment causes CO_2 accumulation and, in turn, leads to the activation of respiratory drive. These sleep disturbances lead to recurrent awakening at night and poor quality of sleep, resulting in decreased overall performance and memory impairment, predisposes people to other high-altitude illnesses [19].

1.3.4 Impact of High Altitude on Hematological Parameters

In the initial days, Hb and hematocrit are increased due to a decrease in effective circulatory volume mediated by hypoxia-induced renal bicarbonate diuresis. This is followed by increased erythropoiesis secondary to increased production of erythropoietin by the kidneys, which stimulates the bone marrow to produce more red blood cells, resulting in increased hematocrit and hemoglobin concentration. There is an increase in 2,3-bisphosphoglycerate levels, leading to a shift of the oxygen–hemoglobin dissociation curve (ODC) to the right. This leads to reduced affinity of oxygen for Hb and thus facilitates efficient oxygen unloading in tissues despite low ambient oxygen tension [20].

1.3.5 Modulating Redox Stress in High-Altitude Terrains

Enhanced oxidative enzyme activity in mitochondria improves aerobic metabolism. This involves hypoxia-induced factors (HIFs). HIF-1α and HIF-2α activate hypoxia response elements (HREs) in the nucleus, thus upregulating the genes involved in angiogenesis, erythropoiesis, glycolysis, and mitochondrial function [21, 22].

HIFs promote mitochondrial biogenesis and enhance the efficiency of the electron transport chain (ETC) by inducing the gene peroxisome proliferator-activated receptor gamma coactivator-1 alpha (PGC-1α), which regulates mitochondrial biogenesis and thus increases mitochondrial density [23]. Also, the AMP-activated protein kinase (AMPK) pathway gets activated in response to increased AMP/ATP ratio, which indicates cellular energy stress often encountered in hypoxic environments. AMPK pathway enhances glucose uptake and reduces fatty acid oxidation as glucose oxidation yields more ATP per mole of oxygen consumed as compared to fatty acid. It also inhibits anabolic functions, thus conserving energy (ATP) and optimizing ATP production. There is upregulation of ETC components (complex I-IV, cytochrome C), and ATP synthase improves the efficiency of electron transport and ATP production [24].

1.3.6 Impact of High Altitude on Exercise and Functional Capacity

Is of greatest concern for soldiers posted at high altitudes. Exercise and functional capacity of an individual depend on the stage of high altitude and the type of exercise performed. Anaerobic exercises for a brief duration, like sprinting, are improved due to reduced wind resistance and hypobaric conditions, while for exercises requiring prolonged anaerobic and aerobic respiration, exercise capacity is decreased substantially due to the low ambient partial pressure of oxygen in inspired air, and due to an unacclimatized cardiovascular system. However, due to physiological adaptations and increased oxygen-carrying capacity, submaximal exercise capacity is improved over time [25].

1.4 Altitude Sickness and Diseases

1.4.1 High-Altitude Headache (HAH)

Headache is the most common symptom experienced by almost half of individuals ascending >5000 m. It can be the only symptom, or it can accompany other symptoms of acute mountain sickness. It is characterized by the development of a headache within 24 hours of ascent to high altitude, generally >5000 m, and the patient develops a bilateral or fronto-temporal headache, which is dull, aching, or a feeling of pressing/constricting. It ranges from mild to moderate intensity and increases with activities, coughing, and bending forward. The exact cause of HAH is not known, but it is postulated to be due to cerebral hypoxia leading to oxidative endothelial damage and microcirculatory changes. Dehydration due to increased hyperventilation, exaggerated by insufficient intake of fluids and increased physical activity, can be confounding factors. It generally resolves within 8–10 hours of descent, along with adequate hydration and over-the-counter (OTC) analgesics [26].

1.4.2 Acute Mountain Sickness (AMS)

AMS is a self-limiting condition that generally happens in non-acclimatized individuals presenting for the first time at high altitude. Its incidence increases with height and can affect 50% of travelers at 6000 m [27]. It is characterized by a headache that is very similar to HAH, with varied intensity, along with light-headedness. It is also accompanied by gastrointestinal symptoms (poor appetite, nausea, vomiting), generalized fatigue, weakness, and sleep disturbances. It occurs within 8–10 hours and resolves within 72–96 hours.

Risk factors include rapid ascent and improper acclimatization at lower altitudes, pre-existing comorbid conditions, physical exertion, use of alcohol or any other substance, and prior history of AMS [28].

The exact pathogenesis of AMS is poorly understood, it is thought to be associated with hypoxia leading to arteriolar dilation which overwhelms.

Diagnosis is based on the history of recent travel to high altitudes with typical symptoms of headache, fatigue, and nausea/vomiting in adults, and nonspecific symptoms are seen in children, such as decreased activity, excessive crying, lethargy, and excessive sleep. The symptoms of AMS can vary in intensity from mild to severe, as categorised by the Lake Louise AMS score, a self-assessment tool used by patients to evaluate the severity of the disease: 3–5: Mild AMS, 6–9 Mod AMS, >10: Severe AMS. This classification can be used as a guide for deciding the urgency of treatment [29].

Mild cases can be treated symptomatically with analgesics and antiemetics, along with adequate hydration and by avoiding further ascent. However, severe cases need rapid descent to lower altitudes and administration of acetazolamide 250 mg BD [30], and in severe cases, dexamethasone 4 mg every 6 hours [31]. Supplemental oxygen can relieve symptoms and act as an interim measure for avoiding descent [32].

1.4.3 High-Altitude Pulmonary Edema (HAPE)

HAPE is a life-threatening condition in high altitude, characterized by the accumulation of fluid inside alveolar sacs due to a non-cardiogenic cause. It is of two types: Classical HAPE, which occurs during initial ascent, and reentry HAPE, which occurs on re-exposure of healthy individuals to high altitude. Early symptoms include shortness of breath (dyspnea) on exertion, cough, and decreased exercise tolerance. These progress over time to dyspnea at rest, cough-producing frothy or blood-tinged sputum, chest tightness, and cyanosis (bluish discoloration of the skin due to lack of oxygen).

Hypoxia induces vasoconstriction in the pulmonary arterial, leading to increased pulmonary arterial pressure. Elevated pressure causes stress on the pulmonary capillaries, leading to leakage of fluid into the alveoli (air sacs in the lungs). The fluid accumulation in the alveoli leads to impaired diffusion of gases across the respiratory epithelium, resulting in dyspnea. It occurs without left ventricular failure, distinguishing it from other forms of pulmonary edema.

Uneven hypoxic vasoconstriction causes some regions of the lung to receive more blood flow than others, leading to overperfusion and increased capillary pressure in these areas. Hypoxia can damage the endothelial cells lining the blood vessels, increasing permeability and promoting fluid leakage [33].

Diagnosis of HAPE is clinical, based on a recent history of ascent and classical symptoms of dyspnea with features of hypoxia, demonstrated by decreased SpO_2 associated with bilateral crackles on chest examination. Immediate descent to a lower altitude is the most effective treatment. Supplemental oxygen can help alleviate symptoms by increasing arterial oxygen saturation. Medications like calcium channel blockers (e.g., nifedipine) can reduce pulmonary artery pressure. Phosphodiesterase inhibitors (e.g., sildenafil*)* can also be used. Continuous positive airway pressure (CPAP) can be beneficial in severe cases. It increases transpulmonary pressure and helps in the resolution of pulmonary edema.

1.4.4 High-Altitude Cerebral Edema (HACE)

HACE is a continuum of severe AMS, seen in 0.1–2% of individuals at heights >13,000 ft. It is characterized by the development of ataxia, severe headache, speaking difficulty, and impaired consciousness, finally leading to coma and death. Initial symptoms are nonspecific, like lethargy, and are often missed and attributed to exhaustion. The exact pathogenesis revolves around hypoxia and cerebral edema [34]. Exposure to a hypoxic environment leads to hypoxia-induced cerebral vasodilation as a compensatory mechanism, which leads to overperfusion of microvascular beds and increased hydrostatic pressure in capillary beds. Increased hydrostatic pressure and increased expression of vascular endothelial growth factor (VEGF) and other cytokines increase the vascular permeability of cerebral vessels, resulting in vasogenic edema of brain parenchyma, if untreated herniation develops. On the other hand, enlarged brain parenchyma further causes venules and venous sinus compression, leading to microhemorrhages. Furthermore, hypoxia leads to increased free radical generation, which causes failure of ionic pumps, leading to astrocyte swelling and subsequently cytotoxic edema [35].

HACE is a medical emergency, and early recognition is key, and treatment is the utmost priority. Rapid descent to an altitude lower than 7000 ft is the goal;

however, even a decrease in altitude of 1000 m is associated with improved outcomes [36]. Hyperbaric chambers can be used in situations where descent is not possible. These chambers mimic atmospheric conditions of 2500 m and help relieve symptoms [37]. Supplemental oxygen should be given at high flow, as required, to keep target saturation $SpO_2 > 90\%$ [38]. Dexamethasone helps reduce the symptoms of HACE and should be administered immediately at a dose of 8–12 mg iv/i.m. stat, followed by 4 mg every 6 hourly until symptoms subside and descent is completed [39]. Medications like acetazolamide are only used as adjunctive therapy to dexamethasone and should not be used alone in patients with HACE.

1.4.5 Cold Injuries

Cold injury, sometimes referred to as a cold weather injury, is damage to the body brought on by exposure to low temperatures. Two primary forms of cold injuries can be distinguished: freezing (e.g., frostbite) and non-freezing (e.g., frostnip, chilblain, and trenchfoot). Cold injuries are a serious problem because they have been known to result in major casualties, especially during military expeditions in extreme cold weather conditions. Also, mountaineers and naiive explorers can get exposed to extreme cold and are more susceptible to frostbite. A number of physiological alterations resulting from cold stress might exacerbate perfusion problems and cold injuries. Hypothermia, increased metabolism in contracting muscles, and decreased blood flow are some of these alterations. The complicated effects of cold damage on blood parameters are demonstrated by the changes in erythrocyte counts, mean corpuscular hemoglobin concentration, hematocrit, ice-crystal formation, increase in cytokines production, and thromboxane formation, which subsequently may lead to amputation if left untreated. In summary, cold injury is a serious problem that could significantly affect operations, particularly those related to the military [40].

1.4.6 Preexisting Illness Aggravation by High Altitude

A thorough examination for comorbid conditions is done before going to high altitude, as these conditions tend to deteriorate at these altitudes, resulting in worse outcomes. People with a history of hypertension can experience raised blood pressure, despite adequate compliance with medication. Coronary artery disease is usually a contraindication for ascent to higher altitudes as they tend to worsen due to hypoxia and tachycardia-induced cardiac workload. Respiratory illnesses like COPD and asthma are exacerbated due to the low availability of oxygen and cold-induced bronchoconstriction. Headaches are common due to dehydration and hypoxia and cannot be distinguished from migraine, which are easily precipitated by hypoxia and cold winds.

1.5 Mitigating High-Altitude Illness (HAI) Through Established Protocols

1. *Preinduction medical examination:* Determination of pre-existing illness mainly cardiopulmonary disease is important as asymptomatic individuals become symptomatic at high altitudes due to poor cardiovascular reserve. Proper adherence to medications is necessary alongside avoidance of exertional activities.
2. *Proper acclimatization*: Following a proper 5-day acclimatization schedule is important in reducing the chances of high-altitude illness. Gradual ascent with night stays at lower altitudes helps in proper physiological changes. Gradual increase in physical activity including slow walking for initial days followed by brisk walking and jogging decreases the incidence of HAPE.
3. *Prevent dehydration:* An adequate intake of water is required to compensate for the insensible loss due to hyperventilation and renal water loss and prevents the occurrence of HAH and AMS.
4. *Monitoring and follow-up of risk case:* Individuals showing symptoms of headache, lethargy, and sleep disturbances during the initial days should be monitored for symptom aggravation. These individuals, if not identified, are at high risk for the progression to severe form of altitude-related illness, including AMS, HAPE, and HACE.
5. *Prophylactic medications:* Preinduction administration of acetazolamide in prophylactic doses, continued for 48 hours, aids in acclimatization and symptom alleviation. It increases renal excretion of bicarbonate and counteracts alkalosis caused by hyperventilation.
6. *Standard operating procedures (SOP) for various illnesses*: A ready-made SOP in a medical facility with detailed information on treatment protocols and sequential actions required for the treatment of the concerned disease is useful for managing HAI.
7. *Practicing casualty evacuation drills*: Regular practice of causality evacuation drills enhances preparedness, and improve efficiency and coordination among medical echelons, thus improving overall patient outcomes.

1.6 Impact of Environmental Conditions on Military Expeditions

1.6.1 Extremes of Weather

Extreme weather has a substantial effect on military expeditions at high altitudes and can have an impact on several different components of such operations. Maintaining territorial integrity and ensuring national security often necessitates deployment of troops in challenging and harsh environmental regions with harsh, dry, and frigid weather conditions, with winter temperatures varying between −35 °C to −55 °C in certain areas, altitude ranging from 8000 to 23,000 ft. Low humidity levels, hypobaric hypoxia, and elevated solar UV radiation, together with their associated issues, intensify the challenges encountered by the soldiers in these weather circumstances [41]. The important components affecting military operations are as follows:

Extreme cold temperature: The primary environmental conditions that define the Western Himalayas are exceptionally low winter time temperatures, which can drop to as little as −60 °C in some areas; year-round, severely low relative humidity, which generally stays below 25%; and high winds reaching 120 knots in some areas, which can have a hazardous chilling effect. It is possible to compute ambient air temperatures at high elevations using conventional techniques. Understanding of the critical weather thresholds impacting the operational systems, personnel, and activities is essential for effective mission planning and success [41].

High density altitude: A state of the atmosphere that, at a certain altitude, lowers an aircraft's performance capabilities below normal levels. This crucial parameter establishes if an aircraft is capable of producing enough lift and performing well enough to take off from high-altitude regions. Weapons, fuel, and passenger loads may be restricted at high-density altitudes [42]. High-density altitude can decrease the performance of the aircraft and unmanned aerial vehicles (UAVs) thus affecting military operations in high-altitude regions. A few factors that affect the performance of the aircraft are: (i) reduced lift (high-density altitude diminishes the air's capacity to provide an upward force on the airfoils, leading to a reduction in lift and an increase in peak speeds), (ii) decreased engine power of an aircraft in high-density altitude region (engine power production is reduced at high density altitudes due to the air's capacity to sustain combustion is diminished), (iii) impaired propeller efficiency of the aircraft in high-density altitude regions (propeller efficiency and thrust are decreased because of the decreased air density, which makes it more difficult for propellers to produce thrust), (iv) decrease in service ceiling, etc. [43].

Increased level of snowfall in high-altitude regions affects military operations: Increased snowfall affects aerosystems performance, which can directly impact military operations in high-altitude regions. The aerodynamics of an aerosystem is impacted by excessive snow on lifting surfaces; even a small amount of snow can cause serious issues. This is especially crucial for military operations at high altitudes where snow conditions are more frequent [43]. At the ground level, high altitude and snow-covered conditions place special demands on ground forces, which can have a significant negative influence on military performance. The inclement weather conditions, such as hypoxia, wind, cold, high UV and solar radiation, make it extremely difficult to sustain and survive [5]. Altitude can cause negative symptoms that impair cohesiveness and morale, such as altitude sickness, tachycardia, and dyspnea [44]. Decision-making and cognitive function can be negatively impacted by mood swings and impaired neuropsychological functioning. Localized cold injuries often affect hands, feet, nose, and ears and frequently result in irreparable damage [40, 44]. Sustained physical activities such as weight bearing, battle, and patrolling, are most significantly impacted, resulting in early fatigue and reduced operational effectiveness [45].

Lower visibility and higher precipitation in high-altitude regions: Snow and rain, even in moderate quantities, can deteriorate the quality of data obtained from photographic and infrared collection devices and cause optical impairments. Heavier rainfall can produce background electronic noise, which lowers the effectiveness of ground surveillance radar (GSR). High altitudes with high densities contribute to increased air interference, leading to signal deterioration and distortion, ultimately diminishing GSR performance.

1.6.2 Terrain Analysis

Terrain analysis is a vital part of battlefield visualization and is necessary for efficiently organizing and carrying out military actions. It aids in the commander's comprehension of how natural and artificial elements, like weather and climate, affect the battlefield. Military engineers are in charge of finding and assessing the potential of different terrain characteristics on the battlefield to provide commanders with terrain evaluation and analysis. By giving information on terrain in three dimensions, geospatial engineering integrates technical capabilities and activities that support a thorough knowledge of the physical environment. Enabling mission planning and decision-making entails creating, maintaining, evaluating, and sharing positionally accurate terrain information [46].

1.7 Tactical Considerations and Disruptive Technologies in Military Operations at High-Altitude

1.7.1 Challenges of Altitude Logistics

High-altitude regions have unique logistics and supply chain challenges because of severe weather and topography. Military operations in these locations are greatly impacted by procurement delays. The health and hygiene of soldiers stationed in these locations may suffer as a result of the delays, which may cause severe shortages of essentials, clothes, ammunition, and other medical facilities [47]. Lack of proper logistics can result in the inability of soldiers to carry out their tasks efficiently and can disrupt proper military operations due to a shortage of necessary supplies. Adversaries may take advantage of high-altitude regions' susceptibility to supply chain interruptions; hence, it is imperative for the military to maintain timely and dependable logistics systems. The troops' morale may suffer as a result of the delays if they feel that their requirements are not being addressed, which might lower their morale and overall effectiveness.

1.7.2 High-Altitude Communication System

High-altitude platform systems (HAPS), another name for high-altitude communication systems, are unmanned aircraft that operate at altitudes of approximately 59,000–65,000 ft. in the stratosphere (an altitude that is normally higher than where normal passenger aircraft flies but lower than satellite orbits). These solar-powered aircraft are intended to fill the void left by traditional satellites in orbit and unmanned aerial vehicles (UAVs) operating at lower altitudes. HAPS constitute a network of aerial stations. They have emerged as a crucial component of upcoming wireless networks, spurred by technical advancements in the domains of communication systems, solar power systems, antenna arrays, and autonomous flying [48]. They provide a range of military services and applications as follows:

Emergency communication services: HAPS can offer dependable and effective communication services, especially in places with inadequate satellite coverage. They are useful for data transfer, tactical communications, and even emergency communication. Search, rescue, and recovery activities during a catastrophic event cannot be carried out without emergency communication infrastructure, which can be seriously disrupted by terrorist attacks and natural catastrophes. After disastrous occurrence, telecommunications equipment placed on high-altitude platforms might be utilized to assist these vital communication operations. Additionally, they can offer monitoring in the event of a disaster or natural calamity. HAPS is a great way to cover high-risk regions for emergencies before major disasters happen since they can be on station all the time [49].

Intelligence, surveillance, and reconnaissance (ISR) through high-altitude communication systems: The HAPS module has all the necessary pieces of equipment required to deal with military operations in high-altitude regions, which includes modern sensors, high-definition optical and infrared cameras, and other sophisticated payloads to enable ISR tasks including target tracking, maritime surveillance, and border patrols by armed forces [49].

When compared with normal airplanes and satellites, which also provide an air-defense system, HAPS stands one step ahead as it is an unmanned system with a number of benefits with very minimal infrastructure and being more flexible in operations.

In many circumstances where the other air-defense systems and satellites remain out of reach and restrict the coverage and search operations, they are especially helpful for expeditionary operations, out-of-area deployments, and disaster relief scenarios [50].

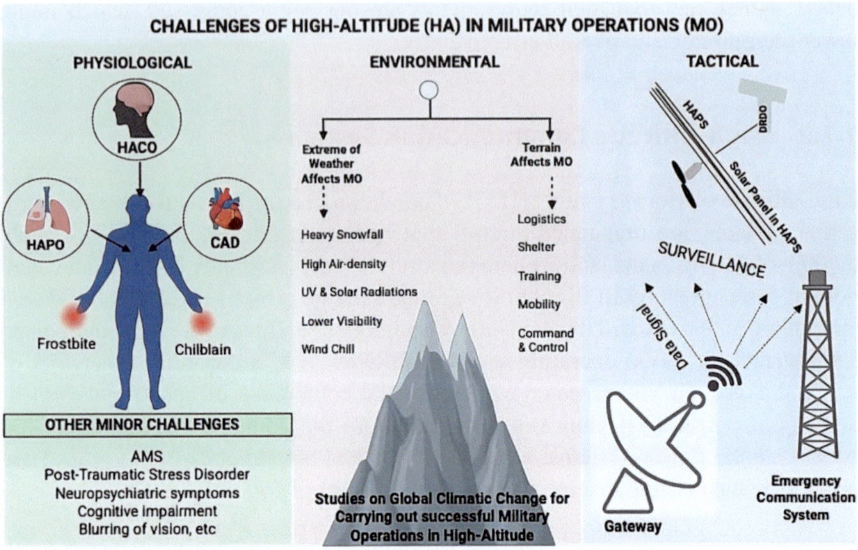

Fig. 1.2 Schematic representations of Challenges of high altitude on military operations

1.7.3 Aerosystems

In tactical decision-making for military operations, aircraft and Unmanned Aerial Vehicles (UAVs) plays a critical role by supplying airpower that may be employed to accomplish tactical and strategic goals. Thus, aviation plays a vital role in assisting military operations in high-altitude regions. It performs combat, surveillance, and reconnaissance duties to support ground soldiers from the air during emergencies and exercises. Air-to-air combat, airlift missions, and close air support are a few examples of this support. Aviation also plays an important role in the transportation of soldiers, supplies, equipment, and medical facilities such as causality evacuations and medical relief services. Sustaining the flow of materials and assisting military operations depends on efficient air transport operations. To ensure safe and efficient operation of air traffic and minimize interruptions and delays, effective coordination and cooperation between military and civilian aviation bodies are vital. This cooperation may entail cooperative training sessions, information and resource sharing, and coordinated emergency response. To make tactical decisions, situational awareness must be maintained, and aviation is vital to this process. Situational awareness such as proper and consistent information gathering is required to perform critical military operations. To accurately depict the conditions on the battlefield, this awareness is required for identifying targets and dangers [51, 52].

1.7.4 Role of Soldier Training and Acclimatization

The efficiency of military operations in high-altitude areas can be augmented through proper training and acclimatization strategies designed specially by considering multiple factors affecting altitude sickness. It is not advisable to send unacclimatized personnel quickly to high-altitude areas as it can have detrimental consequences on operational capabilities, including poor physical performance at work and altitude sickness-related health problems for the force [53].

Soldiers are well trained to provide support during natural disasters and also to their colleagues as well, and army units situated at high altitudes and other remote locations can be supported by telemedicine systems, which can help further lessen the demand for on-site medical facilities required by the native civilian population and military personnel posted in such regions. To ensure overall well-being of soldiers in high-altitude and remote areas, DRDO has proposed specialised ration scale to provide wholesome and balanced meals, further complemented by the specialised yoga and meditation programs designed for safeguarding the overall physical and mental health of troops and ensuring better performance under extreme conditions. These initiatives include dietary adjustments for flatulence, use of biodegradable packaging material, use of insulated, portable, lightweight, and modular antifreeze containers for fresh supply of food materials without compromising its overall quality. These holistic interventions underscore a dedicated commitment to enhance the soldier welfare and operational effectiveness in challenging and extreme environments.

1.7.5 Directed Energy Weapons (DEWs)

The Directed Energy Weapons (DEWs) are a class of ranged weapons that effectively neutralize and disable its targets using concentrated electromagnetic energy in the form of high energy laser beams. In contrast to the conventional kinetic weapons used, DEWs offer several operational advantages including high precision target engagement and low operational cost. These attributes makes them effective against low-radar cross section (RCS) threats, such as unmanned aerial vehicles, swarm drones, and incoming projectiles. Recent advancements have showcased the ability of these systems to accurately detect, track, and neutralize

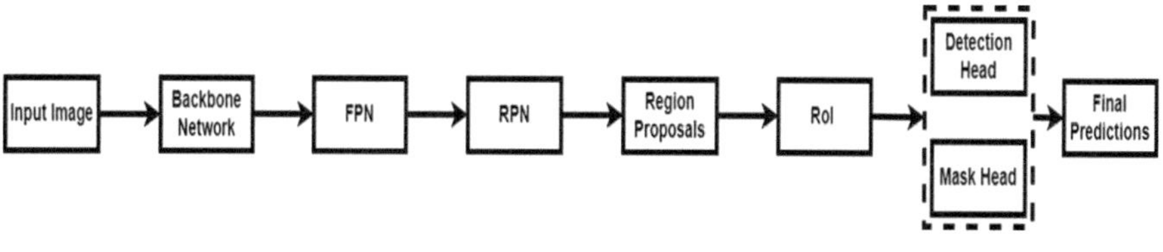

Fig. 1.3 Architecture of mask R-CNN

both aerial and ground-based targets, including the electro-optical sensors and communication systems. Based on their power output, DEWs are categorized into following types: Low energy lasers (<1kW) employed in weapon simulation systems involved in training exercises, jamming of communication sensors, or utilized in antipersonal mode against human eye. Medium energy lasers (10 - 100 kW) deployed for disabling optical and optoelectronic devices in both ground and space based systems, and High energy lasers (>100 kW) used in anti-aerial, drones, and missile combat scenarios, particularly offering strategic advantages in defence system. In parallel with the laser technology, the radiofrequency (RF) and microwaves systems form the backbone of the modern military communication and electronic warfare. Advanced RF systems, particularly the software-defined radios (SDRs) offer high level of reconfigurability, adaptability, and real-time signal processing capabilities, making them ideal for secure communications, and remote command and control operations. Moreover, it supports complex applications, essential for maintaining electromagnetic situational awareness on the battlefield. Microwave communication systems on the other hand provides additional benefits by offering secure, high-throughput data and voice transmission, utilizing air as the medium to the military forces even in the extreme environments. The ability to rapidly establish the robust communication link under both extreme environmental and logistical conditions, makes these systems ideal for both strategic deployments and other military operations. In addition, microwave weapons are also being tested by several countries [1, 54]. Together these advancements in directed energy viz., Lasers, RF, and microwave technologies, signify a paradigm shift in defence strategy, thereby offering efficient and precise tools to address and counteract emerging threats in military operations [55, 56]. However, the use of directed energy weapons is fraught with its own limitations and dangers to human operators as the deleterious bioeffects might be linked to small, yet inevitable exposures.

Various challenges that obstruct military operations are depicted in Fig. 1.2.

1.8 Deep Learning-Based Model as a High Altitude Ailment Diagnostic Tool Using Instant Segmentation: A Future Approach

Artificial intelligence (AI) and machine learning (ML) are revolutionizing the healthcare sector by dramatically enhancing disease detection, diagnosis, and treatment at the point of care. AI can help with diagnosing illnesses and treatment in high-altitude conditions. High-altitude diseases like high-altitude cerebral edema (HACE) and high-altitude pulmonary edema (HAPE) can be detected using AI-powered technologies that evaluate patient data and diagnostic imaging [21, 22]. These technologies analyze clinical images, such as X-rays, MRIs, and CT scans, with unprecedented precision and speed—often outperforming human capabilities by utilizing sophisticated algorithms and large datasets. They facilitate prompt interventions that are essential for improving patient outcomes by enabling early detection of ailments like cancer, cardiovascular diseases, and neurological disorders. AI-driven diagnostic technologies ensure precision care tailored to each patient's specific needs by analyzing medical data, proposing possible diagnosis, and recommending individualized treatment options. Healthcare professionals may improve diagnosis accuracy, decrease chances of miscellaneous error, and deliver more efficient, patient-centered treatment on-site, especially in remote locations such as high-altitude regions by incorporating AI into clinical procedures [57–59].

Machine learning, especially using deep learning-based algorithms, streamlines important processes like disease diagnosis in clinical image analysis [60]. For segmentation problems, a variety of neural network designs are used, such as FC-DenseNet [58, 59], Fully Convolutional Network (FCN) [61], SegNet [62], DeepLab [63], Atrous Convolution, and Atrous Spatial Pyramid Pooling (ASPP). In particular, the Mask R-CNN deep learning model is widely used in medical imaging for disease identification [64, 65]. For instance segmentation tasks, the Mask R-CNN architecture is quite effective since it adds another branch to the Faster R-CNN design that is specific for predicting segmentation masks in addition to bounding boxes and class labels. To create feature maps, an input image is first subjected to feature extraction using a backbone network, usually ResNet or ResNeXt. A feature pyramid network (FPN), which creates a hierarchy of feature maps at multiple scales to improve object detection across a range of sizes, further refines these maps. Next, a region proposal network (RPN) suggests prospective object regions termed anchor boxes, which refines these suggestions using classification and regression layers to produce accurate region recommendations. Mask R-CNN uses a region of interest (RoI) alignment to handle these approaches, which guarantees precise spatial information extraction without any quantification problems associated with RoI pooling. This technique produces fixed-size maps of characteristics essential for later tasks for every region. The detection head and the mask head are the two paths that the model branches into. The detection head uses regression to improve bounding box coordinates and fully connected layers for item cat egorization. Convolutional and deconvolutional layers leading to a sigmoid activation are simultaneously employed by the mask head, a small fully convolutional network, to generate pixel-level probability maps for item existence. An architecture of mask R-CNN is depicted in Fig. 1.3. The integration of Internet-of-Things (LoT) technology into military operations, particularly via advanced biometric wearable technology and AI analysis, marks a significant advancement in sustaining soldier performance under extremes of environment. In the context of soldier monitoring, the integration of LoT facilitates real-time collection and communication of valuable information, such as regarding soldiers location, health metrics (heart rate, respiration capacity, body temperature, cognitive status), and environmental conditions, to the command centre, prompting rapid response to any risk and significantly enhancing the battlefield awareness. Smart wearable such as ruggedized smartwatches, smart clothing, and augmented reality (AR) helmets, equipped with biometric sensors and AI capabilities to provide continuous health monitoring, thereby enabling early detection of potential medical issues are becoming indispensable tools in military operations [66]. Additionally, these devices overlay crucial information onto a soldier's field of vision, aiding real-time decision making, and improving battlefield situational awareness. The LoT-based Soldier Status Monitoring System (SSMS) integrates multiple smart sensors, e.g, gas sensors, motion sensors, biometric sensors, and metal detectors directly within soldier's gear, relaying this information to a centralised command via GPS communication system [67]. Such a comprehensive approach, empowered by AI for data analysis and predictive insights, might play a crucial role in safeguarding soldier health and overall well-being, by providing immediate real-time status updates, location tracking and alerts, contributing to the success and safety of futuristic modern military missions.

Instance Segmentation for High-End Disease Detection In medical imaging, instance segmentation is vital for accurately identifying diseases in the human body. In contrast to traditional approaches that simply demonstrate the boundaries of objects that have been spotted, instance segmentation performs exceptionally well by precisely defining every single instance of interest in medical images. With the use of this technology, medical personnel may accurately locate and measure anomalies down to the pixel level in addition to detecting them, such as tumors or lesions [68]. Through the provision of comprehensive segmentation masks for every recognized occurrence, instance segmentation significantly improves diagnosis accuracy, streamlines treatment planning, and provides continuous illness monitoring. The efficacy of medical imaging in clinical settings can be greatly increased by using deep learning models such as Mask R-CNN, which include sophisticated convolutional neural networks and guarantee robust and trustworthy segmentation results.

To produce precise and efficient results, several technical processes are involved in employing Mask R-CNN for segmentation detection. The first step of the procedure is data preparation, which involves assembling a varied and thoroughly annotated dataset. This dataset consists of comprehensive annotations on medical images that show areas of interest (ROIs) associated with anomalies or diseases. The use of augmentation techniques like rotation, flipping, and scaling increases the robustness and diversity of the dataset. Similarly, using AI–ML tools, various high-altitude ailments can be manifested, which can be beneficial in making quick decisions and managing treatment in such remote locations.

1.9 Conclusion

High altitude environment presents a complex interplay of physiological, psychological, environmental, and operational challenges significantly impacting military performance and operation effectiveness. The human body undergoes several adaptations to cope with these conditions. The reduced partial pressure of oxygen leads to hypoxia-induced conditions which include AMS, HAPE, and HACE, all of which require timely recognition and management to safeguard troops health. Acclimatization is a key strategy in enhancing physiological adaptation, while the use of herbal adaptogens, neutraceuticals, and advanced biomedical technologies offer additional protection. The present chapter has explored the impact of high-altitude environment on military operations, shedding the light on the human response and tailoring interventions to mitigate the problems faced by the warfighter in such extreme scenarios. Managing these challenges requires a multifaceted approach, combining physiological adaptation strategies, environmental preparedness, and state-of-the-art technological support. The particular difficulties presented by high-altitude environments, such as terrain analysis, logistics, weather, staff recruitment and instruction, operational adaptability, and integration with other C4ISR (Command, Control, Communications, Computers, Intelligence, Surveillance, and Reconnaissance) capabilities, must be taken into consideration while taking tactical decisions. Moreover, the integration of AI and ML models will contribute to predictive health surveillance, allowing for prediction, early detection, diagnosis, and development of personalized acclimatization and treatment strategies to support troops under such extreme conditions. Further, integration of AI will drive innovation in drug discovery tailored to high-altitude physiology, thereby enabling faster development of targeted and personalized prophylotherapeutics in near future. Therefore, addressing these challenges demands for an integral framework that combines human physiological understanding with modern medical technologies. The advancements in predictive analytics, personalized prophylotherapeutics, and adaptation strategies is likely to further strengthen the ability to sustain the overall operational readiness and ensure warfighter safety in high-altitude regions. These innovations underscore the importance of strategic planning, adaptability, and provisioning of cutting-edge solutions in mitigating the challenges and risks encountered by military personnel upon deployment at extreme high-altitude.

Acknowledgments The authors (RBY, PK, RA) would like to profusely thank Director, DIPAS for his encouragement and unflinching support. The authors sincerely thank Dr. Niroj Shetty and Dr. Gopinath Bhaumick, Scientists at DIPAS, for going through the manuscript and for their useful suggestions. RBY, PK and VKY are thankful to DRDO for support in the form of a fellowship. AKY, VK, and AS are grateful to Commandant, Army Hospital R&R, New Delhi for encouragement and support in multifarious ways. The authors are thankful to Dr. Upendra Kumar Singh, Director General - Soldier Support System (SSS), and Secretary, Department of Defence R&D and Chairman, DRDO for encouragement, support and facilitation in multifarious ways.

Disclaimer The content of this chapter is of unclassified nature and is purely academic/scientific in nature. The views expressed by the authors are personal and do not necessarily reflect the views of their respective organizations and may be construed as such.

References

1. Muza SR. Military applications of hypoxic training for high-altitude operations. Med Sci Sport Exerc. 2007;39(9):1625–31.
2. Fighting Tactics. https://www.cassindia.com/fighting-tactics. Assessed on 13th June 2024.

3. Grocott M, Montgomery H, Vercueil A. high-altitude physiology and pathophysiology: implications and relevance for intensive care medicine. Crit Care. 2007;11(1):203.
4. Institute of Medicine (US) Committee on Military Nutrition Research. 2, Committee on military nutrition research recommendations and conclusions. In: Marriott BM, Carlson SJ, editors. Nutritional needs in cold and in high-altitude environments: applications for military personnel in field operations. Washington (DC): National Academies Press (US); 1996. Available from: https://www.ncbi.nlm.nih.gov/books/NBK232862.
5. McLaughlin CW, Skabelund AJ, George AD. Impact of high altitude on military operations. Curr Pulmonal Rep. 2017;6:146–54.
6. Jindal AK, editor. The highest battlefield of the world : medical problems and solutions. Med J Arm Forces India. 2009;65(2):170–2. https://doi.org/10.1016/S0377-1237(09)80135-4.
7. Lt. General (Retd). Katoch P.C. Exoskeleton for Armed Forces, 2025.
8. Lt Col (Retd) Tripathi N. Future Combat Soldiers: navigating the Exoskeleton Challenge, 2024.
9. Ralph H, Kellogg. La pression barométrique: Paul Bert's hypoxia theory and its critics. Resp Physiol. 1978;34(1):1–28. https://doi.org/10.1016/0034-5687(78)90046-4.
10. Ploszczyca K, Czuba M, Chalimoniuk M, Gajda R, Baranowski M. Red blood cell 2,3-Diphosphoglycerate decreases in response to a 30 km time trial under hypoxia in cyclists. Front Physiol. 2021;15(12):670977.
11. Luo Y, Wang Y, Lu H, Gao Y. 'Ome' on the range: update on high-altitude acclimatization/adaptation and disease. Mol Biosyst. 2014;10(11):2748–55.
12. Netzer NC, Rausch L, Eliasson AH, Gatterer H, Friess M, Burtscher M, Pramsohler S. SpO2 and heart rate during a real hike at altitude are significantly different than at its simulation in normobaric hypxia. Front Physiol. 2017;13:81.
13. Guyenet Patrice G, Bayliss Doughlas A. Neural control of breathing and CO2 homeostasis. Neurons. 2015;87:946–61. Cell Press.
14. Dunham-Snary KJ, Wu D, Sykes EA, Thakrar A, Parlow LRG, Mewburn JD, Parlow JL, Archer SL. Hyoxic pulmonary vasocontriction: from molecular mechanisms to medicine. Chest. 2017;151(1):181–92.
15. Bhutta BS, Alghoula F, Berim I. StatPerals [Internet]. Treasure Island: StatPearls Publishing; 2024. Available from: https://www.ncbi.nlm.nih.gov/books/NBK482316/.
16. King J, Lowery DR. StatPerals [Internet]. Treasure Island: StatPearls Publishing; 2023. Available from: https://www.ncbi.nlm.nih.gov/books/NBK470455/.
17. Berry CA, Shaywitz SE, Shaywitz BA. Girls with attention deficit disorder: a silent minority? A report on behavioral and cognitive characteristics. Pediatrics. 1985;76(5):801–9.
18. Liu J, Clermont AC, Gao BB, Feener EP. Intraocular hemorrhage causes retinal vascular dysfunction via plasma kallikrein. Invest Opthalmol Vis Sci. 2013;54(2):1086–94.
19. Badr MS, Dingell JD, Javaheri S. Central sleep apnea: a brief review. Curr Pulmonol Rep. 2019;8(1):14–21.
20. Patel S, Jose A, Mohiuddin SS. Physiology, oxygen transport and carbon dioxide dissociation curve. In: StatPearls [Internet]. Treasure Island: StatPearls Publishing; 2024. Available from: https://www.ncbi.nlm.nih.gov/books/NBK539815/.
21. Wei CY, Chen PN, Lin SS, Huang TW, Sun LC, Tseng CW, Lin KF. Using machine learning to determine the correlation between physiological and environmental parameters and the indiction of acute mountain sickness. BMC Bioinform. 2022a;31(22):628.
22. Wei X, Hou Y, Long M, Jiang L, Du Y. Molecular mechanisms underlying the role of hypoxia-indicible factor-1 α in metabolic reprogramming in renal fibrosis. Front Endocrinol (Lausanne). 2022b;25(13):927329.
23. Li L, Pan R, Li R, Niemann B, Aurich AC, Chen Y, Rohrbach S. Mitochondrial biogenesis and peroxisome proliferator-activated receptor-γ coactivator-1α (PGC-1α) deacetylation by physical activity: intact adipocytokine signaling is required. Diabetes. 2011;60(1):157–67.
24. Hardie DG. AMP-activated protein kinase: an energy sensor that regulates all aspects of cell function. Genes Dev. 2011;25(18):1895–908. https://www.faa.gov/air_traffic/publications/atpubs/aim_html/chap4_section_5.html. Accessed on 13th June 2024.

25. Khodaee M, Grothe HL, Seyfert JH, VanBaak K. Athletes at high altitude. Sport Health. 2016;8(2):126–32.
26. Alizadeh R, Ziaee V, Aghsaeifard Z, Mehrabi F, Ahmadinejad T. Characteristics of headache at altitude among trekkers; a comparison between acute mountain sickness and non-acute mountain sickness headache. Asian J Sport Med. 2012;3(2):126–30.
27. Meier D, Collect TH, Locatelli I, Cornuz J, Kayser B, Simel DL, Sartori C. Does this patient have acute mountain sickness?: the rational clinical examination systematic review. JAMA. 2017;318(18):1810–9.
28. Taylor AT. High-altitude illnesses: physiology, risk factors, prevention, and treatment. Rambam Maimonides Med J. 2011;2(1):e0022.
29. Imray C, Wright A, Subudhi A, Roach R. Acute mountain sickness: pathophysiology, prevention, and treatment. Prog Cardiovasc Dis. 2010;52(6):467–84.
30. Grissom CK, Roach RC, Sarnquist FH, Hackett PH. Acetazolamide in the treatment of acute mountain sickness: clinical efficacy and effect on gas exchange. Ann Intern Med. 1992;116(6):461–5.
31. Luks AM, Auerbach PS, Freer L. Wilderness medical society clinical practice guidelines for the prevention and treatment of acute altitude illness. Wilderness Environ Med. 2019;30(4):S3–S-18.
32. Bartsch P, Bailey DM, Berger MM, Knauth M, Baumgartner RW. Acute mountain sickness: controversies and advances. High Alt Med Biol. 2004;5(2):110–24.
33. Paralikar SJ. High altitude puloonary edema-clinical features, pathophysiology, prevention and treatment. Indian J Occup Environ Med. 2012;16(2):59–62.
34. Reis J, Giroud M, Kokubo Y. Environmental risk factors for stroke and cardiovascular disease. In: Encyclopedia of cardiovascular research and medicine. Elsevier; 2018. p. 238–47.
35. Michinaga S, Koyama Y. Pathogenesis of brain edema and investigation into anti-edema drugs. Int J Mol Sci. 2015;16(5):9949–75.
36. Hackett PH, Roach RC. High altitude Cerebral Edema. High Alt Med Biol [Internet]. 2004 May [cited 2025 Jun 30];5(2):136–46. Available from: http://www.liebertpub.com/doi/10.1089/1527029041352054
37. Bartsch B, Merki D, Hofstetter M, Maggiorini B, Kayser O, Oelz. Treatment of acute mountain sickness by simulated descent: a randomised controlled trial. BMJ. 1993;306(6885):1098–101. https://doi.org/10.1136/bmj.306.6885.1098
38. Christopher DP, Hackett. Advances in the prevention and treatment of high altitude illness. Emerg Med Clin N Am. 2017;35(2):241–60. https://doi.org/10.1016/j.emc.2017.01.002
39. Ferrazzini G, Maggiorini M, Kriemler S, Bartsch P, Oelz O. Successful treatment of acute mountain sickness with dexamethasone. Br Med J (Clin Res Ed). 1987;294(6584):1380–2.
40. Yadav RB, Yadav A, Pathak D, Arora R. Cold injury prevention and management in high altitude extreme environments: pharmacological and therapeutical interventions. Def Life Sci J. 2024;9(1):106–13.
41. Jayaswal R, Sivadas P, Mishra SS. Health and performance of military personnel in cold climatic environment of the western Himalayas. Med J Armed Forces India. 2001;57(4):322–5.
42. Gosling Clarke A. Military challenges from climate change. Contemp Issue Air Space Power. 2023;1(1):bp36471855.
43. John ML. Weather operations in the transformation era. 2003. https://apps.dtic.mil/sti/tr/pdf/ADA414781. Assessed on 14th June 2024.
44. Mekjavic IB, Norheim AJ, Friedl KE. Human performance and medical treatment during cold weather operations: synthesis of a symposium. Int J Circumpolar Health. 2023;82(1):2246666.
45. Chatterjee T, Bhattacharyya D, Pramanik A, Pal M, Majumdar D, Majumdar D. Soldier's load carriage performance in high mountains: a physiological study. Mil Med Res. 2017;4(6):2017.
46. Nair Y. https://www.usiofindia.org/publication-journal/terrain-as-a-force-multiplier-in-operational-planning-role-of-the-engineers.html. 2011. Accessed on 14th June 2024.
47. Nikbakhsh E, Farahani RZ. Humanitarian logistic planning in disaster relief operations. In: Logistics operations and management. Elsevier; 2011. p. 291–332.

48. Kurt Kurt G, Khoshkholgh MG, Alfattani S, Ibrahim A, Darwish TSJ, Alam MS, Yanikomeroglu H, Yongacoglu A. A vision and framework for the high altitude platform station (HAPS) network of the future. IEEE Commun Surv Tutor. 2021;23(2):729–79.
49. Di Vito P, Fischer D, Spada M, Rinaldo R, Duquerroy L. HAPs operations and service provision in critical scenarios. SpaceOps Conferences. Marseille, France; 2018. https://doi.org/10.2514/6.2018-2504
50. Mershad K, Dahrouj H, Sarieddeen H, Shihada B, Al-Naffouri T, Alouini MS. Cloud-enabled high-altitude platform systems: challenges and opportunities. Front Commun Netw. 2021;2:716265.
51. Biswas K. Military aviation principles. IntechOpen; 2020. Available from https://doi.org/10.5772/intechopen.87087
52. Mohsan SAH, Othman NQH, Li Y, Alsharif MH, Khan MA. Unmanned aerial vehicles (UAVs): practical aspects, applications, open challenges, security issues, and future trends. Intell Serv Robot. 2023;16:109–37.
53. Kyröläinen H, Pihlainen K, Vaara JP, Ojanen T, Santtila M. Optimising training adaptations and performance in military environment. J Sci Med Sport. 2018;21(11):1131–8.
54. Chen B, Wu Z, Huang X, Li Z, Wu Q, Chen Z. Effect of altitude training on the aerobic capacity of athletes: a systematic review and meta-analysis. Heliyon. 2023;9(9):e20188.
55. McFall S. Microwaves in military. Microw J. 2007;50:42–6.
56. Lazov L, Teirumnieks E, Ghalot RS. Applications of laser technology in army. J Defense Manag. 2021;11(4):1–8. https://doi.org/10.35248/2167-0374.21.11.210
57. Ahsan MM, Luna SA, Siddique Z. Machine-learning-based disease diagnosis: a comprehensive review. Healthcare (Basel). 2022;10(3):541.
58. Wang A, Zhang Q, Han Y, Megason S, Hormoz S, Mosaliganti KR, Lam JCK, Li VOK. A novel deep learning-based 3D cell segmentation framework for future image-based disease detection. Sci Rep. 2022a;12:342.
59. Wang Z, Li X, Yao M, Li J, Jiang Q, Yan B. A new detection model of microaneurysms based on improved FC-DenseNet. Sci Rep. 2022b;12(1):950.
60. Rana M, Bhushan M. Machine learning and deep learning approach for medical image analysis: diagnosis to detection. Multimed Tools Appl. 2023;82:26731–69.
61. Nie D, Cao X, Gao Y, Wang L, Shen D. Estimating CT image from MRI data using 3D fully convolutional networks. In: Deep learning and data lableling for medical applications. Springer; 2016. p. 170–8.
62. Xing W, Zhu Z, Dongni H, Yue Y, Dai F, Li Y, Tong L, Song Y, Ta D. CM-SegNet: a deep learning-based automatic segmentation approach for medical images by combining convolution and multilayer perceptron. Comput Biol Med. 2022;147:105797.
63. Yang J, Tu J, Zhang X, Yu S, Zheng X. TSE DeepLab: an efficient visual transformer for medicine image segmentation. Biomed Signal Process Control. 2023;80:104376.
64. Anantharaman R, Velazquez M, Lee Y. Utilizing mask R-CNN for detection and segmentation of oral diseases. In: IEEE International Conference on Bioinfomatics and Biomedicine (BIBM). Madrid: IEEE; 2018. p. 2197–204.
65. Chaio JY, Chen KY, Liao KYK, Hsieh PH, Zhang G, Huang TC. Detection and classification the breast tumors using mskr R-CNN on sonograms. Medicine. 2019;98(9):e15200.
66. Sandeep KN, Balaji B, Ramachandran. A review on smart wearable devices for soldier safety during battlefield using WSN technology. Mater Today Proc. 2020:334578–85. https://doi.org/10.1016/j.matpr.2020.08.191
67. Revathi NS, Ali V, Navya A, Navdeep G, Murugaperumal S, Krishnamoorthy. AI-driven predictive health monitoring and early warning systems for enhanced soldier safety in IoT-enabled wearable devices. E3S Web of Conferences 61903003. 2025; https://doi.org/10.1051/e3sconf/202561903003
68. De RL, Wen XY, Jullian Y, Karuppasamy S, Rajesh C. A deep learning approach to the screening of malaria infection: automated and rapid cell counting, object detection and instance segmentation using Mask R-CNN. Comput Med Imaging Graph. 2021;88:101845.

Further Reading

1. Rajesh AB, Singh KV, Peter. New age herbals resource quality and pharmacognosy DRDO Herbal Technologies: military and civil applications. Singapore: Springer Singapore. p. 359–408.
2. Richa RP, Geetha S, Lilly S, Ganju. A pharmacological investigation of Hippophae salicifolia (HS) and Hippophae rhamnoides turkestanica (HRT) against multiple stress (C-H-R): an experimental study using rat model. Cell Stress Chaperones. 2015;20(5):821–31. https://doi.org/10.1007/s12192-015-0603-2

Cardiopulmonary Responses at High Altitude (HA): Susceptibility to HA Maladies

2

Rajinder K. Gupta, Sachin Kumar, Abhishek Bharadwaj, and Pooja Chaudhry

Abstract

Cardiopulmonary responses at high altitude (HA) and acclimatisation have been studied in great detail; however, the pathophysiology of HA maladies like acute mountain sickness (AMS), high altitude pulmonary oedema (HAPO), HA cerebral oedema (HACO) and thromboembolic disorders is not clear. Rapid induction of healthy individuals to the same altitude (3500 m) produces heterogeneous symptoms as a majority suffer from AMS while a few suffer from HAPO and HACO which could be fatal if not treated promptly. This chapter explains how modulation of various cardiopulmonary responses at HA can lead to the development of hypoxia susceptibility and the occurrence of HA maladies which are primarily vascular. Previous studies, including ours, have shown that hypoxia susceptibility is associated with endothelium dysfunction and vascular remodelling, leading to reduced vascular compliance during hypoxia. B-type natriuretic peptide (BNP) is a known marker of heart failure but also indicates endothelial dysfunction and vascular remodelling at a lower level. Predicting hypoxia susceptibility is important, as the number of inductions to HA is increasing and limited medical facilities are available in difficult terrain. Currently, there is no marker for predicting HA maladies; however, a subclinical elevation of BNP (>15 pg/mL) has been established as a novel marker for predicting HA maladies.

Keywords

Acclimatisation · Hypoxic pulmonary vasoconstriction · High-altitude pulmonary oedema · B-type natriuretic peptide · Hypoxia susceptibility

R. K. Gupta (✉) · S. Kumar · A. Bharadwaj · P. Chaudhry
Defence Institute of Physiology and Allied Sciences (DIPAS), Defence Research and Development Organization (DRDO), Timarpur, Delhi, India
e-mail: rajindergupta@gov.in

S. Srivastava et al. (eds.), *Health and Sustenance in Extreme Environment*,
https://doi.org/10.1007/978-981-96-8642-1_2

2.1 Introduction

Human exposure to high altitude (HA) is common among tourists, mountaineers, sojourners, trekkers or as a part of army tenure posting. Low oxygen, cold, low humidity, UV radiation, etc. are common stressors encountered at HA. In order to combat hypoxia stress, various modifications and adjustments occur at systemic and cellular levels, called acclimatisation, which becomes important above 2500 m [1]. The cardiovascular and pulmonary system has an important role to play in sustaining life at HA by providing oxygen to peripheral tissues. Pulmonary vasculature constricts and systemic vasculature dilates in response to hypoxia. Fluid shifts from intravascular to extravascular space occurs within minutes of HA induction, leading to a rise in haematocrit, increasing the oxygen content of haemoglobin as a part of acclimatisation. One of the physiological responses to hypoxia is hypoxic pulmonary vasoconstriction (HPV). The magnitude of HPV determines the development of physiological (acclimatisation) or pathological (high-altitude pulmonary oedema—HAPO) conditions associated with either normal or exaggerated HPV responses, respectively [2, 3]. This chapter highlights various cardiopulmonary responses to hypoxia and how modulation of these responses can lead to susceptibility to hypoxia and HA maladies. The rate of ascent, altitude attained and individual susceptibility are the primary determinants for the development of HA maladies and individuals with a previous episode of HAPO (HAPO susceptible—HAPO-S) if ascend to more than 4500 m within 22 h have more than 60% prevalence of developing HAPO, while nifedipine prophylaxis, slow ascent and no past episode of HA maladies significantly reduce the prevalence of HA maladies [4]. Thus, susceptibility to hypoxia is associated with predisposing factors at subclinical levels, which precipitate as HA maladies under hypoxic stress. It is interesting that exposure to HA shows a heterogeneous response, while the majority remain asymptomatic, some develop acute mountain sickness (AMS), with a prevalence of 30% at 3500 m on rapid ascent, and very few develop HAPO and high-altitude cerebral oedema (HACO). Acute HA maladies like AMS, HAPO and HACO are hydrostatic oedema and the heterogeneous pattern of HA disease on acute exposure indicates the existence of mild, moderate and severe subclinical vascular pathology among hypoxia susceptible which precipitates under hypoxic stress [5]. Predicting hypoxia susceptibility is important as health facilities are limited in HA areas due to the difficult terrain and can decrease morbidity and mortality, especially when a large number of individuals were inducted at HA, like army troops for a longer duration of stay.

2.2 Effect of HA Exposure on Cardiovascular System

Hypoxia directly produces vasodilation of systemic arterioles via ATP-sensitive potassium channels while pulmonary vasoconstriction is produced by oxygen-sensitive potassium channels at Type 1 glomus cells in the carotid body. Glomus cells are also present in the aortic body (peripheral chemoreceptors) whose afferent activity rises hyperbolically with an increase in hypoxia, causing an increase in ventilatory and sympathetic activity demonstrated by a rise in urinary and plasma concentration of catecholamines and skeletal muscle sympathetic activity [6]. Acute hypoxia stimulates respiration, leading to carbon dioxide washout causing inhibition of medullary chemoreceptors via respiratory alkalosis during the initial few days of HA induction, while prolonged hypoxia causes increased sensitivity of the peripheral chemoreceptors, augments the ventilatory drive leading to ventilatory acclimatisation. Increased sympathetic activity documented by microneurography at extreme altitudes and elevated catecholamines in urine and plasma indicates severe hypoxia causes anorexia and weight loss on prolonged stay in healthy individuals [7]. The direct effects of hypoxia on systemic resistance vessels and those mediated by the carotid body in both the systemic and pulmonary circulation were antagonistic. During the initial few hours of HA stay, hypoxic vasodilation tends to oppose sympathetic vasoconstriction in the systemic circulation, causing no effect on systemic blood pressure; however, prolonged hypoxia exposure at extreme altitude causes a rise in blood pressure and systemic peripheral resistance. A rise in blood pressure is a part of acclimatisation at extreme altitudes and is not reverted by oxygen inhalation [8], alpha or beta blockers [9], suggestive of structural changes. In contrast, HAPO-S showed an exaggerated systemic and pulmonary vasoconstriction on hypoxia exposure at sea level or at HA [10]. Reduced vascular compliance in response to hypoxia is a pathogonomic feature of hypoxia susceptibility [5].

Stroke volume reduces during the first few weeks of HA stay as a consequence of diuresis and reduction in plasma volume, which decreases by nearly 20% at 4000 m, causing a reduction in left ventricular dimensions and filling pressure [11]. A reduction in plasma volume with the passage of time during a high-altitude stay is different since the initial reduction of plasma volume in part may be mediated by chemoreceptors, atrial natriuretic peptide and a decrease in aldosterone synthesis [12], whereas the further reduction of plasma volume occurs without a net loss of body water by a fluid shift from extracellular to intracellular space.

Cardiac contractility and submaximal cardiac output for a given oxygen uptake are increased initially at HA. Above an altitude of 1500 m, maximum cardiac output remains unchanged and maximum oxygen consumption declines by 1% per 100 m [13]. Cardiac output increases initially at HA induction due to a rise in HR and may be offset by reduced stroke volume. Stroke volume falls during the initial few days at HA and then tends to stabilise [14]. Prolonged HA stay causes a fall in maximum cardiac output and the maximum rate of oxygen consumption remains low [15].

Acute HA exposure causes an increase in heart rate (both resting and exercise), myocardial contractility and cardiac output. Hypoxia-induced increase in heart rate (HR) is related to increased sympathetic activity and vagal withdrawal. Acclimatisation-mediated enhanced parasympathetic neural activity at HA causes a lowering of the heart rate during submaximal exercise, while a reduction in cardiac output on prolonged stay causes a decreased maximum work capacity in order to conserve limited oxygen at HA. HR that can be achieved at maximum exercise at sea level is reduced in parallel with reduced maximum oxygen consumption at HA [16]; however, HR is greater at altitude for a given level of exercise. Glycopyrrolate-mediated vagal blockage completely restores the maximum heart rate to sea level values, whereas cardiac output did not increase at HA even while raising the inspiratory partial pressure of oxygen to sea level values. Prolonged stay at HA causes a decrease in oxygen consumption thus reduces work load on left ventricle but causes increase in work load on right ventricle due to pulmonary hypertension thus acclimatisation gradually increase the physical endurance while susceptible shows an exaggerated response causing left ventricular diastolic dysfunction [17]. Left ventricular systolic function was not affected even during severe exercise in marathon runners, although pulmonary hypertension associated with transient right ventricular dysfunction can occur at extreme altitudes [18]. Myocardial oxygen supply and left ventricular function are maintained in healthy individuals during maximal exercise at extreme altitude, while maximal heart rate is reduced by 20%, and cardiac output as well as exercise capacity are reduced by 40–50% [19]. Left ventricular diastolic volume reduces as a result of right ventricular dilation in response to increased pulmonary vascular resistance in HAPO-S [20]. Left ventricular diastolic dysfunction is overcome by enhanced atrial contraction during acute hypoxia in healthy individuals, whereas chronic hypoxia may cause diastolic dysfunction of the left ventricle in HAPO-S [20].

2.3 Effect of HA Exposure and Exercise on Pulmonary Vascular Response

The basic response of pulmonary vasculature to hypoxia is an increase in pulmonary artery resistance. The magnitude of hypoxic pulmonary vasoconstriction (HPV) is variable and response increases with susceptibility to HA maladies. Lung perfusion scan with a fluorescent microsphere showed that HPV is normally heterogeneous due to heterogeneous distribution of smooth muscle around the pulmonary arteriole [21]. Heterogeneity increases with the magnitude of HPV in HAPO-S [22]. Pulmonary haemodynamic response during exercise in HAPO-S showed a higher pulmonary artery pressure (Ppa) as well as elevated pulmonary artery occlusion pressure [23]. The elevated pulmonary artery occlusion pressure (PAOP) is responsible for a 50% increase in exercise-related increase in Ppa. In contrast, maximal Ppa and PAOP values for the control group remained lower than HAPE-S [23]. Greater baseline pulmonary vascular resistance in HAPO-S, greater pulmonary pressure reactivity to hypoxia and greater PAOP reactivity to exercise are due to vascular modelling, i.e. residual structural changes in response to elevated HIF 1 alpha [3]. The baseline elevated Ppa and exaggerated rise in Ppa in response to hypoxia and exercise is due to vascular remodelling and is a pathogonomic feature of HAPE-S [24]. Heavy exercise at HA causes a rise in Ppa and PAOP in both control and HAPE-S; however, maximum Ppa and PAOP values were low in the control group compared to HAPO-S. Ppa rise alone cannot increase the pulmonary capillary pressure, but a simultaneous rise in Ppa and PAOP can result in augmented pulmonary capillary permeability with enhanced fluid movement. PAOP rises rapidly once the mean Ppa > 30 mmHg; however, in control Ppa never exceeds 35 mmHg during heavy exercise at HA, so the mean pulmonary capillary pressure does not rise beyond 20 mmHg (threshold for the development of HAPO), in contrast to HAPO-S. The rise in Ppa and PAOP is high enough during severe exercise in control to cause reversible interstitial pulmonary oedema [23].

2.4 Factors Predisposing to HA Maladies

HA maladies all acute (AMS, HAPO, HACO), subacute mountain sickness and chronic maladies are associated with excessively elevated Ppa and thus primarily vascular. HAPO is a noncardiogenic hydrostatic oedema which occurs at an altitude of >2500 m [1]. As PaO_2 falls below 60 mmHg above 2500-m, acclimatisation mechanisms become important while susceptible fail to acclimatise and scum to HA illness. Hypoxia susceptibility is associated with reduced vascular compliance during hypoxia and exercise [5]. Exaggerated increase in sympathetic activity with reduced vascular compliance leads to congested cerebral blood vessels on HA exposure in hypoxia susceptible and can manifest as headache, nausea and vomiting features of AMS. In severe cases, remodelling of pulmonary arterioles and veineoles due to cellular hypoxia resulted in a simultaneous rise in Ppa and PAOP, leading to a pulmonary capillary rupture in those HAPO-S who developed HAPO [25].

There is considerable overlap in Ppa between HAPO-S and control individuals at HA, suggesting the role of mechanisms other than Ppa contribute to edema formation in the lungs. A past study showed that an acute increase in Ppa was associated with early diastolic septal flattening with a consecutive LV shape change that appeared responsible for prolonged isovolumic relaxation showed a positive correlation between a rise in Ppa and LV diastolic dysfunction. Also, during heavy exercise at HA, the rising Ppa and right ventricular volume could further worsen LV diastolic dysfunction and lead to elevated PAOP observed in HAPO-S and control [20]. Clinical conditions which aggravate the rise in PAOP during exercise at HA are pulmonary AV malformations and patent foramen ovale predispose to HA maladies.

Patent foramen ovale (PFO) is present in 25% of the general population and is four times more frequent in HAPO-S [26]. A high prevalence of PFO has been reported in the disease state known to be associated with pulmonary hypertension, such as COPD and sleep apnoea syndrome, suggesting that some foramen may reopen under conditions associated with elevated pulmonary vascular resistance and an increase in right-sided cardiac pressure. A reopened foramen oval may contribute to systemic arterial desaturation from right to left intracardiac shunting under any condition that results in high PAOP which occurs during exercise at HA. Therefore, a vicious cycle may be operated at HA in HAPO-S with PFO where hypoxic pulmonary hypertension will initiate right to left shunting, further worsening hypoxia and rise in PAOP.

Prolonged stay at HA can cause a rise in protoporphyrin indicating iron deficiency [27]. Individuals with a lower serum iron level showed an exaggerated rise in Ppa in response to hypoxia as iron is required for proteosomal degradation of HIF 1 alpha. Iron deficiency has been identified as a risk factor for ischaemic strokes in patients with pulmonary AV malformations [28]. Interestingly, strokes occurring at

HA are ischaemic in origin and not associated with risk factors. Long-term stay at HA is associated with a high prevalence of stroke 13.7/1000 compared to sea level. Also, various other risk factors for HAPO like reduced HbA fraction [3] and forced vital capacity [29] are associated with decreased blood oxygen levels, leading to vascular remodelling and hypoxia susceptibility.

2.5 Heterogeneous Pattern of HA Maladies and Vascular Remodelling

AMS is generally associated with headache, anorexia, nausea, vomiting, dizziness and fatigue. The prevalence of AMS on a passive ascent from 3000 to 3500 m is 25–40%, and when an unacclimatised individual ascends more than 500 m/day to an altitude of 4500–6000 m, the incidence of AMS can vary from 40% to 90% [30]. In most cases, symptoms resolve spontaneously after 18–36 h without requiring a descent to low altitude. The incidence of HACO in the general population below 2500 m is <0.01%, but between 4200 and 5500 m, it is 0.5–1% [31].

Slow ascent in >3 days to an altitude of 4000–5000 m has shown a lower prevalence of HAPO compared to 7% on rapid ascent <1 day to the same altitude among mountaineers. Army troops showed a significant increase in HAPO incidence (2.5% vs 15.5%) when an altitude of >5000 m was climbed by trekking over 4–6 days as opposed to air dropped. Also, the prevalence of HAPO rose to >60% in HAPO-S on rapid ascent (climbing 4559 m within <24 h). Up to 50% of HAPO patients have concomitant AMS and up to 14% have concomitant HACO.

A direct correlation has been found between symptom severity and cerebral blood flow velocity in AMS cases. Increased intracranial pressure has been found in moderate-to-severe AMS. MRI showed increased extracellular fluid, leading to an increase in brain volume, i.e. cytotoxic edema and no evidence of vasogenic oedema [32], while increased lumber cerebrospinal fluid pressure, truncal ataxia, decreased consciousness and severe AMS is the clinical hallmark of HACO. SaO_2 and PaO_2 are lower than AMS in HACO cases and MRI showed a leakage in the blood–brain barrier as evidenced by hemosiderin persisting over years in the corpus callosum [33].

Early symptoms of HAPO include exertional dyspnoea, cough and reduced exercise performance followed by severe cough, breathlessness at rest and orthopnoea as the disease progress. Pulmonary hypertension precedes oedema formation in HAPO and dyspnoea on exertion (DOE) is common in both HAPO and pulmonary hypertension, while reduction of Ppa and PAOP caused significant relief in DOE by attenuating J receptor stimulation [34]. A rise in the pulmonary capillary pressure stimulates J receptor which explains the intense fatigueness on exertion in HAPO patients due to inhibition of stretch reflex [34]. Diagnosis of HAPE is clinical and often supported by radiographic evidence showing a patchy, peripheral distribution of oedema on the chest radiograph [35]. Elevated red cell count and protein concentration in the lavage fluid is a pathogonomic feature found within a day of ascent to extreme altitude with no evidence of inflammatory cells and pro-inflammatory mediators in diagnosed HAPO cases [36]. Leakage of fluid in HAPO is debated due to a difference of opinion on whether it occurs due to ruptures of basement membranes [35], or non-traumatic, dynamic, pressure-sensitive stretching of pores or opening of fenestrae [37]. Studies have shown that post-pulmonary capillary resistance was elevated in HAPO-S with HAPO pointing towards a rupture of the basement membrane [38]. Heterogeneous arteriolar vasoconstriction resulting in increased capillary pressure in areas with high blood flow has been suggested in a previous study [39]. Similar findings were reported from animal studies using fluorescent microbeads [21] and from human lung perfusion analysis by magnetic resonance imaging [22]. Hypoxic venoconstriction could account for increased capillary pressure in HAPO, but this mechanism could not explain the often patchy radiographic appearance of early HAPO on lung imaging unless there is regional heterogeneity of hypoxic venoconstriction [40]. However, due to the heterogeneous distribution of smooth muscle in pulmonary arterioles and veineoles, vascular remodelling determines the pulmonary vascular haemodynamic response to slow or rapid ascent to HA and explains the lack of similarity of radiographic appearance between two separate episodes of HAPO in the same HAPO-S individual. Right heart catheterisation studies have shown beyond doubt that post-pulmonary capillary resistance in HAPO-S with HAPO was elevated but not in HAPO-S without HAPO at 4559 m [38].

2.6 BNP as a Predictive Marker for Hypoxia Susceptibility

HA maladies are subclinical vascular pathology which manifest under severe hypoxic stress. Acute HA maladies like AMS, HAPO and HACO are hydrostatic Oedema associated with endothelial dysfunction. Heterogeneous symptoms in acute HA maladies ranging from mild headaches in AMS to severe pulmonary or cerebral oedema are a consequence of vascular remodelling. Endothelial dysfunction and vascular remodelling occur in response to cellular hypoxia mediated by HIF1alpha, causing a baseline high pulmonary artery pressure (Ppa) and exaggerated rise in Ppa in response to hypoxia as Ppa fall is not seen during normoxia at HA [24, 25]. Elevated HIF1 alpha under normoxia has been shown in HAPE-S to cause subclinical elevation of BNP (>15 pg/ml) indicative of hypoxia susceptibility [5, 24]. Endothelial dysfunction has been implicated in HAPE susceptibility as a baseline and HA levels of endothelin-1 (ET-1) were elevated with reduced endothelial nitric oxide levels in HAPO-S mountaineers [41]. HIF 1 alpha acts on the promotor site of the BNP gene and the highest BNP level for a healthy individual was 18.4 pg/mL. BNP is a marker of vascular remodelling and individuals who did not develop HAPO upon repeated exposure to HA had normal PVR (Ppa < 38 mmHg at Fio2 0.12) and have BNP $\leq$ 15 pg/mL [5, 25]. Endothelial function is very precisely regulated by a large number of biological pathways; however, BNP has been found to be an independent predictor of endothelial dysfunction [42]. Individuals with healthy endothelial function (BNP $\leq$ 15 pg/mL) when exposed to extreme altitudes >5000 m were normal compared to individuals with subclinical elevation in baseline BNP (>15 pg/mL) level who suffered severe AMS at same altitude [43]. Studies have shown no change in BNP levels on acute severe hypoxia exposure ($SpO_2 = 62.3\%$) in hypoxia-resistant individuals (BNP $\leq$ 15 pg/mL), while hypoxia-susceptible individuals showed a rise in BNP levels [5]. Stabilisation of HIF 1 alpha has also been implicated in development of micro-vasculopathy, a characteristic feature of lifestyle disease, and a rise in BNP post-hypoxia (Fio2 0.15) can screen subclinical lifestyle disease, thus having the potential to be a marker for increasing life expectancy [44]. Moreover, vascular remodelling has also been associated with heterogeneous phenotype seen in COVID-19 patients presented as asymptomatic, silent hypoxia (mild-to-moderate disease) to lesser cases of acute respiratory distress syndrome (severe disease). Individuals positive for COVID-19 with lower BNP responded to treatment compared to the worst prognosis in those with BNP > 100 pg/mL at the time of admission [45]. COVID-19 patients with solid organ transplants on immunosuppressant calcineurin inhibitors (HIF 1 alpha inhibitor) did not suffer severe disease, suggesting endothelial dysfunction might be the primary pathophysiology of COVID-19 disease [46].

2.7 Conclusion

Reduced vascular compliance during hypoxia is a feature of hypoxia susceptibility. Subclinical elevated BNP (>15 pg/mL) indicates endothelial dysfunction contributing to hypoxia susceptibility; thus, prior estimation of BNP is an early and useful step towards personal protection before going to HA areas.

Acknowledgement The authors would like to express deep gratitude to Dr. M. P. K. Reddy for his support in writing the chapter and to Sri Jagbir Singh for his efforts while the field studies were undergoing on hypoxia susceptibility.

References

1. Luks AM, Swenson ER, Bärtsch P. Acute high-altitude sickness. Eur Respir Rev. 2017;26:160096. https://doi.org/10.1183/16000617.0096-2016.
2. Grover RF, Johnson RL, McCullough RG, McCullough RE, Hofmeister SE, Campbell WB, Reynolds RC. Pulmonary hypertension and pulmonary vascular reactivity in beagles at high altitude. J Appl Physiol. 1988;65:2632–40. https://doi.org/10.1152/jappl.1988.65.6.2632.
3. Soree P, Gupta RK, Singh K, Desiraju K, Agrawal A, Vats P, Bharadwaj A, Baburaj TP, Chaudhary P, Singh VK, Verma S, Bajaj AC, Singh SB. Raised HIF1α during normoxia in high altitude pulmonary edema susceptible non-mountaineers. Sci Rep. 2016;6:26468. https://doi.org/10.1038/srep26468.
4. Bartsch P, Gibbs JSR. Effect of altitude on the heart and the lungs. Circulation. 2007;116:2191–202. https://doi.org/10.1161/CIRCULATIONAHA.106.650796.
5. Khatri R, Gupta RK, Vats P, Bansal V, Yadav AK, Reddy PK, Bharadwaj A, Chaudhary P, Sharma S, Bajaj AC, Deskit P, Dass D, Baburaj TP, Singh SB, Kumar B. Subclinical elevated B-type Natriuretic Peptide (BNP) indicates endothelial dysfunction contributing to hypoxia susceptibility in healthy individuals. Life Sci. 2020;260:118408. https://doi.org/10.1016/j.lfs.2020.118408.

6. Mazzeo RS, Bender PR, Brooks GA, Butterfield GE, Groves BM, Sutton JR, Wolfel EE, Reeves JT. Arterial catecholamine responses during exercise with acute and chronic high-altitude exposure. Am J Phys. 1991;261:E419–24.
7. Hansen J, Sander M. Sympathetic neural overactivity in healthy humans after prolonged exposure to hypobaric hypoxia. J Physiol. 2003;546:921–9.
8. Bender PR, Groves BM, McCullough RE. Oxygen transport to exercising leg in chronic hypoxia. J Appl Physiol. 1988;65:2592–7.
9. Wolfel EE, Selland MA, Cymerman A, Brooks GA, Butterfield GE, Mazzeo RS, Grover RF, Reeves JT. O_2 extraction maintains O_2 uptake during submaximal exercise with beta-adrenergic blockade at 4,300 m. J Appl Physiol. 1998;85:1092–102.
10. Dehnert C, Mereles D, Greiner S, Albers D, Scheurlen F, Zügel S, Böhm T, Vock P, Maggiorini M, Grünig E, Bärtsch P. Exaggerated hypoxic pulmonary vasoconstriction without susceptibility to high altitude pulmonary edema. High Alt Med Biol. 2015;16:11–7.
11. Grover RF, Bärtsch P. Blood. In: Hornbein TF, Schoene RB, editors. High altitude: an exploration of human adaptation, vol. 161. New York: Marcel Dekker Inc; 2001. p. 493–523.
12. Swenson ER. Renal function and fluid homeostasis. In: Hornbein TF, Schoene RB, editors. High altitude: an exploration of human adaptation, vol. 161. New York: Marcel Dekker Inc; 2001. p. 525–68.
13. Wagner PD, Gale GE, Moon RE, Torre-Bueno JR, Stolp BW, Saltzman HA. Pulmonary gas exchange in humans exercising at sea level and simulated altitude. J Appl Physiol. 1986;61:260–70.
14. Klausen K. Cardiac output in man at rest and work during and after acclimatization to 3800 m. J Appl Physiol. 1966;21:609–16.
15. Calbet JA. Chronic hypoxia increases blood pressure and noradrenaline spillover in healthy humans. J Physiol. 2003;551:379–86.
16. Reeves JT, Groves BM, Sutton JR, Wagner PD, Cymerman A, Mal-conian MK, Rock PB, Young PM, Houston CS. Operation Everest II: preservation of cardiac function at extreme altitude. J Appl Physiol. 1987;63:531–9.
17. Boushel R, Calbet JA, Rådegran G, Sondergaard H, Wagner PD, Saltin B. Parasympathetic neural activity accounts for the lowering of exercise heart rate at high altitude. Circulation. 2001;104:1785–91.
18. Dávila-Román VG, Guest TM, Tutcur PG, Rowe WJ, Ladenson JH, Jaffe AS. Transient right but not left ventricular dysfunction after strenuous exercise at high altitude. J Am Coll Cardiol. 1997;30:468–73.
19. Sutton JR, Reeves JT, Wagner PD, Groves BM, Cymerman A, Malconian MK, Rock PB, Young PM, Walter SD, Houston CS. Operation Everest II: oxygen transport during exercise at extreme simulated altitude. J Appl Physiol. 1988;64:1309–21.
20. Allemann Y, Rotter M, Hutter D, Lipp E, Sartori C, Scherrer U, Seiler C. Impact of acute hypoxic pulmonary hypertension on LV diastolic function in healthy mountaineers at high altitude. Am J Physiol Heart Circ Physiol. 2004;286:H856–62.
21. Hlastala MP, Lamm WJE, Karp A, Polissar NL, Starr IR, Glenny RW. Spatial distribution of hypoxic pulmonary vasoconstriction in the supine pig. J Appl Physiol. 2004;96:1589–99.
22. Hopkins SR, Garg J, Bolar DS, Balouch J, Levin DL. Pulmonary blood flow heterogeneity during hypoxia and high altitude pulmonary edema. Am J Respir Crit Care Med. 2005;171:83–7.
23. Eldridge MW, Podolsky A, Richardson RS, Johnson DH, Knight DR, Johnson EC, Hopkins SR, Michimata H, Grassi B, Feiner J, Kurdak SS, Bickler PE, Wagner PD, Severinghaus JW. Pulmonary hemodynamic response to exercise in subjects with prior high-altitude pulmonary edema. J Appl Physiol. 1996;81:911–21.
24. Gupta RK, Himashree G, Singh K, Soree P, Desiraju K, Agrawal A, Ghosh D, Dass D, Reddy PK, Panjwani U, Singh SB. Elevated pulmonary artery pressure and brain natriuretic peptide in high altitude pulmonary edema susceptible non-mountaineers. Sci Rep. 2016;6:21357. https://doi.org/10.1038/srep21357.
25. Maggiorini M. High altitude-induced pulmonary oedema. Cardiovasc Res. 2006;72:41–50. https://doi.org/10.1016/j.cardiores.2006.07.004.

26. Allemann Y, Hutter D, Lipp E, Sartori C, Duplain H, Egli M, Cook S, Scherrer U, Seiler C. Patent foramen ovale and high-altitude pulmonary edema. JAMA. 2006;296:2954. https://doi.org/10.1001/jama.296.24.2954.
27. Reynafarje C, Lozano R, Valdivieso J. The Polycythemia of high altitudes: iron metabolism and related aspects Metabolism. Blood. 1959;14(4):433–55. https://doi.org/10.1182/blood.V14.4.433.433.
28. Shovlin CL. Iron deficiency, ischaemic strokes, and right-to-left shunts: from pulmonary arteriovenous malformations to patent foramen ovale? Intractable Rare Dis Res. 2014;3(2):60–4.
29. Gupta RK, Soree P, Desiraju K, Agrawal A, Singh SB. Subclinical pulmonary dysfunction contributes to high altitude pulmonary edema susceptibility in healthy non-mountaineers. Sci Rep. 2017;7:14892. https://doi.org/10.1038/s41598-017-14947-z.
30. Schneider M, Bernasch D, Weymann J, et al. Acute mountain sickness: influence of susceptibility, preexposure, and ascent rate. Med Sci Sports Exerc. 2002;34:1886–91.
31. Bartsch P, Roach RC. Acute mountain sickness and high-altitude pulmonary edema. High altitude: an exploration in human adaptation. In: Hornbein TF, Schoene RB, editors. Lung biology in health and disease, vol. 161. New York: Marcel Dekker; 2001. p. 731–76.
32. Sagoo RS, Hutchinson CE, Wright A. Magnetic resonance investigation into the mechanisms involved in the development of high-altitude cerebral edema. J Cereb Blood Flow Metab. 2016; https://doi.org/10.1177/0271678X15625350.
33. Schoch HJ, Fischer S, Marti HH. Hypoxia-induced vascular endothelial growth factor expression causes vascular leakage in the brain. Brain. 2002;125:2549–57.
34. Anand A, et al. Early-symptom relief a er valvulotomy in mitral stenosis indicates role of lobeline-sensitive intrapulmonary receptors. Respir Physiol Neurobiol. 2009;169:297–302.
35. West JB, Tsukimoto K, Mathieu-Costello O, Prediletto R. Stress failure in pulmonary capillaries. J Appl Physiol. 1991;70:1731–42.
36. Swenson ER, Maggiorini M, Mongovin S, et al. Pathogenesis of high-altitude pulmonary edema: inflammation is not an etiologic factor. JAMA. 2002;287:2228–35.
37. Neal CR, Michel CC. Openings in frog microvascular endothelium induced by high intravascular pressures. J Physiol. 1996;492:39–52.
38. Maggiorini M, Christian M, Pierre S, Pfeiffer F, Greve I, Sartori C, Lepori M, Hauser M, Scherrer U, Naeije R. High-altitude pulmonary edema is initially caused by an increase in capillary pressure. Circulation. 2001;103:2078–83.
39. Hultgren HN. High altitude pulmonary edema. In: Staub NC, editor. Lung water and solute exchange. New York: Marcel Dekker; 1978. p. 437–64.
40. Zhao Y, Packer CS, Rhoades RA. Pulmonary vein contracts in response to hypoxia. Am J Phys. 1993;265:L87–L92.C.
41. Sartori VL, Löffler BM, Delabays A, Nicod P, Bärtsch P, Scherrer U. Exaggerated endothelin release in high-altitude pulmonary edema. Circulation. 1999;99:2665–8. https://doi.org/10.1161/01.CIR.99.20.2665.
42. Pauriah M, Khan F, Lim TK, Elder DH, Godfrey V, Kennedy G, Belch JJF, Booth NA, Struthers AD, Lang CC. B-type natriuretic peptide is an independent predictor of endothelial function in man. Clin Sci. 2012;123:307–12. https://doi.org/10.1042/CS20110168.
43. Woods DR, Begley J, Stacey M, Smith C, Boos CJ, Hooper T, Hawkins A, Hodkinson P, Green N, Mellor A. Severe acute mountain sickness, brain na- triuretic peptide and NT-proBNP in humans. Acta Physiol. 2012;205:349–55. https://doi.org/10.1111/j.1748-1716.2012.02407.x.
44. Khatri R, Gupta R, Vats P, Bansal V, Yadav AK, Reddy PK, Bharadwaj A, Chaudhary P, Sharma S, Bajaj AC, Singh SB, Varshney R. Hypoxic pulmonary vascular response can screen subclinical lifestyle disease in healthy population. Microvasc Res. 2023;145:104454. https://doi.org/10.1016/j.mvr.2022.104454.

45. Liu Y, Liu D, Song H, Chen C, Lv M, Pei X, Hu Z, Qin Z, Li J. Clinical features and outcomes of 2019 novel coronavirus-infected patients with high plasma BNP levels. MedRxiv. 2020; https://doi.org/10.1101/2020.03.31.20047142.
46. Hage R, Steinack C, Benden C, Schuurmans MM. COVID-19 in patients with solid organ transplantation: a systematic review. Transplantology. 2020;1:1–15. https://doi.org/10.3390/TRANSPLANTOLOGY1010001.

Cardiopulmonary Diseases and High Altitude: Unresolved Issues

3

Dolly Sharma, Anjali Yadav, Anu Kumari, Chhavi Rai, Rajeev Varshney, and Ramesh C. Meena

Abstract

Medical sciences have always shown keen interest in studying the interaction between cardiopulmonary health and high-altitude environments. One such condition that arises under extreme environmental conditions, significantly affecting human cardiovascular and pulmonary systems, is hypobaric hypoxia. At high altitudes, individuals undergo physiological acclimatization, such as increased ventilation, shifts in fluid regulation, elevated heart rates, and vascular adaptations. These conditions result in hypertension, high-altitude pulmonary edema, ischemic heart disease, and heart failure. The present chapter aims to explore unresolved issues concerning the interface between cardiopulmonary diseases and high altitude. The primary focus areas include environmental variables faced at high altitude, cardiopulmonary pathophysiology, possible influencing factors, and treatments available for managing these health conditions that ultimately result in hypobaric hypoxia-induced illnesses. Through filling these knowledge voids, prospective practitioners can gain a deeper understanding of how to reduce the risks associated with high-altitude exposure in susceptible demographics, ultimately improving their quality of life during altitude-related pursuits.

Keywords

High-altitude physiology · Hypobaric hypoxic stress · Cardiopulmonary response · Cardiovascular diseases · Vascular remodeling

D. Sharma · A. Yadav · A. Kumari · C. Rai · R. Varshney · R. C. Meena (✉)
Defence Institute of Physiology and Allied Sciences (DIPAS), Defence Research and Development Organization (DRDO), Timarpur, Delhi, India
e-mail: rcmeena.dipas@gov.in

S. Srivastava et al. (eds.), *Health and Sustenance in Extreme Environment*,
https://doi.org/10.1007/978-981-96-8642-1_3

3.1 Introduction

The relationship between cardiopulmonary diseases and high altitude (HA) presents a complex and multifaceted challenge in research and medical sciences. The HA environments, defined as regions above 2500 m, impose unique physiological strains on the human body, predominantly affecting the cardiovascular and respiratory systems. For individuals with preexisting cardiopulmonary conditions, these stressors can exacerbate their symptoms and complicate disease management. Despite substantial progress in understanding the overall effects of HA on healthy individuals, numerous unresolved questions persist regarding its impact on people with cardiopulmonary diseases. The reduced partial pressure of oxygen at HA leads to hypobaric hypoxia, triggering a cascade of physiological responses to optimize oxygen delivery and its proper utilization. However, these adaptive mechanisms may be insufficient in patients with cardiopulmonary diseases, such as chronic obstructive pulmonary disease (COPD), pulmonary hypertension, and heart failure. It can also lead to severe health implications, ranging from exacerbated symptoms to life-threatening complications.

3.1.1 Human Physiology of Cardiopulmonary Systems and High Altitude

The beauty and adventurous opportunities of the mountains attract millions of visitors to different HA destinations worldwide. In addition, thousands of climbers, trekkers, and skiers worldwide climb them at an ascent rate that exceeds an individual's ability to acclimatize. Such a rapid climb places non-acclimatized traveler at risk of developing high-altitude illness (HAI). The average ambient temperature at HA is reduced by 2 °C for every 300 m rise in altitude. At any given hour, the "physiologic" temperature could drop 20 °C with changes in solar load and wind. In a relatively short time, a person could easily go from a sweating situation with several layers of clothing to one in which there could be a freezing injury. This harsh and fluctuating environment is further complicated by atmospheric conditions specific to different altitudes. Within the troposphere, barometric pressure falls as altitude (vertical height above sea level) increases. Since the concentration of oxygen in the air remains constant, the barometric pressure decreases in proportion to the oxygen pressure. This condition is referred to as "hypobaric hypoxia" (HH). For instance, the partial pressure of oxygen at the altitude of Everest Base Camp (5300 m altitude) is about one-half of the sea level. The peak of Mount Everest is the highest point above sea level on the Earth's surface, at 8850 m altitude, and has a partial pressure of oxygen about one-third of the sea-level values [1].

For environmental triggers/stressors, there are several adaptive mechanisms and responses that exist in the human body. Respiratory adaptations include increased ventilation to enhance oxygen uptake while alveolar gas exchange efficiency decreases, leading to increased pulmonary blood flow. These adaptations are essential for maintaining adequate oxygen levels in the blood, although they can cause respiratory alkalosis, where reduced carbon dioxide levels increase blood pH. Building on these respiratory changes, cardiovascular responses to HA involve an initial increase in cardiac output and heart rate, ensuring sufficient oxygen delivery to tissues. However, chronic exposure to HA can lead to pulmonary hypertension, resulting from hypoxic vasoconstriction of the pulmonary arteries. This condition places additional strain on the right ventricle, potentially leading to right heart failure (Fig. 3.1) if not managed properly. In addition to cardiovascular changes, hematological adaptations at high altitudes include increased erythropoiesis, stimulated by hypoxia-inducible factors. This process boosts the production of red blood cells (RBCs) and oxygen-carrying capacity but raises blood viscosity, which can pose cardiovascular risks such as thrombosis and stroke. These alterations are vital for long-term adaptations at HA but require careful monitoring, especially in individuals with preexisting/predisposed conditions.

Beyond the hematological adaptations, cellular and metabolic responses also play a crucial role. Mitochondria become more efficient, enhancing the capacity for oxygen utilization and anaerobic metabolism. Additionally, the kidneys make renal adjustments to excrete bicarbonate, compensating for respiratory alkalosis and stabilizing blood pH. These adaptations help to maintain proper cellular functions and energy production despite the challenging extreme environmental conditions. These physiological deviations are particularly significant for people with preexisting conditions such as COPD, pulmonary hypertension, and cardiovascular diseases. The decreased oxygen availability and increased physiological demands complicated

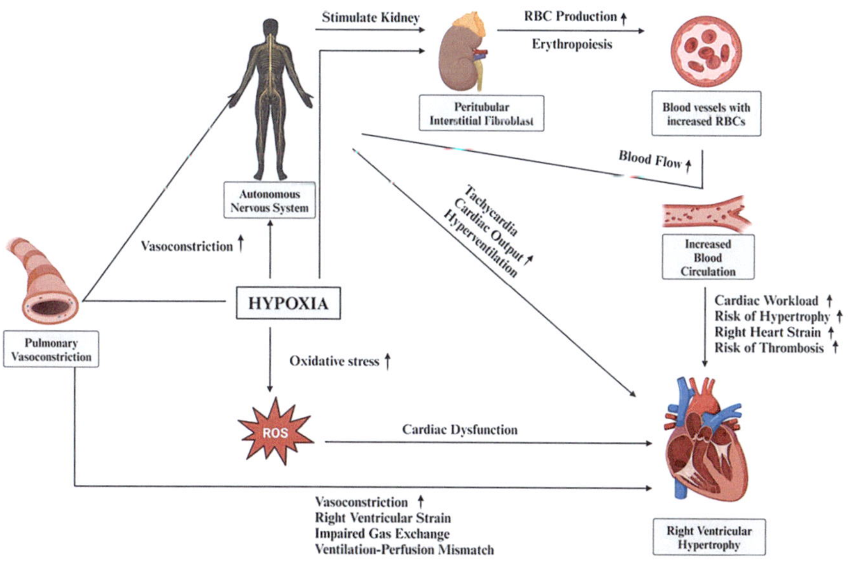

Fig. 3.1 The impact of hypoxia on cardiopulmonary dysfunction with key pathways including the production of ROS, an increase in red blood cell (RBC) count, blood flow changes, vasoconstriction, and the involvement of the autonomic nervous system

disease management, necessitating tailored medical approaches and gradual acclimatization to mitigate risks. Supplemental oxygen and specific medications may be required to support volunteers during ascent and/or stay at HA locations and conditions.

Considering the significant impact of physiological adaptations on individuals' well-being, the quality and quantity of sleep at altitude emerge as crucial factors determining successful acclimatization or gradual deterioration. Disturbances in sleep are related to the terrestrial elevation and the length of exposure. Many studies at very high and extreme altitudes confirm that sleep arousals up to 30–40 per night can occur. These arousals are probably caused by increased episodes of periodic breathing. There is also a concurrent 50% reduction in total sleep time and a fivefold reduction in rapid eye movement. Less-disturbed sleep is observed at lower altitudes, but the phenomena may still persist for weeks. Individuals commonly report "being awake half the night," "not being able to sleep," and having "frequent disturbing dreams." Problems with sleeping can contribute to mood changes during the day, loss of appetite, and daytime drowsiness [2].

Adding to the complexity, fluctuations in ambient temperature and decreased barometric pressure further challenge physiological adaptations at HA. The drop in barometric pressure leads to hypobaric hypoxia, where the partial pressure of oxygen decreases proportionately with altitude. Understanding these adaptations is crucial for managing health risks and improving the quality of life for those who are exposed to various heights of altitudes.

3.1.2 Impact of Hypobaric Hypoxia on Cardiovascular System

The impact of HH on cardiovascular diseases is multifaceted and significant. Hypobaric hypoxia leads to decreased oxygen availability generating challenges to cardiovascular functions. Oxygen is crucial for cellular metabolism, including the myocardium's energy requirements. The hypoxic conditions encountered at HA affect various physiological functions, particularly cardiovascular dynamics [3]. Initially, the autonomic nervous system adapts to low oxygen concentration, resulting in tachycardia reduction to protect the myocardium against excessive energy consumption. However, prolonged exposure to HH induces erythropoiesis (Fig. 3.1), a process aimed at increasing oxygen-carrying capacity through the production of RBCs. While this adaptation is beneficial in improving oxygen delivery, excessive erythropoiesis can lead to chronic mountain sickness, a condition associated with pulmonary hypertension and heart failure (Fig. 3.1). Therefore, individuals with preexisting cardiovascular diseases are prone to higher risk in HA environments due to the compounded effects of HH on heart functions. Understanding these interactions is essential for managing cardiovascular health in individuals exposed to HA and underscores the importance of tailored medical approaches and preventive strategies in such settings.

3.1.2.1 Hypotrophy in Hypoxia

Common adaptive reactions to trauma and stress include cardiac hypertrophy and fibrosis, ultimately resulting in heart failure. It is characterized by larger cardiomyocytes, increased protein synthesis, reactivated fatal genes, cytoskeleton remodeling, and fibrosis. On the other hand, extended hypertrophy is linked to a higher risk of cardiovascular death [4]. Cardiomyocytes proliferate during development, but after birth they lose their capacity to proliferate, and the principal mechanism by which the heart expands is hypertrophy of cardiomyocytes. Both pathological and physiological circumstances result in cardiac hypertrophy. Pathological hypertrophy can be caused by cardiomyopathy, myocardial infarction, or hypertension caused by ventricular remodeling, which in turn causes interstitial fibrosis, systolic and diastolic dysfunction, and other negative consequences like heart failure. The heart contains a variety of cell types, such as smooth muscle cells, cardiomyocytes, endothelial cells, and fibroblasts. The majority of the cells in the heart are endothelial cells, which line the inner layer of the blood vessels. Cardiac vascular rarefaction is connected with pathological cardiac hypertrophy and heart failure. The key transcriptional factors that mediate cellular responsiveness to oxygen levels are called hypoxia-inducible factors, or HIFs. HIF-PHD1-3 (Prolyl-4 hydroxylases 1-3) delicately regulates the α subunit of HIF. Prolyl-4 hydroxylases, or PHDs, hydroxylate HIF-α (mostly HIF-1α and HIF-2α) under normoxic circumstances. Under hypoxic conditions, activities of prolyl-4 hydroxylases are inhibited, leading to the stabilization of HIF-α proteins, which in turn activate downstream target gene expression [5].

3.1.2.2 Hypoxia-Induced Coronary Arterial Disease

Hypoxia is a significant factor in the development of coronary artery disease (CAD) [6]. Central to this process is hypoxia-inducible factor 1 (HIF-1), a key regulator of oxygen homeostasis that stimulates the expression of over 100 genes aiding cellular adaptation to low oxygen levels [7]. HIF-1's role includes enhancing cardiovascular health by promoting the formation of collateral blood vessels and preconditioning mechanisms, which offer protective effects in conditions like ischemia [6]. These genes influence critical biological functions such as iron metabolism, angiogenesis, erythropoiesis, energy metabolism, cell survival, and free radical scavenging. Moreover, manipulating HIF-1 pathways could provide new therapeutic approaches for vascular diseases.

In the context of CAD under hypoxic conditions, several pathological mechanisms have been observed. Hypoxia causes endothelial dysfunction by decreasing the production of nitric oxide (NO), a vasodilator, which results in vasoconstriction and heightened thrombosis risk [10]. The inflammatory response is amplified as hypoxia upregulates inflammatory cytokines and adhesion molecules, contributing to arterial thickening by facilitating the migration of inflammatory cells to the endothelium. Oxidative stress is exacerbated in hypoxic environments due to an imbalance between the synthesis and elimination of reactive oxygen species (ROS), accelerating the development of atherosclerotic plaques and causing endothelial cell damage [9]. Metabolically, hypoxic conditions force cells to switch from aerobic to anaerobic pathways, leading to a drop in ATP levels and a rise in lactate production, resulting in cardiac ischemia and cell death [6]. Furthermore, prolonged hypoxia induces angiogenesis and astrogenesis, leading to the formation of new and collateral blood vessels. Although initially beneficial, these newly formed vessels may be permeable and inadequate for sustaining proper blood flow [11]. Individuals who have long-term hypoxia may exhibit more severe CAD symptoms, including myocardial infarction, angina pectoris, and heart failure. Since conventional therapeutic techniques are complicated by the underlying oxygen deprivation, hypoxia-induced CAD is frequently more difficult to cure.

3.1.2.3 Heart Failure Associated with Hypoxic Stress

The link between hypoxia and heart failure is complex and involves a series of interconnected mechanisms that exacerbate cardiac dysfunction and lead to heart failure. In heart failure, reduced cardiac output leads to inadequate oxygen delivery to tissues, which in turn weakens the myocardium and exacerbates the heart's inability to pump blood efficiently. This creates a vicious cycle where decreased oxygen availability worsens cardiac function, leading to more severe heart failure. At the cellular level, hypoxia activates hypoxia-inducible factors (HIFs), primarily HIF-1 and HIF-2, which are central to the cellular response to low oxygen. Under normoxic conditions, HIFs are degraded, but hypoxia stabilizes these factors, allowing them to enter the nucleus and activate genes involved in angiogenesis, erythropoiesis, energy metabolism, and cell survival [12]. Specifically, HIFs stimulate the production of vascular endothelial growth factor (VEGF) to promote new blood vessel formation and enhance erythropoietin production to increase red blood cell count [6]. Although these responses aim to improve oxygen delivery, the new vessels may be immature and leaky, potentially failing to meet the oxygen demands of cardiac tissue [11].

Hypoxia also induces significant metabolic shifts in cardiac cells, compelling them to rely more on anaerobic pathways for ATP production due to insufficient oxygen for oxidative phosphorylation. This metabolic shift from aerobic to anaerobic glycolysis results in decreased ATP availability and increased lactate production, which can impair cellular functions and exacerbate ischemic conditions [6]. Concurrently, hypoxia increases the generation of ROS, which, coupled with diminished antioxidant defenses, leads to oxidative stress. The accumulation of ROS damages lipids, proteins, and DNA, further impairing mitochondrial function and promoting cell death [8, 9]. The oxidative stress also activates inflammatory pathways, contributing to chronic inflammation and tissue remodeling [10].

At the systemic level, hypoxia-induced endothelial dysfunction reduces the production of nitric oxide (NO), which exacerbates heart failure by raising the afterload and impairing coronary circulation. Additionally, hypoxia activates compensatory neurohormonal responses, including the sympathetic nervous system and the renin–angiotensin–aldosterone system (RAAS), which initially aim to enhance oxygen delivery by increasing the heart rate, blood pressure, and fluid retention [8, 9]. However, these responses eventually add to the heart's workload, further deteriorating cardiac function.

Furthermore, hypoxia promotes chronic inflammation by upregulating inflammatory cytokines and adhesion molecules, attracting inflammatory cells to cardiac tissue [10]. This inflammatory response leads to fibrosis, where the extracellular matrix thickens and stiffens the cardiac tissue, impairing its compliance and contractility [8]. The prolonged hypoxic state also induces vascular remodeling, characterized by the formation of new and collateral blood vessels, a process known as angiogenesis and astrogenesis [11]. While initially beneficial, these new vessels can be fragile and insufficiently integrated into the existing vascular network, potentially failing to provide adequate blood flow and further complicating heart failure.

3.1.3 Effect of Hypoxia on Pulmonary System

3.1.3.1 Hypoxia and Pulmonary Hypertension

When the body acclimatizes to HA, it undergoes immediate and long-term physiological changes to adapt to the reduced oxygen availability. Initial responses include hyperventilation, fluid loss due to a diminished thirst drive, an increased heart rate, and a slight reduction in stroke volume. Over time, the body adjusted by excreting alkali through urine, increasing hematocrit levels (a condition known as polycythemia), enhancing capillary density in skeletal muscles, raising myoglobin levels, and remodeling blood vessels. Vascular remodeling involves hypoxic pulmonary vasoconstriction, where the body constricts pulmonary arteries to redirect blood flow to well-oxygenated lung areas (Fig. 3.1). This adaptation helps maintain efficient oxygen utilization [13]. However, if hypoxia persists, these initial adaptations can become maladaptive. Chronic hypoxia can lead to sustained pulmonary artery constriction, resulting in increased pressure within these vessels, a condition known as "pulmonary hypertension" (PH). PH is a severe condition characterized by a mean pulmonary artery pressure exceeding 20 mmHg, a pulmonary artery occlusion pressure of 15 mmHg or less, and a pulmonary vascular resistance surpassing 3 Wood units [14]. This condition can ultimately progress to heart failure if not managed effectively [15].

PH is classified into five distinct groups: pulmonary arterial hypertension (group 1), pulmonary hypertension due to left heart disease (group 2), pulmonary hypertension associated with lung diseases or hypoxia (group 3), chronic thromboembolic pulmonary hypertension (group 4), and pulmonary hypertension attributed to unclear or multifactorial mechanisms (group 5). Among these, pulmonary hypertension from left heart disease and chronic lung disease (groups 2 and 3) have the largest patient count globally [15].

At the molecular level, hypoxia-induced PH results from an imbalance between vasoconstrictive and vasodilatory factors. Hypoxia causes endothelial dysfunction, leading to the reduced production of vasodilators like nitric oxide (NO) and prostacyclin, and increased release of vasoconstrictors such as endothelin-1 [16]. It also elevates growth factors like platelet-derived growth factor (PDGF) and transforming growth factor-beta (TGF-β), which stimulate smooth muscle cell (SMC) proliferation shown in Fig. 3.2 [17]. VEGF is another crucial factor; it promotes

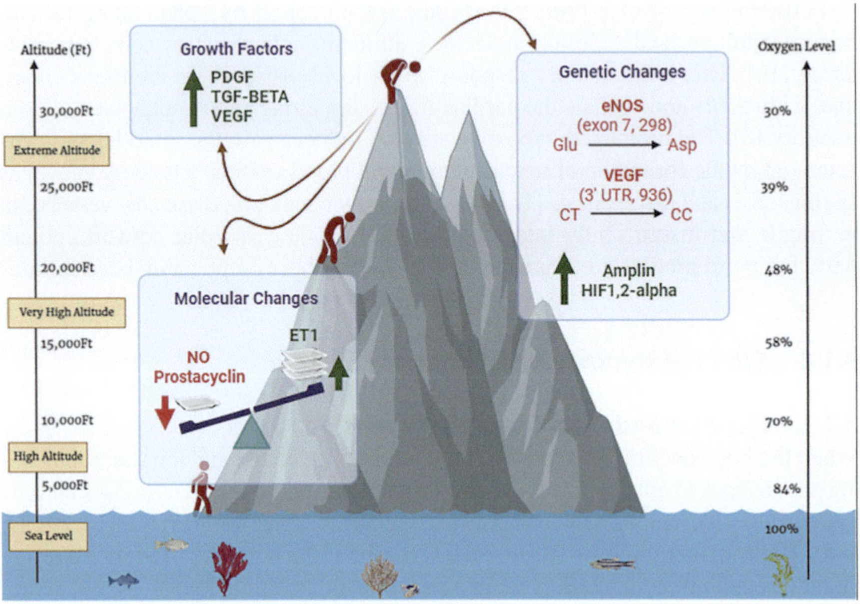

Fig. 3.2 The key genetic and molecular alterations that enable organisms to cope with reduced oxygen availability in high-altitude environments

angiogenesis but also increases capillary permeability and contributes to endothelin overproduction. This leads to calcium influx in vascular smooth muscle cells, activating pathways that promote contraction and proliferation [18].

Additionally, hypoxia influences blood clotting factors, such as α2AP, which regulates plasmin activity and contributes to endothelial repair and angiogenesis. Genetic factors are also significant; for example, mutations in the eNOS and VEGF genes influence the development of PH, with specific polymorphisms linked to altered serum VEGF levels and endothelial NO synthase activity (Fig. 3.2) [19]. The persistent elevation in pulmonary pressure due to hypoxia strains the heart, leading to right ventricular hypertrophy and potential heart failure [20]. Continuous pressure increases and compromised blood flow can also damage pulmonary arteries, contributing to COPD and pulmonary fibrosis [21].

3.1.3.2 Chronic Obstructive Pulmonary Disease (COPD) and Hypoxic Environment

COPD is a lung condition characterized by irreversible damage, posing a significant global health burden. According to the American Thoracic Society, it includes chronic bronchitis and emphysema. Structural and functional alterations in pulmonary circulation are evident in the early stages of COPD. Chronic bronchitis manifests with continuous coughing and sputum production, while emphysema leads to a chronic condition of breathlessness due to lung tissue destruction and enlarged air spaces [22]. The inflammatory process involves increased infiltration of T cells, B cells, neutrophils, dendritic cells, and macrophages. This leads to the upregulation of proinflammatory cytokines, resulting in the deterioration of lung tissue and a subsequent decline in pulmonary function. At the molecular level, hypoxia in COPD leads to significant pathological changes. Hypoxia stabilizes and activates hypoxia-inducible factors (HIF-1α) and (HIF-2α) (Fig. 3.2). These transcription factors regulate the expression of genes involved in angiogenesis, metabolism, and erythropoiesis [23]. In COPD, increased HIF-1α activity is associated with enhanced inflammation, fibrosis, and remodeling of the pulmonary blood vessels [24]. Additionally, hypoxia increases the oxidative stress by increasing the production of ROS. ROS can damage cellular components, including DNA, proteins, and lipids, contributing to the pathogenesis of COPD. This increased oxidative stress under hypoxic conditions further impairs lung functioning and accelerates disease progression [25].

Moreover, hypoxia promotes the release of pro-inflammatory cytokines such as IL-6, IL-8, and TNF-α. These cytokines amplify the inflammatory response in the lungs, leading to chronic inflammation and tissue damage [26]. Elevated levels of these cytokines are commonly observed in COPD patients, particularly during exacerbations and hypoxic episodes. Hypoxia also upregulates the expression of matrix metalloproteinases (MMPs), particularly MMP-9. These enzymes degrade extracellular matrix components, contributing to tissue remodeling and destruction in the lungs [27]. Increased MMP activity is a hallmark of COPD and is closely associated with the development of emphysema and airway remodeling [28]. These molecular changes highlight the complex and detrimental impact of hypoxia on the development of COPD.

3.1.3.3 Hypoxia and Aasthma

Conversely, hypoxia, or low oxygen levels, is a significant concern for asthma patients at HA. The reduced oxygen availability can increase the respiratory rate and effort, potentially triggering asthma symptoms and attacks. Asthma patients may experience more frequent and severe episodes of bronchoconstriction due to the additional stress placed on their respiratory system in a hypoxic environment [29]. At HA environmental conditions, the body compensates for lower oxygen levels by increasing the breathing rate. This hyperventilation can lead to respiratory alkalosis, a condition that can worsen asthma symptoms. The effort to breathe more deeply and rapidly can also tire the respiratory muscles, further complicating the management of asthma [30]. High-altitude environments are often colder and drier, conditions that can also aggravate asthma. Cold air is a known trigger for bronchospasm in asthma patients, and the dry air can dry out the mucosal lining of the airways, making them more susceptible to irritation and inflammation [31]. For asthma patients considering travel to or residence at high altitudes, it is crucial to consult with healthcare professionals. Pre-trip evaluations and an action plan tailored to manage asthma under these unique conditions are recommended. Medications should be adjusted appropriately, and individuals should be prepared for the possibility of increased symptoms [29].

Treatment for PH often involves a multifaceted approach tailored to the underlying cause and severity of the condition which basically involves the reversal of vasculopathy. Administering supplemental oxygen remains an important treatment for hypoxia-induced PH. This helps alleviate hypoxia, reducing pulmonary vasoconstriction and lowering pulmonary artery pressures [32]. Treprostinil, an inhaled prostacyclin analogue, has shown efficacy in patients with PH associated with interstitial lung disease (ILD), improving exercise capacity and quality of life [33]. Use of Vasodilators in treatments such as endothelin receptor antagonists, phosphodiesterase-5 inhibitors, and soluble guanylate cyclase stimulators have shown benefits in pulmonary arterial hypertension (PAH); their effectiveness in hypoxia-induced PH (group 3 PH) has been less conclusive [32]. These help manage fluid retention and reduce the volume load on the heart, easing symptoms associated with right heart failure commonly seen in PH [33].

3.2 Unresolved Issues of Cardiopulmonary Problems

Cardiopulmonary problems at HA environmental conditions are connected in many ways that are still not fully understood. This connection is especially important for people with heart or lung issues who might travel to or live at HA locations/areas. Here are some key unresolved cardiopulmonary system-related issues/problems at extreme environmental conditions.

3.2.1 Genetic Factors

A person's sensitivity to HA is greatly influenced by their genetic constitution. Important genetic markers help to adapt to low-oxygen environments. People who have resided at HA for many generations, like Tibetans, have inherited changes that help them to breathe more efficiently. Results suggest that certain genes, like EPAS1 in Tibetans, helped the body's response to low oxygen levels and improved their ability to tolerate extreme environmental conditions such as HA [34].

3.2.2 Preexisting Health Status

An individual's reaction toward HA is significantly affected by their preexisting illnesses. Clinical/comorbidity conditions, e.g., anemia, COPD, and cardiovascular diseases, can worsen the effect of low oxygen levels. For example, people with heart problems may struggle more with reduced oxygen availability, leading to severe cases of acute mountain sickness (AMS) or high-altitude pulmonary edema (HAPE) [35].

3.2.3 Degree of Acclimatization

The process by which the body steadily adapts to HA is known as acclimatization, and it varies from person to person. Physiological changes in the body that increase oxygen transport to tissues, improve ventilation, and enhance the production of

RBCs are essential for effective acclimatization. The rate and degree of these adaptations may vary according to a number of factors such as fitness level, past exposure to HA, and variability in physiological stress responses at HA [49].

3.3 Predictive Biomarkers for Cardiopulmonary Diseases

Achieving optimal patient outcomes through the classification of risk and specific treatment involves the identification of significant prediction markers for adverse effects in patients with cardiopulmonary disease at HA. A variety of different physiological stressors linked with HA, such as hypobaric hypoxia and altitude pulmonary artery pressure, exacerbate the preexisting cardiovascular disorder.

3.3.1 Hypoxemia

Low blood oxygen levels are an important marker of adverse outcomes. Identification of individuals susceptible to acute hypoxic problems can be assisted through monitoring arterial blood gases (ABG) and arterial oxygen saturation (SpO_2) [30]. Hypoxemia is defined by the American Thoracic Society as a partial pressure of oxygen in arterial blood (PaO_2) of less than 80 mm Hg. It can occur in a variety of diseases, such as acute respiratory distress syndrome (ARDS), interstitial lung disease, congenital heart disease, pneumonia, and COPD [36]. Hypoxemia can cause cyanosis (bluish color of the skin or lips), shortness of breath, rapid breathing, and in extreme situations, loss of consciousness. For the diagnosis and treatment of hypoxemia, ABG analysis is considered the standard parameter. It measures blood pH, partial pressure of oxygen (PaO_2), and carbon dioxide ($PaCO_2$). The SpO_2 of hemoglobin in the blood can be measured noninvasively using pulse oximetry; normal readings usually fall around 95–100%. Impaired oxygen exchange in the lungs, caused by diffusion impairment, hypoventilation, or ventilation-perfusion mismatch, is the root cause of hypoxemia. For instance, airway obstruction can result in hypoventilation and lower blood oxygen levels in diseases like COPD and asthma [37]. The complications from chronic hypoxemia include pulmonary hypertension, right heart failure, and dementia. The root cause and success rate of treatment have an impact on the prognosis of causes like hypoxemia [38].

3.3.2 Cardiac Biomarkers

Increased cardiac biomarkers, such as troponin and B-type natriuretic peptide (BNP), are significant indicators of myocardial stress and possible heart failure. Because of the special stress conditions at HA and their effect on the cardiovascular system, these biomarkers are particularly specific for monitoring individuals who have been exposed to hypobaric hypoxia.

3.3.2.1 B-type Natriuretic Peptide

The cardiovascular system releases a hormone called B-type natriuretic peptide (BNP). When heart failure begins or becomes worse, changes in arterial pressure induce its release. BNP is released by the heart muscle whenever exposed to stress or overwork, helping in controlling blood pressure and fluid balance. When cardiomyocytes are stretched, the heart's ventricles release an excessive amount of BNP. There is a direct link between HA BNP levels and ventricular volume with pressure overload, which is common in diseases such as heart failure. HA exposure can result in hypoxic pulmonary vasocontraction, which increases pulmonary arterial pressure. This condition may cause right ventricle contraction and increase the BNP levels, which could indicate cardiac strain or pulmonary edema, a disease in which fluid overflow occurs in the lungs. According to studies, individuals who reside at HA have noticeably higher BNP levels, which suggest that the physiological stressors and adaptations are caused due to the hypoxic environment conditions [39].

3.3.2.2 Troponin

Troponin is a type of protein that occurs in the heart muscle and goes into the bloodstream when the heart is injured. It acts as a very specific biomarker for diseases like myocardial infarction or heart attacks, which are symptomatic of myocardial damage. An important diagnostic tool for doctors is troponin, which is increased in the blood and indicates injury to the heart muscle. Mechanism of troponin release: when cardiac cells sustain injury either from an infarction (heart attack) or ischemia (reduced blood flow), troponin is released into the blood. Increased troponin levels are used to diagnose acute coronary syndrome and other heart diseases because they signify mitochondrial damage. Troponin at HA: due to hypobaric hypoxic conditions at HA, the heart may suffer ischemia, which increases troponin levels. Recently, it has been shown that following acute exposure to extremely high altitudes, even healthy individuals show elevated troponin levels [40]. This indicates that cardiac cell damage occurs during the body's adaptation to the decreased supply of oxygen [41].

3.3.3 Genetic Markers

Individuals' responses to conditions with lower oxygen levels at HA are mostly influenced by their genetic predisposition. The EPAS1 gene is one of the major genetic markers linked to HA adaptation. The synthesis of HIFs is triggered by hypoxia, or low oxygen levels, and these factors, in turn, regulate the expression of downstream target genes [30]. Genetic variations in the EPAS1 gene have been found in populations that have lived for generations at high altitudes, such as Tibetans, providing them with unique adaptations that enhance their ability to survive under hypoxic conditions. EPAS1 variations in Tibetans are strongly associated with reduced hemoglobin levels under hypoxic conditions. These variations could help in the overproduction of RBCs, which results in consequences of persistent mountain sicknesses [50].

3.3.4 Risk Stratification and Management Strategies

Identifying such markers and individuals can be categorized into different risk groups. Compared to individuals with fewer symptoms and normal pressures, those with substantial hypoxemia and increased pulmonary artery pressures may require more frequent monitoring and early therapies. Specific management strategies can be developed depending on the risk analysis. The methods used for patients at high risk include preventive drugs to avoid high-altitude sicknesses, including acetazolamide [42]. Further oxygen treatment may be used to keep oxygen levels at a suitable level [43].

3.4 Long-Term Adaptation at High Altitude

It is still not clear how long-term exposure to high altitude will affect an individual who has heart problems. This is very essential since low oxygen availability at HA environments can aggravate preexisting cardiopulmonary problems, hence creating new difficulties.

3.4.1 Acclimatization and Physiological Adaptation

Acclimatization is used to describe the physiological changes caused by low oxygen levels or hypoxia at HA. These changes include better utilization of oxygen, improved respiration, and increased oxygen transport capacity. Acclimatization can decrease the adverse effect of exposure to HH at HA in healthy individuals, including AMS and HAPE [44]. The acclimatation process can be insufficient for individuals with cardiopulmonary disorders, which could result in increased health risk [45].

3.4.2 Cardiopulmonary Responses to HA

In order to get enough oxygen into tissues, the heart and lungs have to work harder at HA. The cardiovascular system may be strained by this increased effort, particularly in those who already have heart disease, COPD, or pulmonary hypertension. Results show that prolonged hypoxia-induced pulmonary vasoconstriction caused by hypoxic exposure leads to pulmonary hypertension and right heart failure [46].

3.4.3 Chronic Complications

Extended exposure may cause chronic hypoxemia, which can cause chronic altitude sickness, a long-term consequence that causes fatigue. Although additional research is required to validate this, there is concern that repeated or prolonged exposure to HA may cause cumulative adverse effects on the heart and lungs [47].

3.5 Therapeutic Approaches

3.5.1 Erythropoiesis Stimulating Agents (ESAs)

These drugs increase the blood's ability to carry oxygen by enhancing the production of RBCs. Erythropoietin (EPO) is a naturally occurring hormone that can be supplemented to increase the production of RBCs. Results suggest that ESAs increase oxygen transport to tissues. Further studies are being carried out to evaluate their ability to improve performance and endurance at HA [47].

3.5.2 Phosphodesterase-5 or PDE-5 Inhibitor

Drugs that inflate blood vessels, such as tadalafil (Cialis) and sildenafil (Viagra), may help to enhance blood flow in the lungs and lower pulmonary hypertension, a typical problem at HA. By increasing oxygenation and lowering pulmonary artery pressure, these drugs have shown potential in preventing HAPE [48].

3.5.3 Antioxidants and Anti-Inflammatory Agent

HA-related illnesses/disorders are significantly affected by oxidative stress and inflammation. There is a possibility that anti-inflammatory drugs like dexamethasone and antioxidants like N-acetylcysteine (NAC) can reduce these effects and enhance overall cardiopulmonary function under hypoxic conditions [51].

3.6 Technological Approaches

Oxygen is obtained from the surrounding air by portable oxygen concentrators that provide an ongoing supply of concentrated oxygen; as they are battery-operated, lightweight, and portable, they can be used in HA settings where normal oxygen tanks are not feasible [52]. Enhancing the production of breathable devices, such as rebreathers and masks designed to optimize CO_2 removal and oxygen delivery, is essential. Cutting-edge materials used in contemporary designs enhance functionality and comfort at HA. Heart rate, SpO_2, respiration rate, and other vital indicators should be continuously monitored via wearables with sensors. These gadgets offer useful information that can aid in the early identification of diseases linked to altitude, enabling prompt diagnosis and treatment. Remote monitoring and data analysis are made possible by integration with cloud-based systems and mobile applications [53].

3.7 Conclusions and Future Directions

In conclusion, this chapter provides a comprehensive overview of the intricate relationship between cardiopulmonary health and HA environmental conditions, shedding light on unresolved questions and areas warranting further research. Throughout the chapter, key topics such as the physiological adaptations to hypoxia, the development of pulmonary hypertension, and the impact of genetic factors on cardiopulmonary responses have been meticulously explored. The chapter underscores the immediate and long-term physiological responses that individuals undergo when exposed to HA conditions, including hyperventilation, fluid loss, increased heart rate, and vascular remodeling. While these adaptations serve to optimize oxygen utilization initially, prolonged hypoxia can lead to maladaptive responses, such as pulmonary hypertension, which poses significant risks to cardiovascular health.

Furthermore, the chapter delves into the molecular mechanisms underlying pulmonary hypertension, highlighting the intricate interplay between vasoconstrictive and vasodilatory factors, endothelial dysfunction, and genetic predisposition. By elucidating these mechanisms, the chapter highlights the need for targeted therapeutic interventions to mitigate the adverse effects of chronic hypoxia on cardiopulmonary functions. Despite significant advances in our understanding of cardiopulmonary responses to HA, numerous unresolved issues still persist. The chapter emphasizes the importance of continued research efforts aimed at elucidating the complex pathways involved in cardiopulmonary adaptation and dysfunction at HA. Additionally, interdisciplinary collaborations and innovative methodologies will be crucial for addressing unresolved questions and advancing our understanding of cardiopulmonary systems and diseases, particularly in the context of hypobaric hypoxia at high-altitude environments.

Acknowledgments We are thankful to the Director of the Defence Institute of Physiology and Allied Sciences (DIPAS), for his guidance and financial support for the study. The authors are also thankful to Ms. Harshita Gupta, Technician 'B,' Shri Dhruv Varshney and Ms. Guriya Sahi, Project Trainees, Department of Disruptive and Deterrence Technologies. DS is supported by a fellowship from DIPAS-DRDO, Ministry of Defence, Govt of India, Delhi, India, and AY is supported by a fellowship from UGC-CSIR, Ministry of Human Resources, Govt of India, Delhi, India.

Conflict of Interests The authors declare no conflict of interest.

Funding This work is supported by Defence Institute of Physiology and Allied Sciences, Defence Research and Development Organization, Delhi-110054.

References

1. Grocott M, Montgomery H, Vercueil A. High-altitude physiology and pathophysiology: implications and relevance for intensive care medicine. Crit Care. 2007;11:203. https://doi.org/10.1186/cc5142.
2. Marriott BM, Carlson SJ, editors. Nutritional needs in cold and in high-altitude environments: applications for military personnel in field operations. National Academies Press; 1996. https://www.ncbi.nlm.nih.gov/books/NBK232874/.
3. Doe J, Smith J. The cardiovascular effects of hypobaric hypoxia: from basic mechanisms to clinical perspectives. J Cardiovasc Res. 2020;85(3):215–30. https://doi.org/10.1016/j.jcr.2020.01.015.
4. Brown E, Green M. Cardiac hypertrophy and fibrosis: molecular mechanisms and clinical implications. J Mol Cell Cardiol. 2018;123(4):102–14. https://doi.org/10.1016/j.yjmcc.2018.01.012.
5. Dai Z. Loss of endothelial hypoxia inducible factor-prolyl hydroxylase 2 induces cardiac hypertrophy and fibrosis. J Am Heart Assoc. 2022;10(22):e022077.
6. Kasivisvanathan V, Shalhoub J, Lim CS, Shepherd AC, Thapar A, Davies AH. Hypoxia-Inducible Factor-1 in arterial disease: a putative therapeutic target. Curr Vasc Pharmacol. 2011;9(3):333–49. https://doi.org/10.2174/157016111795495602. Bentham Science Publishers.
7. Semenza GL. Hypoxia-inducible factors: roles in cardiovascular disease progression, prevention, and treatment. Cardiovasc Res. 2023;119(2):371–80. https://doi.org/10.1093/cvr/cvac089.
8. Madamanchi NR, Vendrov A, Runge MS. Oxidative stress and vascular disease. Arterioscler Thromb Vasc Biol. 2005;25(1):29–38. https://doi.org/10.1161/01.ATV.0000150649.39934.13.
9. Libby P, Ridker PM, Maseri A. Inflammation and atherosclerosis. Circulation. 2002;105(9):1135–43. https://doi.org/10.1161/hc0902.104353.
10. Staels B, Maes M, Zambon A. Fibrates and future PPARα agonists in the treatment of cardiovascular disease. Nat Rev Cardiol. 2008;5:542–53. https://doi.org/10.1038/ncpcardio1278.
11. Lopaschuk GD, Ussher JR. Evolving concepts of myocardial energy metabolism: more than just fats and carbohydrates. Circ Res. 2016;119(11):1173–6.
12. Semenza GL. Under normoxic conditions, HIFs are degraded, but hypoxia stabilizes these factors, leading them to enter the nucleus and activate genes involved in angiogenesis, erythropoiesis, energy metabolism, and cell survival. Angiogenesis. 2014;17(1):7–13. https://doi.org/10.1007/s10456-014-9446-2.
13. Srivastava S, Singh SB. High-altitude pulmonary edema. J Occup Health. 2014;56(4):235–43.
14. Hannemann J, Böger R. Dysregulation of the nitric oxide/Dimethylarginine pathway in hypoxic pulmonary vasoconstriction molecular mechanisms and clinical significance. Front Med. 2022;9:835481.
15. Bousseau S, Sobrano Fais R, Gu S, Frump A, Lahm T. Pathophysiology and new advances in pulmonary hypertension. BMJ Med. 2023;2(1):e000137. https://doi.org/10.1136/bmjmed-2022-000137. PMID: 37051026; PMCID: PMC10083754.
16. Srivastava S, Bhagi S, Gupta RK. Positive association of mutations in genes related to vascular endothelial dysfunction in relation to high altitude pulmonary edema (HAPE): a case control study. Adv Biotech Microbiol. 2019;13(4). https://doi.org/10.19080/AIBM.2019.13.555870.
17. Simons M, Gordon E, Claesson-Welsh L. Mechanisms and regulation of endothelial VEGF receptor signalling. Nat Rev Mol Cell Biol. 2016;17:611–25. https://doi.org/10.1038/nrm.2016.87.
18. Crosswhite P, Sun Z. Molecular mechanisms of pulmonary arterial remodeling. Mol Med. 2014;20:191–201. https://doi.org/10.2119/molmed.2013.00165.
19. Reddy PK, Malik SM. Alpha-2-Antiplasmin: a fibrinolytic peptide, could be a potential biomarker for diagnosis of high altitude induced pulmonary hypertension. Def Sci J. 2017;67(6):631.

20. Shi J, Liu S, Guo Y, et al. Association between eNOS rs1799983 polymorphism and hypertension: a meta-analysis involving 14,185 cases and 13,407 controls. BMC Cardiovasc Disord. 2021;21:385. https://doi.org/10.1186/s12872-021-02192-2.
21. Stenmark KR, Mecham RP. Cellular and molecular mechanisms of pulmonary vascular remodeling. Annu Rev Physiol. 1997;59(1):89–144.
22. Vestbo J, Hurd SS, Agustí AG, Jones PW, Vogelmeier C, Anzueto A, Barnes PJ, Fabbri LM, Martinez FJ, Nishimura M, Stockley RA, Sin DD, Rodriguez-Roisin R. Global strategy for the diagnosis, management, and prevention of chronic obstructive pulmonary disease: GOLD executive summary. Am J Respir Crit Care Med. 2013;187(4):347–65. https://doi.org/10.1164/rccm.201204-0596PP. Epub 2012 Aug 9.
23. Semenza GL. Hypoxia, clonal selection, and the role of HIF-1 in tumor progression. Crit Rev Biochem Mol Biol. 2000;35(2):71–103. https://doi.org/10.1080/10409230091169186.
24. Shimoda LA, Laurie SS. HIF and pulmonary vascular responses to hypoxia. J Appl Physiol. 2014;116(7):867–74.
25. Rahman I, Adcock IM. Oxidative stress and redox regulation of lung inflammation in COPD. Eur Respir J. 2006;28(1):219–42. https://doi.org/10.1183/09031936.06.00053805.
26. Barnes PJ. Alveolar macrophages as orchestrators of COPD. COPD: J Chron Obstruct Pulmon Dis. 2004;1(1):59–70. https://doi.org/10.1081/COPD-120028701.
27. Chung KF, Adcock IM. Multifaceted mechanisms in COPD: inflammation, immunity, and tissue repair and destruction. Eur Respir J. 2008;31(6):1334–56.
28. Vignola AM, Bonanno A, Mirabella A, Riccobono L, Mirabella F, Profita M, Bellia V, Bousquet J, Bonsignore G. Increased levels of elastase and α1-antitrypsin in sputum of asthmatic patients. Am J Respir Crit Care Med. 1998;157(2):505–11.
29. Vinnikov D, Khafagy A, Blanc PD, Brimkulov N, Steinmaus C. High-altitude alpine therapy and lung function in asthma: systematic review and meta-analysis. ERJ Open Res. 2016;2(2):00097.
30. West JB. Respiratory physiology: the essentials. 7th ed. Lippincott Williams & Wilkins; 2004.
31. Boulet LP. Influence of comorbid conditions on asthma. Eur Respir J. 2009;33(4):897–906.
32. Nathan SD, Barbera JA, Gaine SP, Harari S, Martinez FJ, Olschewski H, Olsson KM, Peacock AJ, Pepke-Zaba J, Provencher S, Weissmann N, Seeger W. Pulmonary hypertension in chronic lung disease and hypoxia. The Eur Respir J. 2019;53(1):1801914. https://doi.org/10.1183/13993003.01914-2018.
33. Gonzales J, Fraidenburg DR. Pharmacology and emerging therapies for group 3 pulmonary hypertension due to chronic lung disease. Pharmaceuticals. 2023;16(3):418.
34. Beall CM. Two routes to functional adaptation: Tibetan and Andean high-altitude natives. Proc Natl Acad Sci U S A. 2007;104(Suppl 1):8655–60. https://doi.org/10.1073/pnas.0701985104. Epub 2007 May 9. PMID: 17494744; PMCID: PMC1876443.
35. Taylor AT. High-altitude illnesses: physiology, risk factors, prevention, and treatment. Rambam Maimonides Med J. 2011;2(1):e0022. https://doi.org/10.5041/RMMJ.10022. PMID: 23908794; PMCID: PMC3678789.
36. Parshall MB, Schwartzstein RM, Adams L, Banzett RB, Manning HL, Bourbeau J, Calverley PM, Gift AG, Harver A, Lareau SC, Mahler DA, Meek PM, O'Donnell DE, American Thoracic Society Committee on Dyspnea. An official American Thoracic Society statement: update on the mechanisms, assessment, and management of dyspnea. Am J Respir Crit Care Med. 2012;185(4):435–52. https://doi.org/10.1164/rccm.201111-2042ST. PMID: 22336677; PMCID: PMC5448624.
37. Matthay MA, Zimmerman GA, Esmon C, Bhattacharya J, Coller B, Doerschuk CM, Floros J, Gimbrone MA Jr, Hoffman E, Hubmayr RD, Leppert M. Future research directions in acute lung injury: summary of a National Heart, Lung, and Blood Institute working group. Am J Respir Crit Care Med. 2003;167(7):1027–35.
38. Fayyaz AU, Edwards WD, Maleszewski JJ, Konik EA, DuBrock HM, Borlaug BA, Frantz RP, Jenkins SM, Redfield MM. Global pulmonary vascular remodeling in pulmonary hypertension associated with heart failure and preserved or reduced ejection

fraction. Circulation. 2018;137(17):1796–810. https://my.clevelandclinic.org/health/diagnostics/22629-b-type-natriuretic-peptide.
39. Gao M, Wang R, Jiayong Z, Liu Y, Sun G. NT-ProBNP levels are moderately increased in acute high-altitude pulmonary edema. Exp Ther Med. 2013;5(5):1434–8. https://doi.org/10.3892/etm.2013.976. Epub 2013 Feb 26. PMID: 23737894; PMCID: PMC3671802.
40. Cameron AJ, Roberts LM. Troponin release in myocardial injury and hypoxia: mechanisms and implications. J Am Coll Cardiol. 2017;70(10):1204–12. https://doi.org/10.1016/j.jacc.2017.07.012.
41. Boos CJ, Mellor A, Begley J, et al. The effects of exercise at high altitude on high-sensitivity cardiac troponin release and associated biventricular cardiac function. Clin Res Cardiol. 2014;103:291–9. https://doi.org/10.1007/s00392-013-0654-2.
42. West JB. High-altitude medicine. Am J Respir Crit Care Med. 2012;186(12):1229–37. https://doi.org/10.1164/rccm.201207-1323CI. Epub 2012 Oct 26.
43. Luks AM, Auerbach PS, Freer L, Grissom CK, Keyes LE, McIntosh SE, Rodway GW, Schoene RB, Zafren K, Hackett PH. Wilderness medical society clinical practice guidelines for the prevention and treatment of acute altitude illness: 2019 update. Wilderness Environ Med. 2019;30(4S):S3–S18. https://doi.org/10.1016/j.wem.2019.04.006. Epub 2019 Jun 24.
44. Basnyat B, Murdoch DR. High-altitude illness. Lancet. 2003;361(9373):1967–74. https://doi.org/10.1016/S0140-6736(03)13591-X.
45. Grover RF, Weil JV, Reeves JT. Cardiovascular adaptation to exercise at high altitude. Exerc Sport Sci Rev. 1986;14:269–302.
46. Luks AM, Swenson ER. Travel to high altitude with pre-existing lung disease. Eur Respir J. 2007;29(4):770–92. https://doi.org/10.1183/09031936.00052606.
47. Smith KJ, Bleyer AJ, Little WC, Sane DC. The cardiovascular effects of erythropoietin. Cardiovasc Res. 2003;59(3):538–48. https://doi.org/10.1016/S0008-6363(03)00468-1.
48. Burtscher M, Hefti U, Hefti JP. High-altitude illnesses: old stories and new insights into the pathophysiology, treatment and prevention. Sports Med Health Sci. 2021;3(2):59–69. https://doi.org/10.1016/j.smhs.2021.04.001. PMID: 35782163; PMCID: PMC9219347.
49. Saito S, Takeda N, Eto Y. Individual variations in adaptation to high-altitude hypoxia and their mechanisms. In: Cymerman N, Reeves JT, editors. Hypoxia: the adaptations. CRC Press; 1995. p. 233–54.
50. Cynthia. Genetic variation in EPAS1 and its implications for adaptation to high altitude hypoxia in the Tibetan population. J Genet. 2007;86(3):211–8.
51. Zhang Y, Fang J, Dong Y, Ding H, Cheng Q, Liu H, Xu G, Zhang W. High-altitude hypoxia exposure induces iron overload and ferroptosis in adipose tissue. Antioxidants (Basel). 2022;11(12):2367. https://doi.org/10.3390/antiox11122367. PMID: 36552575; PMCID: PMC9774922.
52. Furlanetto KC, Pitta F. Oxygen therapy devices and portable ventilators for improved physical activity in daily life in patients with chronic respiratory disease. Expert Rev Med Devices. 2017;14(2):103–15. https://doi.org/10.1080/17434440.2017.1283981. Epub 2017 Jan 31.
53. Iqbal MH, Aydin A, Brunckhorst O, Dasgupta P, Ahmed K. A review of wearable technology in medicine. J R Soc Med. 2016;109(10):372–80. https://doi.org/10.1177/0141076816663560.

Impact of High Altitude on Skeletal Muscle Health

4

Akshita Kumar, Geetha Suryakumar, Som Nath Singh, and Richa Rathor

Abstract

High-altitude-associated skeletal muscle atrophy is a well-established phenomenon that is further responsible for the decline in physical performance. One of the major stressors of high altitude is hypobaric hypoxia, which contributes to disturbed redox homeostasis that further results in pathophysiological conditions, including skeletal muscle atrophy. Other recent research establishes the fact that hypobaric hypoxia-induced muscle protein loss is regulated via multiple pathways, including oxidative stress, inflammation, ER stress, disturbed calcium homeostasis, RyR1 remodeling, perturbed myokine profiling, altered anabolic pathways, enhanced protein degradation, and apoptosis in the skeletal muscle. Awareness of the molecular mechanism of altitude-linked muscle atrophy opens a way to develop strategies and interventions to ameliorate this protein loss. Nowadays dietary supplementation is gaining interest in amelioration of skeletal muscle atrophy and enhancing physical performance due to its safe profile, easy ingestion, and availability. Therefore, the present chapter sketches the details of the molecular mechanisms responsible for high-altitude-associated muscle protein loss and further sheds light on dietary intervention for managing the skeletal muscle health during stressful conditions.

Keywords

High altitude · Hypobaric hypoxia · Muscle atrophy · Dietary supplement

A. Kumar · G. Suryakumar · S. N. Singh · R. Rathor (✉)
Defence Institute of Physiology and Allied Sciences (DIPAS), Defence Research and Development Organization (DRDO), Timarpur, Delhi, India
e-mail: richa110in.dipas@gov.in

S. Srivastava et al. (eds.), *Health and Sustenance in Extreme Environment*, https://doi.org/10.1007/978-981-96-8642-1_4

4.1 Introduction

Skeletal muscle is a vastly adaptable tissue accounting for 40% of body mass. It acts as a biomechanical device coordinating contraction, force generation, and body movements [1]. It is engaged in managing functional metabolic capacity for lifestyle disease and managing daily physical activity [2]. The maintenance of the strength of muscle mass is considered to be muscle quality (MQ), and various studies concluded a positive correlation between higher MQ and increased muscle strength, function, and physical performance [3]. Muscle-related problems not only decline physical function but also compromise quality of life [4]. Consequently, skeletal muscle mass loss results into frailty and difficulty in carry out daily activities [5].

Skeletal muscle tissue exhibits diverse properties resulting from the combination of two distinct types of muscle fibers. Slow oxidative Type I fibers, enriched with dense capillary supply, elevated mitochondrial, and myoglobin content, and fast glycolytic Type IIB and Type IIX fibers with greater cross-sectional area contribute to the heterogeneity of muscle tissue [6]. Skeletal muscle is capable of handling moderate energy uptake during the event of normal contractions. It efficiently switches to anaerobic metabolism to fulfill the increased energy demands in the case of intense physical exercise. Adenosine triphosphate (ATP) hydrolysis resulting from actin–myosin cross-bridge formation is the source of energy during contractions [7]. However, anaerobic metabolism fulfills energy demands during intense physical exercise. Henceforth, aerobic and anaerobic metabolism both are crucial for the maintenance of energy demand, while approximately 90% of energy demands are encountered through aerobic metabolism [6]. Consequently, disturbance in regular oxygen supply would exacerbate skeletal muscle homeostasis. This situation arises in the case of ascent to high altitude, where the reduced atmospheric pressure is linked with a simultaneous decrease in partial pressure of inspired oxygen, leading to modifications in skeletal muscle metabolism. Previous studies have reported a decline in muscle mass on exposure to hypoxia. The occurrence of hypoxia has also been observed in multiple malignancies, such as diabetes, muscular dystrophy, chronic obstructive pulmonary disease, cachexia, heart failure, anemia, and ischemia. The risk of increased fatigue, higher mortality, and a sedentary lifestyle can result from altered muscle metabolism and decline in muscle mass; thus, a proper understanding of the impact of hypoxia on skeletal muscle mass is required. Moreover, skeletal muscle also performs the role of an endocrine organ by acting as the production and secretion site for bioactive molecules. Thus, maintenance of a healthy musculoskeletal system is a vital requirement for overall body health.

4.2 High-Altitude Environment

An elevation beyond 2500 m is considered high altitude. The existing adverse environmental conditions at high altitudes prove to be detrimental to human health and have a negative impact on physical performance and physiological functions

Table 4.1 Altitude: Their ranges and reported changes

Classification	Altitude	Reported changes	Reference
Intermediate altitude	1500–2500 m	Physiological changes occur due to hypobaric hypoxia; arterial oxygen saturation observed to be above 90%	[8]
High altitude	2500–3500 m	Altitude illness observed	[9]
Very high altitude	3500–5800 m	Arterial oxygen saturation falls below 90% and hypoxaemia is observed	[10]
Extreme altitude	>5800 m	Survival becomes further difficult	[11]

(Table 4.1). An extensive amount of research has been conducted on the consequences of acute and chronic exposure to high-altitude conditions. Multiple pathophysiological responses arise on exposure to high altitudes to combat the adverse conditions. The major stresses posed by high-altitude environments include hypobaric hypoxia, cold, high solar radiation, high-velocity winds, and aridity. The impact of exposure to these conditions varies among populations depending on age and physiological conditions. Such conditions pose a health threat to trekkers, climbers, and soldiers stationed at high altitudes, which might lead to life-threatening illnesses. As human biology requires consistent and efficient oxygen supply to mitochondria, the understanding of the adverse impact posed by high-altitude ascent holds utmost significance.

4.3 Environmental Challenges Posed at High Altitude

4.3.1 Hypobaric Hypoxia

Hypobaric hypoxia is a distinctive stress at high altitudes. Hypobaric hypoxia can be described as a condition that arises when the human body is exposed to diminished levels of oxygen at high altitudes. A sharp decline in barometric pressure along with increasing altitude is observed due to the shorter air column. Nonetheless, the oxygen percentage remains at 21% of barometric pressure [12]. However, due to a reduction in atmospheric pressure, the oxygen partial pressure also decreases. This further changes oxygen delivery to body tissues and organs, posing major health risks and physiological challenges. The natives residing at high altitudes, particularly in the Andean region and Tibetan plateau, exhibited relatively lower occurrence of chronic mountain sickness due to adaptation. Among the physiological responses generated by hypoxia encounters are hyperventilation, increased heart rate, cardiac output, and hemoconcentration [13]. The increase in ventilation during acclimatization to hypoxia facilitates recovery of the reduced alveolar and arterial oxygen partial pressures and arterial oxygen saturation. In the hypoxic high-altitude environment, increased hemoconcentration and cardiac output support oxygen transport to tissues both at rest and during physical exercise. This is in addition to the ventilatory response. The reduced barometric pressure at high altitudes evokes a decline in alveolar oxygen tension and in arterial partial pressure of oxygen [14]. The diminished partial pressures cause reduced oxgen saturation in arterial hemoglobin [15], thus leading to a decline in physical performance and arising physiological challenges.

4.3.2 UV Radiation

The factors influencing the impact of UV solar radiation are variations in altitude, latitude, and atmospheric conditions such as total ozone, cloud coverage, fluctuations in ozone, aerosol amount, and surface coverage. At high altitudes, UV radiation is of higher intensity due to low aerosol amounts and high ground reflectance [16]. The UV radiation reaching high altitudes has less scattering and absorption because it travels through the atmosphere over a shorter distance as compared to radiation at lower altitudes. The escalation in solar radiation along with increasing altitude is termed the altitude effect (AE) which is stated as an increase in irradiance in percent per 1000 m in relation to the valley site, where I_m refers to irradiance at the mountain site, I_v is the irradiance at valley site, Δ refers to the difference in altitude between valley site and mountain site in meters [17]:

$$\mathrm{AE} = \left(\frac{I_\mathrm{m}}{I_\mathrm{v}} - 1\right) \times \frac{\Delta(\mathrm{Altitude})}{1000\,\mathrm{m}} \times 100\%$$

The altitude effect is of high significance for a better understanding of UV radiation in high-altitude environments. The altitude effect fluctuates at different altitudes and is greater for low-altitude sites as compared to high-altitude sites [18].

4.3.3 Cold Conditions

High-altitude areas have relatively lower temperatures due to lower air pressure. People ascending above mean sea level are susceptible to temperatures ranging from mild to extremely cold. A reduction of 2 °C for every 300 m or 984 ft has been reported [19]. Hypothermia or local cold injury, such as frostbite or trench foot, can be caused by exposure to cold conditions [20]. Cold conditions lead to vasoconstriction of the skin and shivering so that metabolic heat could be produced. Managing tissue oxygenation in hypoxia-associated cold environment is similar to hypercapnia in which heat loss happens, as a result ventilation increases to prevent heat loss.

4.3.4 High-Velocity Winds

Wind refers to the flow of air from an area of higher pressure to an area of lower pressure and the flow gets obstructed due to friction at lower altitudes, hence the flow of air slows down. In contrast, at higher altitudes, there is no friction; hence, the flow of air is stronger, resulting in high-velocity winds. Also, with air density being lower at higher altitudes, the wind speed is further increased.

4.4 High Altitude and Skeletal Muscles

Skeletal muscle tissue, being a plastic organ, can undergo modifications in its contractile properties under a variety of stimuli such as environmental stress, hormonal influence, and mechanical stress. Hypobaric hypoxia is capable of triggering structural and metabolic alterations in the skeletal muscle, leading to muscle atrophy. Major changes in skeletal muscle include alterations in muscle fiber size, resting and active force, contractile velocity, and function of the neuromuscular junction.

4.4.1 Muscle Fiber Size

Skeletal muscle fibers are multinucleated syncytia forming the largest cell types across the animal kingdom. The diameter ranges from a few microns to thousands of microns [21]. Aerobic metabolism relying on the transport of oxygen from the blood to the mitochondria and further diffusion of ATP from mitochondria to cellular ATPases is linked with diffusion restrictions acting as a limiting factor for maximal fiber size [22]. The skeletal muscle fiber size is the outcome of the equilibrium between protein synthesis and degradation controlled by specific signaling pathways. The drastic reduction in muscle fiber size and function can be explained by muscle fiber loss. Ascent to high altitude and induction of hypobaric hypoxia manifests in the form of muscle fiber size reduction.

A loss of body mass ranging around 5–10% along with a loss of muscle volume was observed in sojourners on expeditions to the Himalayas [23]. Further, it has also been evidenced that lack of nutrition is not the reason [24]. A 20% decline in the cross-sectional area of vastus lateralis muscle fibers in 14 mountain climbers was noted after 8 weeks of stay at an altitude greater than 5000 m [23]. Another indicator of declining muscle health is a threefold elevation in levels of lipofuscin in muscle fibers as a consequence of lipid peroxidation of mitochondrial membranes resulting from oxidative stress [25]. It has been proposed that muscle loss due to oxidative stress might be occurring as a result of reduced rates of protein synthesis. An in vivo study suggested increased oxidative stress, protein oxidation, and lipid peroxidation resulting in the accumulation of misfolded proteins post 24-hour exposure [26]. Agrawal et al. [27] also reported a decrement in the cross-sectional area (CSA) of skeletal muscle fiber of male Sprague–Dawley rats around 43.3% and 66.8%, respectively, on 07d and 14d chronic hypobaric hypoxia exposure, compared with control rats [27]. The study also reported a reduction in Feret's diameter of myofibers in the case of 14-day chronic hypobaric hypoxia-exposed rats. In accordance with the above results, another study also showed a significant decrease in cross-sectional area and Feret's diameter in rat skeletal muscle on acute and chronic exposure to hypobaric hypoxia [28].

4.4.2 Muscle Atrophy

The damaging effects of hypobaric hypoxia insult on skeletal muscle health are reliant on the duration of exposure and the altitude of the ascent. The reduction in body weight and diminished physical performance are encountered in sojourners. The subjects exposed to hypobaric hypoxia for 2 weeks showed a significant modification in muscle fraction without any change in total protein [29]. A loss in body fat amounting to a 70.5% reduction in body weight on sojourning to high altitude has been observed, whereas on residing at 5400 m, body fat loss was associated only with 27.3% body weight loss [30]. A proportionate decrease was also observed in the arm and leg circumferences on residing above 5400 m. Indian troops residing at 4100 and 3750 m for a period of 10 months revealed a decrease in muscle mass of 1.74 and 1.38 kg, respectively [31]. A disrupted equilibrium between protein synthesis and protein degradation is also attributed to hypobaric hypoxia-induced muscle atrophy. The impaired rates of protein synthesis were reported via measuring markers of protein synthesis such as mTOR [32]. Additionally, an in vivo study also reported disturbed proteostasis in terms of an elevated protein synthesis rate by 1.5-fold with an increased protein degradation rate of 5.0-fold that results in muscle atrophy [33]. A detailed study provided evidence of upregulated calpain and ubiquitin–proteasome pathways as the main cause behind increased rates of protein degradation [34].

4.4.3 Physical Performance

Prolonged exposure to high-altitude-associated hypobaric hypoxia leads to compromised physical performance, disrupted sleep patterns, cognitive impairment, and loss of body weight [35]. High-altitude inhabitants and sea-level natives display a similar maximal oxygen uptake (VO_{2max}) in normoxic conditions, but the high-altitude natives have the capability of achieving a higher VO_{2max} even when oxygen availability is lowered at an altitude of 3600 m [36]. Sea-level natives, after acclimatization to high altitude, display improved physical capacity and exercise performance, though not at similar levels to highlanders [36]. Under the condition when the fraction of inspired oxygen was reduced to 14% (similar to 3200 m altitude), the peak power output and VO_{2max} were both reduced [37]. Additionally, the peak heart rate and maximal oxygen capacity obtained from graded exercise tests also diminished during acute hypoxia exposure [38]. Weckbach et al. [39] displayed a considerable decline in peak power output and power output at reaching the lactic acid threshold on ascent to 2650 m. The lactic acid threshold refers to the intensity of exercise at which lactic acid accumulates and shows an exponential rise in the blood. In a study, 14 healthy individuals performed an exercise test on a cycle ergometer at baseline (521 m) and also at Zugspitze (2650 m). The workload to reach the lactate threshold was reduced by 10–20% at Zugspitze, thus indicating a significant reduction in lactate threshold at high altitudes. Thus, it indicates faster lactic acid buildup during exercise routines at higher altitudes.

4.4.4 Fiber Type

The skeletal muscle fibers with diameters varying from several microns to thousands of microns, rank among the largest cell types of the animal kingdom. They are formed from multinucleated syncytia, and the size of fibers appears to be limited by circulation restrictions linked with aerobic metabolism. Skeletal muscle is highly adaptable to external stimuli such as contractile activity and environmental conditions; for instance, on exposure to hypoxia, a decline in peak power output and peak oxygen uptake is observed in the event of whole-body exercise [40].

Skeletal muscle tissue is made up of heterogeneous fiber types, exhibiting distinctive metabolic and contractile characteristics. According to recent studies, skeletal muscle fibers can be classified based on the isoforms of myosin heavy chain (MHC): Type-I fibers, Type-IIA fibers, and Type-IIX/IIB fibers (Table 4.2).

Early studies conducted in lowland native animals and humans have reported the adaptive response of skeletal muscle tissue to ambient hypoxia. Earlier studies conducted on participants exposed to extreme altitude conditions have reported a decline of Type II and Type I fibers by 25% and 26%, respectively [41]. A study described the increase in slow Type I fibers and reduction in Type IIa fibers, highlighting fast to slow muscle fiber transition on chronic hypoxia exposure of 43 days, out of which 23 days were spent above 5000 m, consequently leading to rapid activation of mitochondrial oxidative metabolism [42].

Several studies have documented the role of the HIF-1 signaling pathway in fiber-type distribution in skeletal muscle with contrasting results. Mason et al. [43] estimated the composition of skeletal muscle fiber type in mice by preparing knockout mice through deletion of skeletal muscle-specific HIF-1α. The proportion of Type IIa fibers was documented to show a declining trend in the soleus muscle of HIF-1α knockout mice in contrast to wild-type mice. Besides this, the untrained HIF-1α knockout mice exhibited higher endurance performance than the wild-type mice. The knockout mice also displayed reduced fiber-type shifting from fast to

Table 4.2 Types of fibers and their properties

Types of fibers	Properties
Type-I fibers Slow-twitch oxidative	High mitochondrial content, high in myoglobin, high capillary supply, fatigue rate slower
Type-IIA fibers Fast oxidative glycolytic	Low oxidative capacity, reduced capillary density, larger cross-sectional areas, intermediate fatigue rate
Type-IIX/IIB fibers Fast-twitch glycolytic	Intermediate properties, high glycogen content, fatigue rate faster

slow in the case of strenuous exercise training and increased oxidative capacity after 6 weeks of training [44]. A recent study explored the effect of prolyl hydroxylase 2 (PHD2) protein terminations on fiber-type distribution in mice. This study established that HIF-1α stabilization in skeletal muscle tissue is driven by the deletion of PHD2 and influences a shift from fast fiber to slow fiber in the soleus and gastrocnemius muscle [45].

4.5 Underlying Mechanisms and Causes

Environmental hypoxia leads to a decline in fat-free mass percentage. Disturbed protein homeostasis, oxidative stress, imbalance in energy metabolism, and activation of the HIF pathway are all involved in the consequences afflicted by high-altitude exposure.

4.5.1 Oxidative Stress

Reactive oxygen species (ROS), such as superoxide, hydroxyl radicals, and hydrogen peroxide, occur naturally through oxidative metabolism in mitochondria. Cell signaling pathways employ these reactive oxygen species as chemical messengers, but an external adverse stimulus triggers excessive production of ROS, which leads to damage to biological macromolecules such as proteins, lipids, and DNA [46]. Nevertheless, protection is provided by antioxidant enzymes such as glutathione peroxidase and superoxide dismutase. The ascent to high altitude leads to formation of excess reactive oxygen species (ROS) and decrease antioxidant defence mechanisms [47]. Elevated and uncontrolled production of reactive oxygen and nitrogen species impairs cellular function and might have serious implications for organ functions. Mountaineering involves physical exercise combined with high-altitude exposure, resulting in oxidative damage [48].

4.5.1.1 Sources of Reactive Oxygen Species in Hypoxia

In living organisms, mitochondria play a crucial role in producing ROS, especially complex I and complex III of the electron transport chain (ETC) are believed to be the primary site of ROS production [49]. Under hypoxic conditions, the amount of oxygen available for reduction to H_2O molecules is reduced. This leads to the accumulation of reducing equivalents in the inner mitochondrial membrane and a decrease in the ability of cells to counteract oxidative stress [50]. This process, which results in ROS formation by auto-oxidation of one or more than one mitochondrial complexes like in ubiquinone-ubiquinol coupling, is termed reductive stress [51].

In the event of hypoxia-linked reductive stress, an augmented $NADH/NAD^+$ ratio has also been registered [52]. NAD^+-regulated proteins such as sirtuins, lactate dehydrogenase, and poly-ADP ribose polymerase are all affected by $NADH/NAD^+$

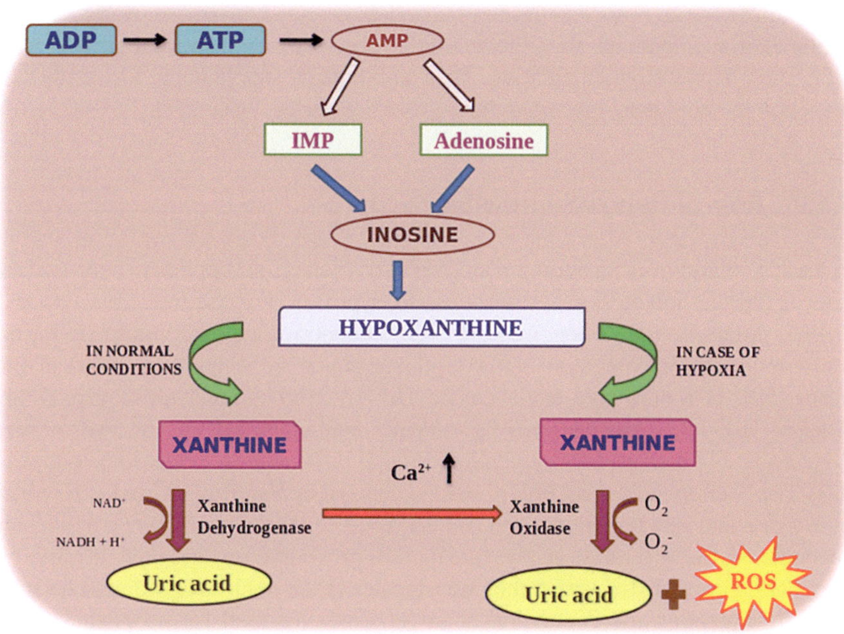

Fig. 4.1 Xanthine oxidase pathway

ratio [53]. At altitudes above 6000 m, cells produce ATP through an adenylate kinase-catalyzed reaction of two ADP molecules, also generating adenosine monophosphate (AMP) breaking down to form hypoxanthine. The conversion of xanthine dehydrogenase to xanthine oxidase occurs in the presence of calcium proteases and in this reaction, NAD^+ acts as an electron acceptor in place of an oxygen molecule [54]. The resulting xanthine production is coupled with superoxide anion formation. Thus, upon hypoxia exposure, xanthine dehydrogenase/oxidase system also acts as a dominant source of ROS production [46]. It has been observed that elevated ROS levels during hypoxia can revert to previous values that existed before hypoxia on reaching normoxia, although there is an upregulated inducible nitric oxide synthase, highlighting disrupted ROS/NO balance due to hypoxia [55] (Fig. 4.1).

4.5.1.2 Impact on Antioxidant Systems

Multiple findings have reported the impact of high altitude on the antioxidant system. Radak et al. [54] have reported a decline in activity and protein content of mitochondrial superoxide dismutase in rat skeletal muscle on exposure to intermittent hypoxia for 6 months. This signifies that intermittent hypoxic exposure amends the transcription of the enzyme as well. These findings also align with another study showing enhanced levels of mitochondrial superoxide dismutase in serum with a reduction in the liver and lung [56]. In another study, residents of high altitude exhibited low levels of glutathione peroxidase activity, which is under the influence of the thiol system. Some studies have also demonstrated the preventive efficacy of antioxidant supplements, thus preventing oxidative damage to biological macromolecules. Upon supplementation, such antioxidant supplements also induce a reduction in ventilatory threshold during exercise performance [57].

4.5.2 Role of Hypoxia-Inducible Factor

In the state of hypoxia, an optimum amount of oxygen is not supplied to body tissues, thus disturbing homeostasis. In response to hypoxia, Hypoxia Inducible Factor 1 (HIF-1) comes into play, acting as the chief regulator of hypoxic impact [58]. This transcription factor exists as an α/β heterodimer, acting as the master regulator of 100 genes [59]. In a normoxic environment, HIF-1β exhibits constitutive expression, whereas HIF-1α is hydroxylated by enzymes containing HIF prolyl hydroxylase domains and is marked for degradation through the ubiquitin–proteasome pathway [60]. The hydroxylated form of HIF-α is recognized by Von Hippel-Lindau E3 ubiquitin ligase and gets ubiquitinylated and degraded. On the contrary, when hypoxia is encountered, the lower oxygen pressure inhibits prolyl hydroxylase domain proteins, thus stabilizing HIF-1α, which then translocates to the nucleus and dimerizes with HIF-1β, promoting the expression of target genes required for adaptation under hypoxia. The understanding of physiological responses in body tissues has been greatly aided by the discovery of oxygen-sensing pathways [61, 62] (Fig. 4.2).

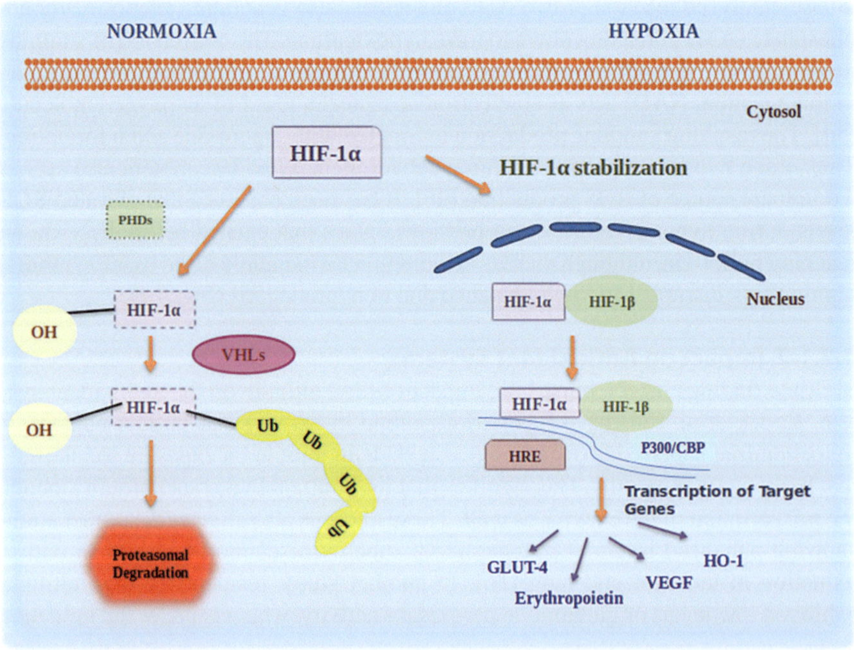

Fig. 4.2 Hypoxia-inducible factor (HIF) pathway

In skeletal muscle, transient hypoxia occurs when, in response to physical exercise, the oxygen pressure declines; under these circumstances, HIF-1α comes into play [63]. In this instance, the oxygen supply is unable to meet oxygen requirements during contraction. HIF-1α stabilizes and translocates to the nucleus in response, inducing VEGF expression [64]. HIF-1α induces angiogenesis in skeletal muscle via upregulation of VEGF [65]. VEGF can trigger angiogenesis not only in exercising muscle but also in resting muscle [66]. Skeletal muscle relies on either nitric oxide-dependent or -independent mechanisms to induce vasodilation to recover oxygen supply for heightened metabolic demands [67]. In harmony with these observations, a correlation between HIF-1α protein content and free radical production was registered in hypoxia-exposed rat muscle [33]. Other target genes regulated by HIF-1α are EPO, HO-1, and GLUT-4. Erythropoietin (EPO), a glycoprotein responsible for the proliferation and activation of vascular endothelial cells, is upregulated in hypoxic conditions. Increased EPO during reduced oxygen supply makes our body produce more RBCs for the augmentation of oxygen delivery [68]. HIF-1α also regulates glucose metabolism in skeletal muscle. The major glucose transporter protein dominantly present in skeletal muscle is GLUT-4 [69]. The stimulation of glucose transport through activation of GLUT-4 translocation in myocytes has been demonstrated through in vitro studies. In a study on healthy humans, GLUT-4 was found to be increased by 30% in the sarcolemmal membrane after a 4-hour hypoxia exposure. Hypoxia also led to a diminished glucose response. Thus, hypoxia may decrease glucose response after a high glycemic meal via increasing GLUT-4 abundance at the sarcolemmal membrane through insulin-independent pathways [70]. Acute hypoxia exposure causes protein misfolding and modifications. In response, heat shock proteins and chaperones get activated. The accumulation of misfolded proteins induces an enhancement in 20S proteasomal activity. Agrawal et al. have reported significant increases in levels of HSP60, HSP70, and GRP78 on hypoxic exposure [26]. Heme oxygenase 1 (HO-1) activation is induced during cellular stress through HIF-1α activation [71]. The increased expression of HO-1 offers cytoprotection and antiapoptotic effects through multiple pathways [72].

4.5.3 Endoplasmic Reticulum Stress

Endoplasmic reticulum (ER) stress is characterized by endoplasmic reticulum dysfunction. Sojourn to high altitude acts as a trigger for inducing oxidative stress, resulting in oxidatively modified proteins, leading to ER stress and progressing into pulmonary hypertension. Under pathological circumstances, unfolded and misfolded proteins accumulate as a result of ER stress. When ER stress occurs, cells produce an unfolded protein response [73]. The unfolded protein response generated by ER stress activates PRK-like ER kinase (PERK), inositol requiring enzyme-1 (IRE1α), and activating transcription factor-6 (ATF6) in the ER membrane [74]. The activation of PERK, IRE1α, and ATF6 is mediated by competitive binding with Bip/GRP78, following which Bip/GRP78 gets dissociated, in return upregulating the proapoptosis transcription factor CHOP, thereby activating apoptosis pathway in ER stress-mediated fashion. IRE1α, acting as an endonuclease post activation, initiates splicing of a 26-base intron from X-box binding protein 1 (XBP1) mRNA, which further activates the expression of chaperones to boost the folding capacity of ER [75]. An earlier study published by our lab stated that ER stress due to chronic hypobaric hypoxia stimulates cardiac hypertrophy and injury in exposed rats [76]. The study registered activated expression of ATF4, phosphorylation of eIF2α, ER chaperone expression, and, lastly, protein degradation. Another study, also published by our lab, demonstrated elevated levels of chaperones and heat shock proteins such as GRP78, HSP70, and HSP60 in the skeletal muscle of hypoxia-exposed rats [26]. Further, the upregulation of caspase 12 is also mediated by the ER stress pathway [77].

Jain et al. [78] observed an increased expression of endoplasmic reticulum (ER) chaperones in the myocardium of rats, such as GRP78, PDI, and calreticulin, thus activating the ER stress response in animals susceptible to hypoxia. The susceptible animals showed a 50% enhancement in ER stress markers compared to the control group. There was no alteration in the expression of ER chaperones in tolerant animals. GRP94 was found to be threefold higher in susceptible animals, whereas in tolerant animals and normal animals, it was found to be twofold higher compared to unexposed animals [78]. Existing evidence suggests oxidative stress and reactive oxygen species are an integral part of ER stress.

4.5.4 Proteostasis (Protein Synthesis and Degradation)

The myofibrillar proteins, a part of skeletal muscle proteins, are characterized by a substantial turnover rate. Variations in the rate of protein synthesis may be influenced by dietary adjustments or the emergence of any pathological disease. The balance between protein production and breakdown is essential for controlling muscle hypertrophy [79]. Anabolic resistance may be posed by acute hypoxia exposure, according to research on the rates of protein synthesis at rest and after exercise in seven healthy men. The subjects were exposed to normobaric hypoxia for 3.5 hours and a decline in muscle protein synthesis was observed, majorly as a result of resistance exercise [80]. Ergo, other studies also were in agreement with these results as disturbed protein turnover rate was reported on ascent to high altitude around 4300–4600 m [81]. An in vivo study published by our lab observed continually increased rates of protein degradation in male Sprague–Dawley rats exposed to an altitude of 7620 m for 3, 7, and 14 days [33].

The IGF-1-Akt-mTOR pathway acts as a positive regulator of muscle mass, while Myostatin-Smad2 and Smad3 pathway is the negative regulator of muscle mass. Muscle atrophy, resulting from an enhancement in the rate of protein degradation and a decline in the rate of protein synthesis, is characterized by the reduction in the cross-sectional area of muscle fibers and a loss of total proteins [82]. The occurrence of acute muscle atrophy has been observed in multiple conditions including cachexia, muscular dystrophy, starvation, aging, cancer, and diabetes, along with exposure to high altitude. There are primarily two mechanisms that cause protein degradation: the ubiquitin–proteasome and the autophagy–lysosomal systems. The musculoskeletal system holds such a large proportion of body mass; thus, loss of muscle mass is of high significance in protein homeostasis.

4.5.4.1 Protein Synthesis Pathway

The musculoskeletal system plays a critical role in the health and well-being of an individual and has a direct relationship with the quality of life. The preservation of skeletal muscle mass requires the maintenance of equilibrium between protein synthesis and protein degradation.

The rate of intracellular protein synthesis is dictated by the efficiency and capacity of translation. Protein synthesis per unit amount of RNA is termed translational efficiency, whereas the total amount of ribosomal content per unit tissue is termed translational capacity [83]. Past research emphasized understanding the signaling pathways leading to the enhancement of translational efficiency, and in recent times, the prime focus has been on investigating the pathways regulating translational capacity [84]. The study by Nader and his team highlighted the inhibitory role of rapamycin in the increment of ribosomal RNA in the case of muscle hypertrophy, suggesting mTOR/TORC1 might have a role in the synchronization of translational efficiency and capacity. The mTOR pathway is the major pathway involved in the regulation of protein synthesis and enhancement of translational efficiency [85].

A downregulation of the Akt/mTOR pathway is also observed after hypoxia exposure in rats [86]. Costes et al. showed that IGF-1-mediated activation of the Akt/mTOR pathway was blocked in C2C12 myotubes under hypoxia conditions [87]. Myostatin, a myokine that is a negative regulator of muscle mass, was found to be increased during hypoxia exposure [88]. The inhibition of myostatin prevented muscle atrophy under hypoxic conditions, thus proving the contribution of myostatin to muscle atrophy [89].

4.5.4.2 Protein Degradation Pathways

Muscle atrophy is caused by elevated levels of protein breakdown, and the rates of protein synthesis and degradation are influenced by a number of diseases and environmental variables. Protein synthesis rates are surpassed by rates of protein breakdown in catabolic circumstances, which results in atrophy. The proteolytic systems in cells are well regulated via transcription and a set of genes are involved in atrophy, which either get upregulated or downregulated.

The protein degradation can occur through four different pathways: the ubiquitin–proteasome pathway, autophagy, the calpain system, and the cysteine aspartate protease system [90, 91]. The ubiquitin–proteasome pathway is the main route responsible for protein breakdown. Ubiquitin, enzymes linked to ubiquitin, and proteasomes comprise the pathway's constituents. Proteins are marked for degradation by the 26S proteasome and tagged for destruction by the 76 amino acid ubiquitin. The reversible nature of ubiquitination allows deubiquitinating enzymes to inhibit the breakdown of target proteins by removing ubiquitin from them. The enzymes involved in ubiquitination are: E1, responsible for ubiquitin activation; E2, ubiquitin conjugating enzyme, E3 ligase, and E4 for ubiquitin chain extension. The E3 ligase determines the specificity and rate of the ubiquitin–proteasome system. The 26S proteasome is involved in the degradation of polyubiquitinated proteins into peptides and is composed of 20S and either one or two 19S proteasomes. 20S recognizes and degrades misfolded or damaged proteins whereas 19S recognizes ubiquitinated proteins. Target proteins get unfolded in order to pass through the 20S core protein. Consequently, substrates get degraded into shorter peptides, and the attached ubiquitin chains are released for reuse.

4.5.4.3 Autophagy–Lysosomal Pathway

Autophagy is a homeostatic mechanism employed for degradation and recycling under different pathological conditions. Earlier findings have suggested the activation of the autophagy–lysosome system in muscle cells under catabolic conditions [92]. Autophagy is induced through both endurance and resistance exercise [93]. It is activated through multiple mechanisms, including ULK1 activation resulting from TORC1 inhibition. An elevated expression of autophagy markers such as BNIP3 and LC3 has been linked with hypoxic exposure [94]. The activation of muscle mitophagy under a hypoxic environment appears to be evident. Nevertheless, the detailed involvement of autophagy in muscle mass regulation needs to be extensively studied.

4.5.5 Mitochondrial Dysfunction

The hypobaric hypoxia-induced decline in partial pressure of arterial oxygen is counteracted by the transformed rate of ventilation, hemoglobin concentration, and capillary density. A significant reduction in mitochondrial volume density has been registered in the skeletal muscle of individuals ascending to high altitudes [95]. Numerous studies have investigated the metabolic response of skeletal muscle tissue to hypoxia exposure. Chronic exposure to high-altitude conditions culminates in a loss of mitochondrial volume density, including a larger loss of subsarcolemmal population [96]. A 50% loss of mRNA of peroxisome proliferator-activated receptor gamma co-activator 1 alpha (PGC1α), along with modified expression of genes and alterations in protein expression levels of metabolic enzymes and regulators has been observed [96].

In vitro studies have supported the claim of hypoxia-inducible factor (HIF) stabilization and altered metabolism in humans at high altitudes and rat models. HIF-1α knockout mice displayed augmentation in exercise capacity and oxidative metabolism in the gastrocnemius muscle of mice. HIF-1α is a key transcription factor responsible for the upregulation of glycolytic enzymes, including aldolase A, glyceraldehyde-3-phospahte dehydrogenase, hexokinase, lactate dehydrogenase, and pyruvate kinase, all of which are involved in the enhancement of the glycolytic capacity of the cell. Lactate dehydrogenase and hexokinase have been reported to be markedly amplified on hypoxia exposure [97]. Furthermore, HIF-1α acts as an inhibitor for oxidative metabolism by limiting the entry of pyruvate into the tricarboxylic acid cycle through the activation of pyruvate dehydrogenase kinase [98].

Pyruvate dehydrogenase enzyme complex catalyzing the conversion of pyruvate into acetyl coenzyme A is inactivated through phosphorylation by pyruvate dehydrogenase kinase. Upregulation of pyruvate dehydrogenase kinase has been observed in the skeletal muscle of rats exposed to 10% oxygen availability [99]. These results were not in accordance with the expression levels of pyruvate dehydrogenase kinase in humans ascended to 4560 m for 7–9 days [32]. Contradictory to this, augmented pyruvate dehydrogenase kinase protein levels were reported after 24 hours of exposure to 4300 m but declined to basal levels after 1 week. The alterations in muscle respiratory function seem to occur at moderate altitudes and are entirely dependent on the duration of exposure. Several studies have described the repression of fatty acid oxidation enzymes on chronic exposure at different altitudes ranging from 4300 to 8848 m [96]. HIF-1α has been demonstrated to cause a reduction in the DNA-binding capability of peroxisome proliferator-activated receptor α (PPARα), which regulates lipid peroxidation [100]. The initiation of oxidative metabolism by impairment of fatty acid oxidation has been reported along with HIF-1α knockout mice displaying enhanced lipid oxidation [44]. Furthermore, hypoxic exposure causes a decline in the protein content of PPARα and PGC-1α and also reduces the transcriptional activity of the PPARα/ PGC-1α complex [100]. Thus, the import and β-oxidation of fatty acids are greatly hindered by hypoxia exposure, as PPARα and PGC-1α are major regulators of fatty acid metabolism. A significant reduction in fatty acid oxidation has been reported after 1 week of hypoxia exposure [101].

4.5.6 Myokines

In past studies, skeletal muscle has been described as an endocrine organ that produces and secretes an array of proteins and peptides. The term "secretome" refers to the collection of biological substances and compounds that cells secrete into the extracellular space [102]. An understanding of the secretome of muscle cells could help us recognize the processes that manage systemic interactions, organization, and synthesis of skeletal muscles, which further could help in identifying proteins involved in several conditions such as adaptation, degeneration, exercises, and aging. Skeletal muscle is capable of secreting several factors facilitating communication with other organs including the brain, adipose tissue, pancreas, gut, liver, and skin [103]. Myokines are defined as cytokines or peptides secreted by skeletal muscle cells and subsequently released into circulation exerting endocrine or paracrine effects in other cells, tissues, and organs [104]. Muscle contraction is the regulator of muscle secretome, which is also coupled with physical exercise [105]. Myokines also display effects on physiological processes, including muscle contraction, cell proliferation, differentiation, growth, and muscle metabolism. Cellular injury activates myokines causing their migration to the site of injury, leading to satellite cell activation. Thus, myokines are crucial for balanced quiescence and skeletal muscle homeostasis [106]. The concept of myokines has opened new avenues for the understanding of molecular consequences of physical exercise for the formation of strategies for interventions and identification of target molecules for moderating the beneficial effects of regular physical activity.

Currently, more than 600 myokines have been reported [107]. The myokines involved in the positive regulation of muscle mass are Irisin, decorin, interleukin-6, interleukin-15, and fibroblast growth factor 21 (FGF-21), while myostatin is involved in the negative regulation of muscle mass and induction of muscle atrophy. As these myokines are involved in the regulation of equilibrium between protein synthesis and degradation, upon ascent to high altitude, the myokine profile is also disturbed, being a cause behind the disturbance.

Śliwicka et al. conducted a human study on individuals in the age group between 23 and 31 years who ascended to an altitude greater than 3000 m and reported a significant decline in irisin and 25-hydroxy vitamin D levels [108]. Accordingly, research was conducted in vivo on male Sprague–Dawley rats in which hypobaric hypoxia exposure was simulated for 1, 3, and 7 days at an altitude of 7620 m in an animal decompression chamber. The study revealed that the hypoxic group had considerably higher levels of myostatin compared to the normoxic group when exposure times were compared. Also, the myogenesis-related markers such as PAX-7, Cyclin D1, and myogenin were found to be downregulated [28]. A recent study was also performed on humans who stayed at high altitudes for acute and chronic periods. The study demonstrated significantly increased irisin levels in the acute exposure group compared to the control group, whereas myostatin levels showed a significant increase in the chronic exposure group [109]. Only a handful number of studies are available describing the impact of high-altitude stay on myokines profile. Further detailed studies are required to unfold the mechanisms behind the muscle atrophy and link with myokines.

4.5.7 microRNA (miRNA)

MicroRNAs are a class of noncoding RNAs with short chain lengths of around 22 nucleotides. They regulate diverse cellular activities like apoptosis, development, cellular growth, and differentiation. miRNAs facilitate the modulation of gene expression via complementary base pairing with target nucleotide sequences. The diverse processes involved in muscle growth and development, like muscle fiber type transition, atrophy, cell proliferation, hypertrophy, and differentiation, are regulated by miRNAs [110]. Research investigations have described alterations in miRNA expressions of hypoxia-exposed murine muscles, focusing on their role in AMPK-mediated energy metabolism [111]. On the other hand, little is known about how the miRNA expression profile changes when skeletal muscle atrophy is brought on by hypobaric hypoxia. In skeletal muscle, hypoxia caused differential expression of nine miRNAs: miR-1306-3p, miR-32-5p, miR-203a-3p, miR-341, miR-326-3p, miR-760-3p, miR-465-5p, miR-499-5p, and miR-205 [112]. Currently, the role of miRNA in skeletal muscle atrophy has become a subject of interest. Under various catabolic conditions, an altered miRNA profile is attributed to skeletal muscle loss and atrophy. An alteration in around 272 miRNA was reported by a study conducted

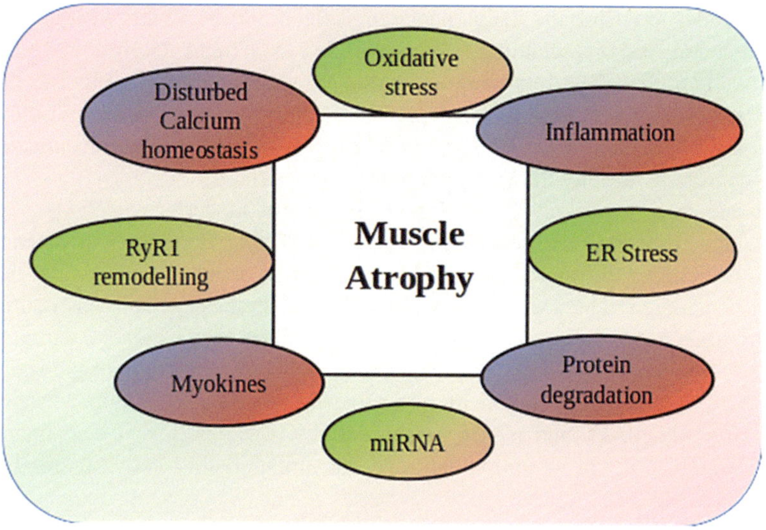

Fig. 4.3 Hypobaric hypoxia-associated skeletal muscle atrophy regulated by multiple signaling pathways

on murine muscles during space flight [113]. In the case of amyotrophic lateral sclerosis, miR-206 was found to be elevated [114]. A pilot study conducted on individuals of two different age groups: 68-72 years and 27-31 years reported the role of miRNA in skeletal muscle atrophy. The results demonstrated elevated expression of miRNA-133a-1, pri-miRNA-133a-2, primiRNA-1–2, and pri-miRNA-1-1 [115]. Our most recent study described the overexpression of miR1 in the case of hypobaric hypoxia-induced skeletal muscle loss. The muscle degradation markers such as FOXO, myostatin, and atrogin were also found to be increased, thus suggesting impaired myogenesis [116]. A detailed study on the role of miRNAs is further required to elucidate the mechanism behind hypobaric hypoxia-induced skeletal muscle atrophy.

In summary, skeletal muscle atrophy linked to hypobaric hypoxia is a complex pathophysiological state controlled by a variety of signaling mechanisms, including altered network, disturbed anabolic signaling, impaired calcium homeostasis, RyR1 remodeling, ER stress, inflammation, apoptosis, and enhanced protein degradation pathway (Fig. 4.3).

4.6 Strategies for Managing Hypobaric Hypoxia-Induced Muscle Atrophy

Managing muscle health is essential for quality of life, especially for personnel who are deployed to and are staying at high-altitude regions. Optimal physical performance of sojourners depends on skeletal muscle health and muscle mass. Thus, preventive strategies and interventions are required for individuals ascending to high altitudes to ensure the maintenance of skeletal muscle health. Multiple health issues encountered on ascending to high altitude can be managed by a well-planned nutritional strategy. The interventions employed should target the precise mechanisms behind high-altitude-induced muscle atrophy and ameliorate muscle loss. The following section describes a few dietary supplements and nutraceuticals that alleviate muscle atrophy induced by hypobaric hypoxia.

Ursolic acid, a pentacyclic triterpene, is present in herbs, vegetables, whole grains, and fruits [117], especially in apples and in the waxy coats of other fruits [118]. Enriched with pharmacological activities and being relatively safe, ursolic acid is considered to be a wonder molecule. Several studies have already confirmed the non-toxic nature of ursolic acid [90, 91, 119]. In cases of oxidative damage and inflammation due to hypobaric hypoxia, ursolic acid can regulate multiple signaling pathways. Ursolic acid supplementation enhances muscle protein concentration by improving Akt, pAkt, and p70S6 kinase activity. Besides this, it also maintains redox homeostasis via its antioxidant and anti-inflammatory activities in skeletal muscle [90, 91].

Carnosine, formed by the amide linkage between β-alanaine and L-histidine, is a naturally occurring dipeptide. It is reported to be found in the skeletal muscle, brain, heart, and gastrointestinal tissues of humans. The potential efficacy of carnosine has already been registered in multiple malignancies such as Alzheimer's disease [120], lung injury [121], atherosclerosis, dementia, Parkinson's [122], and cancer [123]. The biological properties of carnosine include pH buffering ability, antioxidant activity, anti-inflammatory activity, antiaging effect, enhancement of cognitive function, and immunomodulation, thus making it an ideal target for nutraceutical development. Carnosine is able to maintain physiological pH, thus ameliorating the damage caused due to oxidative stress and pH imbalance. Recently, carnosine supplementation was found to be effective in ameliorating hypobaric hypoxia-induced skeletal muscle loss by attenuating ER stress response. The expression levels of ER stress-linked proteins such as PERK, ATF4, and phosphorylated-eIF-2α showed a declining trend in hypobaric hypoxia-exposed rats after carnosine supplementation [124].

Curcumin longa (Turmeric) with a plethora of health benefits has widely been documented for its medicinal properties [125]. In India, it has been used for generations as a traditional spice. Curcumin is a therapeutic plant because of its antibacterial, anti-inflammatory, and antioxidant properties. Surplus human trials have documented its safety and efficacy. A few recent studies have indicated the upregulation of the Nrf2 pathway in skeletal muscle atrophy conditions [126]. The supplementation of curcumin has been shown to promote muscle satellite cell regeneration, elevation of desmin expression, reduced lipid peroxidation, and fasten muscle recovery [127]. It has been observed that curcumin ameliorates chronic hypobaric hypoxia-induced skeletal muscle protein loss through the upregulation of multiple cellular signaling pathways. The administration of carnosine has been noted to elevate the expression of myogenic proteins like MyF5, MyoD, and myogenin needed for proliferation and myogenesis. It also reduced muscle permeability, thus restricting creatine phosphokinase leakage and enhancing the physical performance of chronic hypoxia-exposed rats [128].

Sea buckthorn (*Hippophae rhamnoides L.*), a Himalayan medicinal plant, is renowned for its nutritional and medicinal properties. Majorly found in Europe and Asia, it is a cold-resistant nitrogen-fixing shrub. Our previous work has extensively reviewed the medicinal and therapeutic properties of sea buckthorn [129]. It possesses anti-stress and adaptogenic properties against cold, hypoxia, and restraint [130]. Sea buckthorn boosts hypoxic tolerance and survival when administered orally. In vivo investigations conducted by our lab showed that sea buckthorn extracts increased the expression of hypoxia-responsive proteins, including HIF-1α and HO-1 [131]. Sea buckthorn also worsened NO production by increasing iNOS and eNOS expression. Furthermore, sea buckthorn extract reduced hypoxia-induced ER stress by downregulating CHOP and PERK. Sea buckthorn significantly reduced inflammation and apoptosis in hypoxia-exposed mice via altering proteostasis signaling systems [76, 132].

Conclusively, high-altitude-linked skeletal muscle atrophy is regulated by multiple signaling pathways, including oxidative stress, ER stress, inflammation, distorted anabolic pathways, RyR1 remodeling, calcium imbalance, apoptosis, altered myokine profiles, and excessive protein degradation.

Recent scenarios force to use dietary molecules as an intervention due to safe, easy availability, and palatability. A number of in vitro and in vivo studies have established the protective efficacy of dietary intervention for mitigating hypobaric hypoxia-induced muscle protein loss. Now, it is high time to translate these interventions into clinical studies and establish proof of concept.

Acknowledgments The authors are thankful to Dr. Rajeev Varshney, Director, DIPAS, for his constant support and encouragement.

Conflict of Interest The authors declare no conflict of interest in the contents of this article.

References

1. Mukund K, Subramaniam S. Skeletal muscle: a review of molecular structure and function, in health and disease. Wiley Interdiscip Rev Syst Biol Med. 2020;12(1):e1462.
2. Saltin B, Helge JW. Skelettmuskulatur, körperliche Aktivität und Gesundheit [Skeletal muscles, physical activity and health]. Orthopade. 2000;29(11):941–7. German. https://doi.org/10.1007/s001320050546.
3. Brown JC, Harhay MO, Harhay MN. The muscle quality index and mortality among males and females. Ann Epidemiol. 2016;26:648–53.
4. Riviati N, Indra B. Relationship between muscle mass and muscle strength with physical performance in older adults: a systematic review. SAGE Open Med. 2023;11:20503121231214650.
5. Wang X, Pickrell AM, Rossi SG, et al. Transient systemic mtDNA damage leads to muscle wasting by reducing the satellite cell pool. Hum Mol Genet. 2013;22(19):3976–86.
6. Favier FB, Britto FA, Freyssenet DG, Bigard XA, Benoit H. HIF-1-driven skeletal muscle adaptations to chronic hypoxia: molecular insights into muscle physiology. Cell Mol Life Sci. 2015;72:4681–96.
7. Gherardi G, De Mario A, Mammucari C. The mitochondrial calcium homeostasis orchestra plays its symphony: skeletal muscle is the guest of honor. Int Rev Cell Mol Biol. 2021;362:209–59.

8. Rojas-Camayo J, Mejia CR, Callacondo D, Dawson JA, Posso M, Galvan CA, Davila-Arango N, et al. Reference values for oxygen saturation from sea level to the highest human habitation in the Andes in acclimatized persons. Thorax. 2018;73(8):776–8.
9. Prince TS, Thurman J, Huebner K. Acute mountain sickness. [Updated 2023 Jul 10]. In: StatPearls [Internet]. Treasure Island: StatPearls Publishing; 2024. Available from: https://www.ncbi.nlm.nih.gov/books/NBK430716/.
10. Theunissen S, Balestra C, Bolognési S, Borgers G, Vissenaeken D, Obeid G, De Bels D. Effects of acute hypobaric hypoxia exposure on cardiovascular function in unacclimatized healthy subjects: a "rapid ascent" hypobaric chamber study. Int J Environ Res Public Health. 2022;19(9):5394.
11. West JB. Human responses to extreme altitudes. Integr Comp Biol. 2006;46(1):25–34.
12. Beall CM. Adaptation to high altitude: phenotypes and genotypes. Annu Rev Anthropol. 2014;43:251–72.
13. Bärtsch P, Gibbs JS. Effect of altitude on the heart and the lungs. Circulation. 2007;116:2191–202.
14. Fedele FG, Giaccio B, Isaia R, Orsi G. Ecosystem impact of the Campanian ignimbrite eruption in Late Pleistocene Europe. Quat Res. 2002;57(3):420–4.
15. Peacock AJ. ABC of oxygen: oxygen at high altitude. BMJ. 1998;317:1063–6.
16. Gröbner J, Albold A, Blumthaler M, Cabot T, De la Casiniere A, Lenoble J, Weihs P. Variability of spectral solar ultraviolet irradiance in an Alpine environment. J Geophys Res Atmos. 2000;105(D22):26991–7003.
17. Blumthaler M, Ambach W, Ellinger R. Increase in solar UV radiation with altitude. J Photochem Photobiol B Biol. 1997;39(2):130–4.
18. Schmucki DA, Philipona R. Ultraviolet radiation in the Alps: the altitude effect. Opt Eng. 2002;41(12):3090–5.
19. Cymerman A, Marriott BM, Carlson SJ. The physiology of high-altitude exposure. In: Marriott BM, Carlson SJ, editors. Nutritional needs in cold and high-altitude environments. Washington DC: National Academic Press; 1996.
20. Milledge JS. Altitude medicine and physiology including heat and cold: a review. Travel Med Infect Dis. 2006;4(3–4):223–37.
21. Feher JJ. Quantitative human physiology: an introduction. Academic press; 2017.
22. Jones RD, Hancock JT, Morice AH. NADPH oxidase: a universal oxygen sensor? Free Radic Biol Med. 2000;29:416–24.
23. Hoppeler H, Howald H, Cerretelli P. Human muscle structure after exposure to extreme altitude. Experientia. 1990;46:1185–7.
24. Kayser BENGT, Acheson KEVIN, Decombaz JACQUES, Fern EDWARD, Cerretelli PAOLO. Protein absorption and energy digestibility at high altitude. J Appl Physiol. 1992;73(6):2425–31.
25. Hoppeler H, Vogt M, Weibel ER, Flück M. Response of skeletal muscle mitochondria to hypoxia. Exp Physiol. 2003;88(1):109–19.
26. Agrawal A, Rathor R, Suryakumar G. Oxidative protein modification alters proteostasis under acute hypobaric hypoxia in skeletal muscles: a comprehensive in vivo study. Cell Stress Chaperones. 2017;22(3):429–43.
27. Agrawal A, Rathor R, Kumar R, Suryakumar G, Singh SN, Kumar B. Redox modification of ryanodine receptor contributes to impaired Ca2+ homeostasis and exacerbates muscle atrophy under high altitude. Free Radic Biol Med. 2020;160:643–56.
28. Srivastava S, Rathor R, Singh SN, Suryakumar G. Insight into the role of myokines and myogenic regulatory factors under hypobaric hypoxia induced skeletal muscle loss. Biomarkers. 2022;27(8):753–63.
29. Surks MI, Chinn KS, Matoush LR. Alterations in body composition in man after acute exposure to high altitude. J Appl Physiol. 1966;21(6):1741–6.
30. Boyer SJ, Blume FD. Weight loss and changes in body composition at high altitude. J Appl Physiol. 1984;57(5):1580–5.

31. Bharadwaj H, Prasad J, Pramanik SN, Kishnani S, Zachariah T, Chaudhary KL, Sridharan K, Srivastava KK. Effect of prolonged exposure to high altitude on skeletal muscles of Indian soldiers. Def Sci J. 2000;50:167–76.
32. Viganò A, Ripamonti M, De Palma S, Capitanio D, Vasso M, Wait R, Gelfi C. Proteins modulation in human skeletal muscle in the early phase of adaptation to hypobaric hypoxia. Proteomics. 2008;8(22):4668–79.
33. Chaudhary P, Suryakumar G, Prasad R, Singh SN, Ali S, Ilavazhagan G. Chronic hypobaric hypoxia mediated skeletal muscle atrophy: role of ubiquitin–proteasome pathway and calpains. Mol Cell Biochem. 2012;364(1–2):101–13.
34. Agrawal A, Rathor R, Kumar R, Suryakumar G, Ganju L. Role of altered proteostasis network in chronic hypobaric hypoxia induced skeletal muscle atrophy. PLoS One. 2018;13(9):e0204283.
35. Weil JV, White DP. Sleep. In: Hornbein TF, Schoene RB, editors. High altitude. New York: Marcel Dekker; 2001. p. 707–76.
36. Favier R, Spielvogel H, Desplanches D, Ferretti G, Kayser B, Hoppeler H. Maximal exercise performance in chronic hypoxia and acute normoxia in high-altitude natives. J Appl Physiol. 1995;78:1868–74.
37. Hughson RL, Green HJ, Sharratt MT. Gas exchange, blood lactate, and plasma catecholamines during incremental exercise in hypoxia and normoxia. J Appl Physiol. 1995;79:1134–41.
38. Ozcelik O, Kelestimur H. Effects of acute hypoxia on the estimation of lactate threshold from ventilatory gas exchange indices during an incremental exercise test. Physiol Res. 2004;53(6):653–9.
39. Weckbach LT, Kassem S, Maier F, Hamm W, Schüttler D, Kellnar A, Lackermair K, Brunner S. Impact of acute altitude exposure on lactate threshold. High Alt Med Biol. 2019;20(3):322–3.
40. Sandiford SD, Green HJ, Duhamel TA, Schertzer JD, Perco JD, Ouyang J. Muscle Na-K-pump and fatigue responses to progressive exercise in normoxia and hypoxia. Am J Phys Regul Integr Comp Phys. 2005;289(2):R441–9.
41. MacDougall JD, Green HJ, Sutton JR, Coates G, Cymerman AYP, Houston CS. Operation Everest-II: structural adaptations in skeletal muscle in response to extreme simulated altitude. Acta Physiol Scand. 1991;142:421–7.
42. Doria C, Toniolo L, Verratti V, Cancellara P, Pietrangelo T, Marconi V, Paoli A, Pogliaghi S, Fano G, Reggiani C, et al. Improved VO2 uptake kinetics and shift in muscle fiber type in high-altitude trekkers. J Appl Physiol. 2011;111:1597–605.
43. Mason SD, Howlett RA, Kim MJ, Olfert IM, Hogan MC, McNulty W, Johnson RS. Loss of skeletal muscle HIF-1α results in altered exercise endurance. PLoS Biol. 2004;2(10):e288.
44. Mason SD, Rundqvist H, Papandreou I, Duh R, McNulty WJ, Howlett RA, Johnson RS. HIF-1α in endurance training: suppression of oxidative metabolism. Am J Phys Regul Integr Comp Phys. 2007;293(5):R2059–69.
45. Shin J, Nunomiya A, Kitajima Y, Dan T, Miyata T, Nagatomi R. Prolyl hydroxylase domain 2 deficiency promotes skeletal muscle fiber-type transition via a calcineurin/NFATc1-dependent pathway. Skelet Muscle. 2015;6:1–16.
46. Irarrázaval S, et al. Oxidative stress in acute hypobaric hypoxia. High Alt Med Biol. 2017;18(2):128–34.
47. Bailey DM, Davies B, Young IS, Hullin DA, Seddon PS. A potential role for free radical-mediated skeletal muscle soreness in the pathophysiology of acute mountain sickness. Aviat Space Environ Med. 2001;72(6):513–21.
48. Wozniak A, Drewa G, Chesy G, Rakowski A, Rozwodowska M, Olszewska D. Effect of altitude training on the peroxidation and antioxidant enzymes in sportsmen. Med Sci Sports Exerc. 2001;33(7):1109–13.
49. Brand MD. The sites and topology of mitochondrial superoxide production. Exp Gerontol. 2010;45:466–72.
50. Mohanraj P, Merola AJ, Wright VP, Clanton TL. Antioxidants protect rat diaphragmatic muscle function under hypoxic conditions. J Appl Physiol. 1998;84(6):1960–6.

51. Kanter M. Free radicals, exercise and antioxidant supplementation. Proc Nutr Soc. 1998;57:9–13.
52. Khan S, O'Brien PJ. Modulating hypoxia-induced hepatocyte injury by affecting intracellular redox state. Biochim Biophys Acta. 1995;1269:153–61.
53. Radak Z, Acs Z, Bori Z, Taylor AW, Yang H. The effects of high-altitude exposure on reactive oxygen and nitrogen species. In: Systems biology of free radicals and antioxidants. Berlin Heidelberg: Springer-Verlag; 2014. p. 407–16.
54. Radak Z, Asano K, Inoue M, Kizaki T, Oh-Ishi S, Suzuki K, Taniguchi N, Ohno H. Superoxide dismutase derivative reduces oxidative damage in skeletal muscle of rats during exhaustive exercise. J Appl Physiol. 1995;79:129–35.
55. Gonzalez NC, Wood JG. Leukocyte-endothelial interactions in environmental hypoxia. Adv Exp Med Biol. 2001;502:39–60.
56. Nakanishi K, Tajima F, Nakamura A, Yagura S, Ookawara T, Yamashita H, Suzuki K, Taniguchi N, Ohno H. Effects of hypobaric hypoxia on antioxidant enzymes in rats. J Physiol. 1995;489(Pt 3):869–76.
57. Subudhi AW, Jacobs KA, Hagobian TA, Fattor JA, Muza SR, Fulco CS, Cymerman A, Friedlander AL. Changes in ventilatory threshold at high altitude: effect of antioxidants. Med Sci Sports Exerc. 2006;38:1425–31.
58. Semenza GL, Wang GL. A nuclear factor induced by hypoxia via de novo protein synthesis binds to the human ery thropoietin gene enhancer at a site required for transcriptional activation. Mol Cell Biol. 1992;12:5447–54.
59. Ke Q, Costa M. Hypoxia-inducible factor-1 (HIF-1). Mol Pharmacol. 2006;70:1469–80.
60. Ivan M, Kondo K, Yang H, Kim W, Valiando J, Ohh M, Salic A, Asara JM, Lane WS, Kaelin WG. Jr HIFalpha targeted for VHL-mediated destruction by proline hydroxylation: implications for O2 sensing. Science. 2001;292:464–8.
61. Leu T, Schutzhold V, Fandrey J, Ferenz KB. When the brain yearns for oxygen. Neurosignals. 2019;27:50–61.
62. Zhang Q, Yan Q, Yang H, Wei W. Oxygen sensing and adaptability won the 2019 Nobel prize in physiology or medicine. Genes Dis. 2019;6:328–32.
63. Richardson RS, Noyszewski EA, Kendrick KF, Leigh JS, Wagner PD. Myoglobin O2 desaturation during exercise. Evidence of limited O2 transport. J Clin Invest. 1995;96:1916–26.
64. Ameln H, Gustafsson T, Sundberg CJ, Okamoto K, Jansson E, Poellinger L, Makino Y. Physiological activation of hypoxia inducible factor-1 in human skeletal muscle. FASEB J. 2005;19(8):1009–11.
65. Niemi H, Honkonen K, Korpisalo P, Huusko J, Kansanen E, Merentie M, Rissanen TT, Andre H, Pereira T, Poellinger L, Alitalo K, Yla-Herttuala S. HIF-1alpha and HIF-2alpha induce angiogenesis and improve muscle energy recovery. Eur J Clin Investig. 2014;44:989–99.
66. Tang K, Breen EC, Gerber HP, Ferrara NM, Wagner PD. Capillary regression in vascular endothelial growth factor-deficient skeletal muscle. Physiol Genomics. 2004;18:63–9.
67. Hong KS, Kim K. Skeletal muscle contraction-induced vasodilation in the microcirculation. J Exerc Rehabil. 2017;13:502–7.
68. Watts D, Gaete D, Rodriguez D, Hoogewijs D, Rauner M, Sormendi S, Wielockx B. Hypoxia pathway proteins are master regulators of erythropoiesis. Int J Mol Sci. 2020;21(21):8131.
69. Stuart CA, Yin D, Howell MEA, Dykes RJ, Laffan JJ, Ferrando AA. Hexose transporter mRNAs for GLUT4, GLUT5, and GLUT12 predominate in human muscle. Am J Physiol Endocrinol Metab. 2006;291(5):E1067–73.
70. D'Hulst G, Sylow L, Hespel P, Deldicque L. Acute systemic insulin intolerance does not alter the response of the Akt/GSK-3 pathway to environmental hypoxia in human skeletal muscle. Eur J Appl Physiol. 2015;115(6):1219–31.
71. Lee PJ, Jiang BH, Chin BY, Iyer NV, Alam J, Semenza GL, Choi AK. Hypoxia-inducible factor-1 mediates transcriptional activation of the heme oxygenase-1 gene in response to hypoxia. J Biol Chem. 1997;272(9):5375–81.
72. Dawn B, Bolli R. HO-1 induction by HIF-1: a new mechanism for delayed cardioprotection? Am J Phys Heart Circ Phys. 2005;289(2):H522–4.

73. Marciniak SJ. Endoplasmic reticulum stress in lung disease. Eur Respir Rev. 2017;26:170018.
74. Breckenridge DG, Germain M, Mathai JP, Nguyen M, Shore GC. Regulation of apoptosis by endoplasmic reticulum pathways. Oncogene. 2003;22(53):8608–18.
75. Urra H, Pihán P, Hetz C. The UPRosome–decoding novel biological outputs of IRE1α function. J Cell Sci. 2020;133(15):jcs218107.
76. Jain K, Suryakumar G, Ganju L, Singh SB. Amelioration of ER stress by 4-phenylbutyric acid reduces chronic hypoxia induced cardiac damage and improves hypoxic tolerance through upregulation of HIF-1α. Vasc Pharmacol. 2016a;83:36–46.
77. Nakagawa T, Zhu H, Morishima N, Li E, Xu J, Yankner BA, Yuan J. Caspase-12 mediates endoplasmic reticulum-specific apoptosis and cytotoxicity by amyloid beta. Nature. 2000;403:98–103.
78. Jain K, Suryakumar G, Prasad R, Ganju L. Differential activation of myocardial ER stress response: a possible role in hypoxic tolerance. Int J Cardiol. 2013;168(5):4667–77.
79. Smith GI, Patterson BW, Mittendorfer B. Human muscle protein turnover—why is it so variable? J Appl Physiol. 2011;110(2):480–91.
80. Etheridge T, Atherton PJ, Wilkinson D, Selby A, Rankin D, Webborn N, Smith K, Watt PW. Effects of hypoxia on muscle protein synthesis and anabolic signaling at rest and in response to acute resistance exercise. Am J Physiol Endocrinol Metab. 2011;301(4):E697–702.
81. Preedy VR, Smith DM, Sugden PH. The effects of 6 hours of hypoxia on protein synthesis in rat tissues in vivo and in vitro. Biochem J. 1985;228:179–85.
82. Schiaffino S, Dyar KA, Ciciliot S, Blaauw B, Sandri M. Mechanisms regulating skeletal muscle growth and atrophy. FEBS J. 2013;280(17):4294–314. https://doi.org/10.1111/febs.12253.
83. Millward DJ, Garlick PJ, James WP, Nnanyelugo DO, Ryatt JS. Relationship between protein synthesis and RNA content in skeletal muscle. Nature. 1973;241(5386):204–5.
84. Armstrong DD, Wong VL, Esser KA. Expression of beta-catenin is necessary for physiological growth of adult skeletal muscle. Am J Physiol Cell Physiol. 2006;291(1):C185–8.
85. Nader GA, McLoughlin TJ, Esser KA. mTOR function in skeletal muscle hypertrophy: increased ribosomal RNA via cell cycle regulators. Am J Physiol Cell Physiol. 2005;289(6):C1457–65.
86. Favier FB, Costes F, Defour A, Bonnefoy R, Lefai E, Baugé S, Freyssenet D. Downregulation of Akt/mammalian target of rapamycin pathway in skeletal muscle is associated with increased REDD1 expression in response to chronic hypoxia. Am J Phys Regul Integr Comp Phys. 2010;298(6):R1659–66.
87. Costes F, Gosker H, Feasson L, Desgeorges M, Kelders M, Castells J, Freyssenet D. Impaired exercise training-induced muscle fiber hypertrophy and Akt/mTOR pathway activation in hypoxemic patients with COPD. J Appl Physiol. 2015;118(8):1040–9.
88. de Theije CC, Langen RC, Lamers WH, Gosker HR, Schols AM, Koehler SE. Differential sensitivity of oxidative and glycolytic muscles to hypoxia-induced muscle atrophy. J Appl Physiol. 2015;118(2):200–11.
89. Hayot M, Rodriguez J, Vernus B, Carnac G, Jean E, Allen D, Bonnieu A. Myostatin upregulation is associated with the skeletal muscle response to hypoxic stimuli. Mol Cell Endocrinol. 2011;332(1–2):38–47.
90. Rathor R, Agrawal A, Kumar R, Suryakumar G, Singh SN. Ursolic acid ameliorates hypobaric hypoxia-induced skeletal muscle protein loss via upregulating Akt pathway: an experimental study using rat model. IUBMB Life. 2021a;73(2):375–89.
91. Rathor R, Suryakumar G, Singh SN. Diet and redox state in maintaining skeletal muscle health and performance at high altitude. Free Radic Biol Med. 2021b;174:305–20.
92. Bechet D, Tassa A, Taillandier D, Combaret L, Attaix D. Lysosomal proteolysis in skeletal muscle. Int J Biochem Cell Biol. 2005;37(10):2098–114.
93. Luo L, Lu AM, Wang Y, Hong A, Chen Y, Hu J, Qin ZH. Chronic resistance training activates autophagy and reduces apoptosis of muscle cells by modulating IGF-1 and its receptors, Akt/mTOR and Akt/FOXO3a signaling in aged rats. Exp Gerontol. 2013;48(4):427–36.

94. de Theije CC, Langen RC, Lamers WH, Schols AM, Köhler SE. Distinct responses of protein turnover regulatory pathways in hypoxia-and semistarvation-induced muscle atrophy. Am J Phys Lung Cell Mol Phys. 2013;305(1):L82–91.
95. Howald H, Hoppeler H. Performing at extreme altitude: muscle cellular and subcellular adaptations. Eur J Appl Physiol. 2003;90:360–4. https://doi.org/10.1007/s00421-003-0872-9.
96. Levett DZ, Radford EJ, Menassa DA, Graber EF, Morash AJ, Hoppeler H, Clarke K, Martin DS, Ferguson-Smith AC, Montgomery HE, Grocott MP, Murray AJ, Caudwell Xtreme Everest Research Group. Acclimatization of skeletal muscle mitochondria to high-altitude hypoxia during an ascent of Everest. FASEB J. 2012;26:1431–41.
97. Green HJ, Sutton JR, Wolfel EE, Reeves JT, Butterfield GE, Brooks GA. Altitude acclimatization and energy metabolic adaptations in skeletal muscle during exercise. J Appl Physiol. 1992;73(6):2701–8.
98. Kim J, Tchernyshyov I, Semenza GL, Dang CV. HIF-1 mediated expression of pyruvate dehydrogenase kinase: a metabolic switch required for cellular adaptation to hypoxia. Cell Metab. 2006;3:177–85.
99. De Palma S, Ripamonti M, Vigano A, Moriggi M, Capitanio D, Samaja M, Gelfi C. Metabolic modulation induced by chronic hypoxia in rats using a comparative proteomic analysis of skeletal muscle tissue. J Proteome Res. 2007;6(5):1974–84.
100. Regnault TR, Zhao L, Chiu JS, Gottheil SK, Foran A, Yee SP. Peroxisome proliferator-activated receptor-β/δ,-γ agonists and resveratrol modulate hypoxia induced changes in nuclear receptor activators of muscle oxidative metabolism. PPAR Res. 2010;2010(1):129173.
101. Morash AJ, Kotwica AO, Murray AJ. Tissue-specific changes in fatty acid oxidation in hypoxic heart and skeletal muscle. Am J Physiol Regul Integr Comp Physiol. 2013;305(R534):R541.
102. Daneshmandi L, Shah S, Jafari T, Bhattacharjee M, Momah D, Saveh-Shemshaki N, Laurencin CT. Emergence of the stem cell secretome in regenerative engineering. Trends Biotechnol. 2020;38(12):1373–84.
103. Severinsen MCK, Pedersen BK. Muscle–organ crosstalk: the emerging roles of myokines. Endocr Rev. 2020;41(4):594–609.
104. Pedersen BK, Febbraio MA. Muscles, exercise and obesity: skeletal muscle as a secretory organ. Nat Rev Endocrinol. 2012;8:457–65.
105. Lee JH, Jun HS. Role of myokines in regulating skeletal muscle mass and function. Front Physiol. 2019;10:435092.
106. Furuichi Y, Fujii NL. Mechanism of satellite cell regulation by myokines. J Phys Fitness Sport Med. 2017;6(5):311–6. https://doi.org/10.7600/jpfsm.6.311.
107. Gorgens SW, Eckardt K, Jensen J, Drevon CA, Eckel J. Exercise and regulation of adipokine and myokine production. Prog Mol Biol Transl Sci. 2015;135:313–36.
108. Śliwicka E, Cisoń T, Straburzyńska-Lupa A, Pilaczyńska-Szcześniak Ł. Effects of whole-body cryotherapy on 25-hydroxyvitamin D, irisin, myostatin, and interleukin-6 levels in healthy young men of different fitness levels. Sci Rep. 2020;10(1):6175.
109. Rathor R, Suryakumar G, Singh SN. A pilot study investigating the impact of high altitude on myostatin and irisin levels. Age (yrs). 2024;41(3.124):31–48.
110. Oikawa S, Akimoto T. Functional analysis of MicroRNAs in skeletal muscle. Methods Mol Biol. 2023;2640:339–49.
111. Hadj-Moussa H, Chiasson S, Cheng H, Eaton L, Storey KB, Pamenter ME. MicroRNA-mediated inhibition of AMPK coordinates tissue-specific downregulation of skeletal muscle metabolism in hypoxic naked mole-rats. J Exp Biol. 2021;224(15):jeb242968.
112. Mei T, Hu Y, Zhang Y, Li Y. Hypoxia treatment and resistance training alters microRNA profiling in rats skeletal muscle. Sci Rep. 2024;14(1):8388.
113. Allen DL, Bandstra ER, Harrison BC, Thorng S, Stodieck LS, Kostenuik PJ, Barth JL. Effects of spaceflight on murine skeletal muscle gene expression. J Appl Physiol. 2009;106(2):582–95.

114. Williams AH, Valdez G, Moresi V, Qi X, McAnally J, Elliott JL, Olson EN. MicroRNA-206 delays ALS progression and promotes regeneration of neuromuscular synapses in mice. Science. 2009;326(5959):1549–54.
115. Drunmond MJ, McCarthy JJ, Fry CS, Esser KA, Rasmussen BB. Aging differentially affects human skeletal muscle microRNA expression at rest and after an anabolic stimulus of resistance exercise and essential amino acids. Am J Physiol-Endocrinol Metab. 2008;295(6):E1333–40.
116. Srivastava S, Mondal S, Rathor R, Srivastava S, Suryakumar G. Increased expression of MiRNA-1 contributes to hypobaric hypoxia-induced skeletal muscle loss. Adv Biol. 2024;8(3):2300573.
117. Yamaguchi H, Noshita T, Kidachi Y, Umetsu H, Hayashi M, Komiyama K, Ryoyama K. Isolation of ursolic acid from apple peels and its specific efficacy as a potent antitumor agent. J Health Sci. 2008;54(6):654–60.
118. Frighetto RT, Welendorf RM, Nigro EN, Frighetto N, Siani AC. Isolation of ursolic acid from apple peels by high speed counter-current chromatography. Food Chem. 2008;106(2):767–71.
119. Liao Q, Yang W, Jia Y, Chen X, Gao Q, Bi K. LC-MS determination and pharmacokinetic studies of ursolic acid in rat plasma after administration of the traditional chinese medicinal preparation Lu-Ying extract. Yakugaku Zasshi. 2005;125(6):509–15.
120. Hipkiss AR. Could carnosine or related structures suppress Alzheimer's disease? J Alzheimers Dis. 2007;11(2):229–40.
121. Tanaka KI, Sugizaki T, Kanda Y, Tamura F, Niino T, Kawahara M. Preventive effects of carnosine on lipopolysaccharide-induced lung injury. Sci Rep. 2017;7(1):42813.
122. Bermúdez ML, Seroogy KB, Genter MB. Evaluation of carnosine intervention in the Thy1-aSyn mouse model of Parkinson's disease. Neuroscience. 2019;411:270–8.
123. Gaunitz F, Hipkiss AR. Carnosine and cancer: a perspective. Amino Acids. 2012;43:135–42.
124. Agrawal A, Rathor R, Kumar R, Singh SN, Kumar B, Suryakumar G. Endogenous dipeptide-carnosine supplementation ameliorates hypobaric hypoxia-induced skeletal muscle loss via attenuating endoplasmic reticulum stress response and maintaining proteostasis. IUBMB Life. 2022;74(1):101–16.
125. Hewlings SJ, Kalman DS. Curcumin: a review of its effects on human health. Food Secur. 2017;6(10):92.
126. Sahin K, Pala R, Tuzcu M, Ozdemir O, Orhan C, Sahin N, Juturu V. Curcumin prevents muscle damage by regulating NF-κB and Nrf2 pathways and improves performance: an in vivo model. J Inflamm Res. 2016;9:147–54.
127. Tsai SW, Huang CC, Hsu YJ, Chen CJ, Lee PY, Huang YH, Tung YT. Accelerated muscle recovery after in vivo curcumin supplementation. Nat Prod Commun. 2020;15(1):1934578X20901898.
128. Chaudhary P, Sharma YK, Sharma S, Singh SN, Suryakumar G. High altitude mediated skeletal muscle atrophy: protective role of curcumin. Biochimie. 2019;156:138–47.
129. Suryakumar G, Gupta A. Medicinal and therapeutic potential of sea buckthorn (Hippophae rhamnoides L.). J Ethnopharmacol. 2011;138(2):268–78.
130. Saggu S, Kumar R. Modulatory effect of seabuckthorn leaf extract on oxidative stress parameters in rats during exposure to cold, hypoxia and restraint (C-H-R) stress and post stress recovery. J Pharm Pharmacol. 2007;59(12):1739–45.
131. Rathor R, Sharma P, Suryakumar G, Ganju L. A pharmacological investigation of Hippophae Salicifolia (HS) and Hippophae Rhamnoides Turkestanica (HRT) against multiple stress (C-H R): an experimental study using rat model. Cell Stress Chaperones. 2015;20(5):821–31.
132. Jain K, Suryakumar G, Prasad R, Ganju L, Bala Singh SB. Enhanced hypoxic tolerance by Seabuckthorn is due to upregulation of HIF-1a and attenuation of ER stress. J Appl Biomed. 2016b;14(1):71–83.

Risk of Venous Thrombosis Under Extreme Environment and Situations: Military Perspective

5

Rashi Khare, Sunanda Arya, Swati Srivastava, and Iti Garg

Abstract

Venous thromboembolism (VTE) presents two clinical manifestations: deep vein thrombosis (DVT) and pulmonary embolism (PE), both of which result in serious complications, and if left untreated, it may result in the death of the affected individual. While significant research has been conducted on civilians regarding the risk assessment and management of VTE, there is a scarcity of data regarding military personnel. This chapter investigates the unique challenges faced by military personnel exposed to extreme environmental conditions and stressful situations that may predispose them to VTE.

Key scenarios such as prolonged stasis in submarines, high-altitude postings, extreme temperature exposure, and the microgravity environment encountered during space missions are major highlights of the chapter. The physiological and pathophysiological processes that underlie VTE in these harsh environments are also explained, which include vascular injury, hypercoagulability, and stasis. Combat-related trauma and its role in increasing the risk of VTE are given special attention. While limited data are currently available, the chapter highlights the importance of targeted studies to address these knowledge gaps. Recent research from international space agencies and military healthcare systems can provide potential insights into the mechanisms and preventative measures for VTE. This chapter serves as a foundation for understanding the implications of extreme environments on VTE risk, aiming to improve strategies for preventive measures, clinical outcomes, and operational readiness from a military perspective.

R. Khare · S. Arya · S. Srivastava · I. Garg (✉)
Defence Institute of Physiology and Allied Sciences (DIPAS), Defence Research and Development Organization (DRDO), Timarpur, Delhi, India
e-mail: itigarg.dipas@gov.in

S. Srivastava et al. (eds.), *Health and Sustenance in Extreme Environment*,
https://doi.org/10.1007/978-981-96-8642-1_5

Keywords

Venous thromboembolism · Hypoxia · Deep vein thrombosis · Pulmonary embolism

Abbreviations

DVT	Deep vein thrombosis
HA	High altitude
PE	Pulmonary embolism
VTE	Venous thrombo-embolism

5.1 Overview

Venous thromboembolism (VTE) is a common pathology comprising deep vein thrombosis (DVT) and pulmonary embolism (PE). It results in significant mortality and morbidity and involves substantial healthcare expenditure [1]. The annual incidence of VTE in adult populations has been estimated to be approximately 1 per 1000 annually [2]. VTE has severe outcomes such as recurrence, postthrombotic syndrome, major bleeding due to anticoagulation, and even death, and it has a strong relation with impaired quality of life [3, 4]. Among total cases of VTE, approximately two-thirds of the patients present with DVT and one-third present with PE, which is the primary cause of mortality associated with this disease. It is also the leading cause of maternal mortality and preventable hospital deaths in the United States [5, 6]. The incidence rate of VTE is highly variable across different populations, sex, and age groups. Overall, women have a lower rate of incidence than men except during the reproductive years [7, 8]. Even though a large amount of data is present for civilian VTE patients. It is difficult to access data from a military perspective. In this chapter, we discuss how different environments can potentially affect military personnel physiology and how long durations of stay at places like submarines, aircraft, and high altitudes cause the formation of venous thromboembolism (Fig. 5.1).

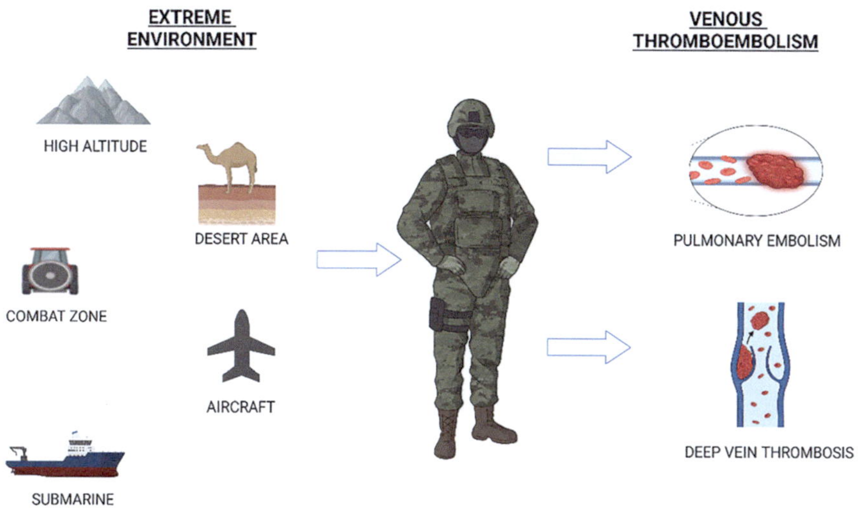

Fig. 5.1 Diagrammatic representation of a variety of extreme conditions affecting soldiers ultimately causing deep vein thrombosis in the lower extremity and pulmonary embolism in the lungs

5.2 Introduction

Army personnel often have to deal with extreme environments and adjust to them as soon as possible to protect their country and its citizens. They are posted to places with extreme environmental conditions which can alter their physiological system. Military personnel also face a variety of extreme situations that test their physical and mental resilience. Most of these environmental conditions and situations are high altitudes, extreme temperatures, combat injury/trauma, confined spaces, etc. Here we will discuss the extreme conditions and combat trauma in detail (Fig. 5.2).

5.2.1 Submarine

The submarine environment is unique due to the confined space, limited access to sunlight, and alternation in atmospheric pressure that people experience. Although submarines were initially not considered in terms of causing damage, submarines nowadays seem to be one of the most critical assets to the naval military, including India, the United States, Russia, China, etc. They enhance a navy's capacity to show power, enhance strategic deterrence, and safeguard maritime security. Keeping in mind that the human body reacts differently in stressful environments. The environment inside the submarine ought to have a potential influence on blood coagulation parameters. Research done on change in coagulation factor for submarine cabin crew is quite niche but few studies show the physiological effect of prolonged missions done inside submarine. The earliest study on blood coagulation variation in

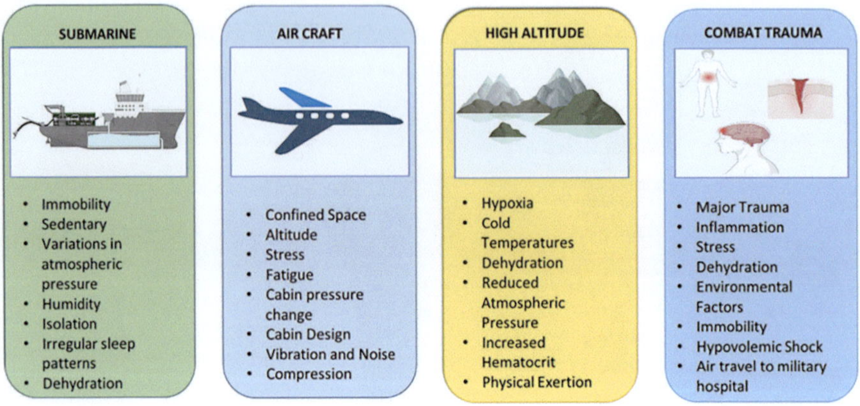

Fig. 5.2 Summarized representation of multiple causes behind the development of venous thromboembolism in extreme conditions

submarine crew was done by Lt Ganson Purcell Jr. in 1965. A group of 10 subjects was chosen in the age range of 23–47 years old. Tests performed weekly on the subjects were as follows: prothrombin time (PT), partial thromboplastin time (PTT), quantitative fibrinogen determination, euglobulin lysis time, and prothrombin consumption time (PCT). These tests were done for 10 weeks on the selected subjects. His study showed all the test results were within the acceptable normal range [9]. A group of animals such as Sprague-Dawley rats and rabbits were used to study the effects of different simulated submarine escape depths. The result showed variation in coagulation factors of animals subjected to 200 and 250 m escape depths. D-dimer value was higher in both groups compared to the normal control group. The D-dimer value was even higher at 250 m depth animals. Results showed that the PT value increased with more escape depth compared to the control group and had a notably higher value in the 250-m escape group than in the control group [10]. Since the D-dimer percentage was higher as the depth increased, we can predict there are chances of venous thrombosis formation in crew members/navy personnel. The conditions experienced by individuals inside a submarine can increase the risk of a thrombotic event. This study opens a window for further investigating the effect of the submarine environment on the increased risk of venous thrombosis.

5.2.2 Aircraft

Just like submarines, aircraft have their own set of features that create stressful conditions for the air force members. Prolonged hours of sitting in a tight space with a gravitation pull putting pressure on the body while dealing with psychologically challenging conditions have their specific consequence on the human body. Focusing on the fact that traveling for long hours can increase the chances of thrombus formation, air force crew have higher chances of developing venous thrombosis. Various countries have conducted studies on coagulation factors and thrombotic risks in military personnel, including those in aircraft. These studies often focus on understanding the physiological impacts of military operations and environments and how to decrease health risks for personnel. In 1989, a case report was done on a US military Air Force pilot developing deep vein thrombosis [11]. Sprague-Dawley model rat was used to study the effect of military flights on coagulation, fibrinolysis, and blood flow. Results from the experiment showed that no clear association between various hypobaric conditions and activation of coagulation was observed [12]. In military combat casualties, the air travel transport of the patients could result in the development of venous thrombosis. Gajic et al. found prolonged air travel before major surgery increased perioperative VTE rates [13].

5.2.3 High Altitude

High altitude has been established as a risk factor for the development of venous thrombosis for almost a decade now. At high altitudes, all three factors of the Virchow triad, that is, venous stasis, hypercoagulability, and vessel-wall injury, all appear to be active [14]. During combat operations, high altitude brings about a multitude of challenges like hypoxia, cold weather, altitude sickness, and cognitive impairment. One of the high-altitude illnesses is venous thromboembolism. The low available oxygen level acts as a gasoline to fuel up the development of venous thromboembolism [6]. At the time of early investigation focusing on high altitude as a risk factor for venous thrombosis, Kumar Shishir's study focused on Indian soldiers posted at high altitude and their risk of developing deep vein thrombosis. The study stated that soldiers posted at high altitudes had a higher relative risk of deep vein thrombosis compared to those posted in the plains [15]. Nair et al. screened 960 healthy male subjects in the plains, of whom 750 ascended to altitudes above 15,000 ft. All the participants were part of the Indian army. Their study showed that 15 subjects developed thrombosis at altitudes >15,000 (12 developed venous thrombosis and 3 developed arterial thrombosis). Hence, it was concluded that high altitude acts as an independent risk factor for the development of venous thromboembolism [16]. Many other studies also consider high altitude as a trigger for the initiation of venous thrombosis [17, 18]. High altitude brings about cardiovascular and hematological changes. Hypoxia triggers erythropoietin release, which increases hematopoiesis to increase oxygen-carrying capacity. Blood viscosity increases due to volume depletion and increased red cell volume. Headache, fatigue, somnolence, and depression result from blood hyperviscosity. These changes can be the reason for the start and progress of venous thrombosis [19]. A 2005 study conducted on Indian soldiers states that high altitude serves as a risk factor for the development of symptomatic portal system thrombosis [19]. Another case study highlights that high altitude acts as a risk factor for the formation of cerebral venous sinus thrombosis (CSVT). In this study, Hussan et al. studied 28 cases of symptomatic CVST at high altitudes between August 2017 and

December 2018. Their results positively correlated CSVT and high altitude [20]. Paliwal et al. presented a compilation of case reports on cerebral venous thrombosis occurring at high altitudes in Indian Armed Forces personnel. Their case series study concluded that even though cerebral venous thrombosis is uncommon in the general population, it's common for military personnel at high-altitude areas because of the varied clinical presentations and other mimickers at high altitudes, diagnosis and treatment are frequently delayed [21].

5.2.4 Military Combat Trauma

The risk of venous thrombosis development in combats has been noticed frequently. Trauma caused by combat is complicated due to thrombus formation. Reported combat trauma VTE rates range from 2.3% to over 23% [22, 23]. Gillern et al. utilized retrospective data from the Combat Trauma Registry (National Naval Medical Center (NNMC), Bethesda, Maryland) and examined a group of 263 patients were treated from March 2003 to December 2007. Under the study, 15 patients were diagnosed with DVT (5.7%) and 15 patients with PE (5.7%). Both DVT and PE were diagnosed in three patients subsequently rate of VTE was 10.3%. Only one death was attributed to an acute PE [24]. At the Walter Reed National Military Medical Center (Bethesda, Maryland) Caruso et al. reported that between November 2010 and November 2012, a single institution rate of VTE occurrence after arrival to the United States was 11.4%. In 2014, Hutchison et al. performed a retrospective review of all US military combat casualties with VTE records from September 2001 to July 2011. Among 26,634 subjects, 587 (2.2%) had a VTE. Out of 587 subjects, 270 patients with deep venous thrombosis (DVT-1.0%), 223 patients with PE (0.8%), and 94 patients with both DVT and PE (0.4%). Out of these, 75 patients had DVT, 70 had PE, and 29 were found to have both a DVT and a PE [13]. While identifying risk factors for military settings, both Hutchison et al. and Gillern et al. result concluded that bilateral lower extremity fracture seems to be a significant risk factor for VTE development [22, 24]. Risk factors for venous thromboembolism development in military trauma specifically injury patterns involving the lower extremities [13]. Even though Virchow's Triad only states three reasons accountable for the development of VTE in trauma patients, research shows there are many more reasons to be added to the list. Five additional independent risk factors were identified by Geerts et al. for the development of VTE in 1994. These factors were older age, blood transfusion, surgery, femur and tibia fractures, and spinal cord injury [25]. Using Hutchison's study, Holley et al. analyzed and found that a subset of patients had blood transfusion predictive of VTE [26].

There have been studies that suggest that pulmonary embolism might not primarily be caused by the dislodging of clots from veins [27, 28]. Lundy et al. noted after reviewing the CT scan of 708 consecutive combat-injured patients in southern Afghanistan from July 1, 2011 to June 30, 2012, that their acute peritraumatic pulmonary thrombus (APPT) rate of 9.3% [29]. Knudson also suggested in the case of civilian pulmonary embolism local inflammation from chest trauma is a fourth risk factor [28]. After soldiers get injured during combat, they are mostly transported to hospital through air travel which has been seen to increase the risk of venous thrombosis development. Not a lot of studies can be found that relate VTE with surgery and flying. Gajic et al. stated in their study that postoperative VTE development was high in patients who had air traveled for prolonged hours before major surgery (4.9% flight vs 0.15% no flight, $p < 0.001$) [14].

In a recent study, Cobler-Lichter et al. did a 20-year retrospective study to evaluate the risk of VTE development for combat casualties requiring damage control laparotomy. Data from 2002 to 2023 were queried for US Military combat casualties which showed that VTE rates were high around 15.8% (DVT: 10.3% and PE 7.1%). In the massively transfused population, it was noticed that the VTE rate was even higher (26.7% vs 10.2%, $p < 0.001$) [30].

In 2016, Hannon et al. determined the risk and time duration for the development of deep vein thrombosis and pulmonary embolism after combat-related major limb amputations. The study population consisted of patients with amputation surgery in Iraq or Afghanistan from 2009 to 2011. The results indicated that 103 patients (28%) were diagnosed with Deep Vein Thrombosis/Pulmonary Embolism whereas 59 subjects were diagnosed with PE (16%) and 59 subjects with DVT (16%) out of 366 patients. It was also observed that a large number of DVT (69%) and PE (66%) cases occurred within 10 days. A spike in ventilator days (odds ratio [OR], 1.97; 95% confidence interval [CI], 1.16–3.37) and the number of blood transfusions (OR, 1.72; 95% CI, 1.11–2.68) were associated with a higher risk of DVT. The incidence of DVT/PE is high following combat-related amputation, with most cases occurring early. Therefore, prophylaxis is recommended [31].

A summary of the following case study/research study can be found in Table 5.1.

Table 5.1 Summary of the studies related to extreme environment

S no.	Study type	Finding	Experimental model	Reference
1.	Submarine Crew	Prothrombin time (PT), partial thromboplastin time (PTT), quantitative fibrinogen determination, euglobulin lysis time, and prothrombin consumption time (PCT) tests were within normal range in all the subjects.	Human subject	Purcell et al. [9]
2.	Animal Model	The prothrombin time value increased with escape depth in both the animal model.	Sprague–Dawley rats and rabbits	Bao et al. [10]
3.	Military Pilot	Incidence of VTE reported	Three US military pilot	Steinhauser et al. [11]

(continued)

Table 5.1 (continued)

S no.	Study type	Finding	Experimental model	Reference
4.	Animal Model	Activation of coagulation factor was observed in both the group of animal exposed at 609 and 7620 m for 2 and 12 h duration but there was no significant difference in parameters of coagulation, fibrinolysis and blood flow.	Sixty Sprague–Dawley male rat	Levkovsky et al. [12]
5.	Indian Soldiers with Deep Vein Thrombosis (high altitude vs. sea level)	Case study revealed that high altitude act as a risk factor for development of deep vein thrombosis.	Twenty-eight cases of deep venous thrombosis	Shishir [15]
6.	Indian Soldiers with Venous Thromboembolism	Incidents of thrombosis were higher in high-altitude subjects compared to sea-level patients.	960 male subjects	Nair et al. [16]
7.	Indian Soldiers with Venous Thromboembolism	High-altitude exposure alters hemostatic systems and contributes to the onset of VTE.	Indian male patient at high altitude = 53 Indian male patient at sea level = 45	Prabhakar et al. [17]
8.	Indian Soldiers with Systemic Portal System Thrombosis	An extended stay at high-altitude area may be a risk factor for development of symptomatic portal system thrombosis.	A total of 37 cases of symptomatic portal system thrombosis from 26 were from high- altitude area.	Anand et al. [19]
9.	Case Report of Male Patients at High Altitude	Cerebral venous thrombosis development was found to be influenced by high altitude.	28 high-altitude patients with cerebral venous thrombosis.	Hassan et al. [20]
10.	Case Study Series	High altitude is a key factor that increases the chances for the development of cerebral venous thrombosis.	Four Indian male patients.	Paliwal et al. [21]

5.3 Microgravity

Earth creates a field of gravitational force that pulls objects within the field with a force that is inversely proportional to the square of the distance between the object's center and the Earth's center. When we measure the acceleration of an object solely under Earth's gravity at the surface, we refer to it as 1 g or one Earth gravity, which is approximately 9.8 m per second squared (m/s^2). An object's mass determines how much it accelerates under a given force, while its weight is the gravitational force exerted on it by Earth. However, microgravity is defined as an environment where the apparent weight of a system is much smaller than its actual weight due to gravity. When discussing how microgravity environments can be created, it's important to remember that many factors influence experienced accelerations, and the quality of the microgravity environment depends on the method used to produce it [32]. Microgravity experienced by astronauts in space can significantly affect the body's physiology, leading to an increased risk of VTE. All three elements of Vichow's triad are disturbed in the following ways:

1. Stasis: On earth due to the gravitational pull on our body, a head-to-foot (Gz) hydrostatic pressure gradient in the body's fluid-filled systems. In space, this Gz pressure is missing which causes abrupt distribution of the body's fluid-filled system. Crew members on the International Space Station (ISS) feel weightlessness, leading to a continuous redistribution of body fluid toward the head. Weightlessness is often associated with stasis which leads to the formation of venous thrombosis in otherwise healthy astronauts [33]. Studies report increased jugular pressure during spaceflight because of microgravity [34, 35].
2. Hypercoagulability: No study evaluating the change in coagulation factor during spaceflight can be found. However, few studies show the change in coagulation pre- and postflight data [36, 37]. In these two studies, the common factor was that their result showed increased levels of fibrinogen. Fibrinogen is a well-established marker of hypercoagulation or acute phase reactant [36, 37] but further studies are needed to get a proper picture.
3. Vascular injury: Laminar flow and high wall shear stress (normal venous flow velocities) promote vessel protective signaling supporting endothelial health in normal physiological states. However, in stasis or reduced flow (and consequently low shear stress), the risk of vessel inflammation, oxidative stress, thrombosis, and stenosis increases [38]. These issues arise mainly because flow disturbances activate various modulators (mechanotransducers, mitogen-activated protein kinases, and effector molecules) that lead to a prothrombotic state. This is mediated through the expression of inflammatory markers (via tumor necrosis factor and endothelial cells) and thrombotic markers (via endothelin 1 and platelet-derived growth factor β) [39–41]. Only one dry immersion study reports that increased endothelial microparticles are associated with vascular distension which might point to enhanced endothelial damage [42].

There is a thrust area for research regarding the study of venous thromboembolism in military personnel experiencing microgravity. We can expect more to come in the upcoming future from many space agencies including ISRO, NASA, Royal Society of Astronomy, etc., which would be helpful for designing new strategies for better management of the disease under extreme conditions as well as under microgravity.

References

1. Cohen AT, Agnelli G, Anderson FA, Arcelus JI, Bergqvist D, Brecht JG, Greer IA, Heit JA, Hutchinson JL, Kakkar AK, Mottier D, Oger E, Samama MM, Spannagl M, VTE Impact Assessment Group in Europe (VITAE). Venous thromboembolism (VTE) in Europe. The number of VTE events and associated morbidity and mortality. Thromb Haemost. 2007;98(4):756–64.
2. White RH. The epidemiology of venous thromboembolism. Circulation. 2003;107:I-4–8.
3. Van Korlaar IM, Vossen CY, Rosendaal FR, Bovill EG, Cushman M, Naud S, et al. The impact of venous thrombosis on quality of life. Thrombi Res. 2004;114:11–8.
4. Kahn SR, Ducruet T, Lamping DL, Arsenault L, Miron MJ, Roussin A, et al. Prospective evaluation of health-related quality of life in patients with deep venous thrombosis. Arch Intern Med. 2005;165:1173–8.
5. Maynard G, Stein J. Preventing hospital-acquired venous thromboembolism: a guide for effective quality improvement. Rockville: (AHRQ Publication No. 08–0075) Agency for Healthcare Research and Quality; 2008.
6. Chang J, Elam-Evans LD, Berg CJ, et al. Pregnancy-related mortality surveillance—United States, 1991–1999. MMWR CDC Surveill Summ. 2003;52:1–8.
7. White R, Zhou H, Murin S, Harvey D. Effect of ethnicity and gender on the incidence of venous thromboembolism in a diverse population in California in 1996. Thromb Haemost. 2005;93:298–30.
8. Heit JA, Silverstein MD, Mohr DN, et al. The epidemiology of venous thromboembolism in the community. Thromb Haemost. 2001;86:452–63.
9. Purcell G Jr. Blood clotting studies in a submarine crew. Serial analysis of the clotting mechanism during prolonged submergence. Arch Environ Health. 1965;11(6):804–9. https://doi.org/10.1080/00039896.1965.10664304.
10. Bao XC, Wang N, Xu J, Ma J, Fang YQ. Effects of different simulated submarine escape depths by free ascent in animal models. Front Physiol. 2023;14:1107782. https://doi.org/10.3389/fphys.2023.1107782.
11. Steinhauser RP, Stewart JC. Deep venous thrombosis in the military pilot. Aviat Space Environ Med. 1989;60(11):1096–8.
12. Levkovsky A, Dardik R, Barazany D, Steinberg DM, Kirichenko MD, Apter S, Peleg E, Silverberg D, Grossman E, Salomon O. The effect of civil and military flights on coagulation, fibrinolysis and blood flow: insight from a rat model. Thromb J. 2020;18:24. https://doi.org/10.1186/s12959-020-00237-8.
13. Fang R, Rodriguez CJ. Venous thromboembolism among military combat casualties. Curr Trauma Rep. 2016;2:48–53. https://doi.org/10.1007/s40719-016-0037-z.
14. Evans CE, Humphries J, Waltham M, Saha P, Mattock K, Patel A, Ahmad A, Wadoodi A, Modarai B, Burnand K, Smith A. Upregulation of hypoxia-inducible factor 1 alpha in local vein wall is associated with enhanced venous thrombus resolution. Thromb Res. 2011;128(4):346–51. https://doi.org/10.1016/j.thromres.2011.05.006.
15. Kumar S. High altitude induced deep venous thrombosis: A study of 28 cases. Indian J Surg. 2006;68:84–8.
16. Velu N, Singh S, Ashraf M, Yanamandra U, Sharma V, Prabhakar A, Ahmad R, Chatterjee T, Behera V, Guleria V, Patrikar S, Gupta S, Vishnoi M, Kalshetty K, Sharma P, Bajaj N, Khaling T, Wankhede T, Bhattachar S, Ganguli L. Epidemiology and pathophysiology of vascular thrombosis in acclimatized lowlanders at high altitude: a prospective longitudinal study. Lancet. 2022:3. https://doi.org/10.1016/j.lansea.2022.05.005.
17. Prabhakar A, Chatterjee T, Bajaj N, Tyagi T, Sahu A, Gupta N, Kumari B, Nair V, Kumar B, Ashraf MZ. Venous thrombosis at altitude presents with distinct biochemical profiles: a comparative study from the Himalayas to the plains. Blood Adv. 2019;3(22):3713–23. https://doi.org/10.1182/bloodadvances.2018024554.

18. McLaughlin CW, Skabelund AJ, George AD. Impact of high altitude on military operations. Curr Pulmonol Rep. 2017;6:146–54. https://doi.org/10.1007/s13665-017-0181-0.
19. Anand AC, Saha A, Seth AK, Chopra GS, Nair V, Sharma V. Symptomatic portal system thrombosis in soldiers due to extended stay at extreme altitude. J Gastroenterol Hepatol. 2005;20(5):777–83. https://doi.org/10.1111/j.1440-1746.2005.03723.x.
20. Hassan WU, Syed MJ, Alamgir W, Awan S, Bell SM, Majid A, Wasay M. Cerebral venous thrombosis at high altitude: analysis of 28 cases. Cerebrovasc Dis (Basel, Switzerland). 2019;48(3–6):184–92. https://doi.org/10.1159/000504504.
21. Paliwal AK, Gupta M, Peeyush P, Sharma V. Imaging the diverse presentations of cerebral venous thrombosis occurring at high altitude area in Armed Forces personnel: case series. Med J Armed Forces India. 2019;75(3):282–7. https://doi.org/10.1016/j.mjafi.2018.03.009.
22. Hutchison TN, Krueger CA, Berry JS, Aden JK, Cohn SM, White CE. Venous thromboembolism during combat operations: a 10-y review. J Surg Res. 2014;187(2):625–30.
23. Caruso JD, Elster EA, Rodriguez CJ. Epidural placement does not result in an increased incidence of venous thromboembolism in combat-wounded patients. J Trauma Acute Care Surg. 2014;77(1):61–6. This paper found no difference in VTE incidence between enoxaparin 40 mg daily in patients with patient-controlled epidural anesthesia versus 30 mg twice daily for VTE prophylaxis in combat-wounded patients. However, 55/484 patients developed VTE during hospitalization despite prophylaxis.
24. Gillern SM, Sheppard FR, Evans KN, Graybill JC, Gage FA, Forsberg JA, et al. Incidence of pulmonary embolus in combat casualties with extremity amputations and fractures. J Trauma. 2011;71(3):607–12.
25. Geerts WH, Code KI, Jay RM, Chen E, Szalai JP. A prospective study of venous thromboembolism after major trauma. N Engl J Med. 1994;331(24):1601–6. https://doi.org/10.1056/NEJM199412153312401.
26. Holley AB, Petteys S, Mitchell JD, Holley PR, Collen JF. Thromboprophylaxis and VTE rates in soldiers wounded in operation enduring freedom and operation Iraqi freedom. Chest. 2013;144(3):966–73. https://doi.org/10.1378/chest.12-2879.
27. Knudson MM, Gomez D, Haas B, Cohen MJ, Nathens AB. Three thousand seven hundred thirty-eight posttraumatic pulmonary emboli: a new look at an old disease. Ann Surg. 2011;254(4):625–32. https://doi.org/10.1097/SLA.0b013e3182300209.
28. Velmahos GC, Spaniolas K, Tabbara M, Abujudeh HH, de Moya M, Gervasini A, Alam HB. Pulmonary embolism and deep venous thrombosis in trauma: are they related? Arch Surg (Chicago, Ill. : 1960). 2009;144(10):928–32. https://doi.org/10.1001/archsurg.2009.97.
29. Lundy JB, Oh JS, Chung KK, Ritter JL, Gibb I, Nordmann GR, Sonka BJ, Tai NR, Aden JK, Rasmussen TE. Frequency and relevance of acute peritraumatic pulmonary thrombus diagnosed by computed tomographic imaging in combat casualties. J Trauma Acute Care Surg. 2013;75(2 Suppl 2):S215–20. https://doi.org/10.1097/TA.0b013e318299da66.
30. Cobler-Lichter MD, Collie BL, Delamater JM, Shagabayeva L, Lyons NB, Bustillos LT, Namias N, Stallings JD, Gross KR, Buzzelli MD, Gurney J, Proctor KG, Wetstein PJ. A 20-year retrospective analysis of deep venous thrombosis and pulmonary embolism among combat casualties requiring damage control laparotomy at US Military Role 2 surgical units. J Trauma Acute Care Surg. 2024; https://doi.org/10.1097/TA.0000000000004405.
31. Hannon M, Tadlock MD, Melcer T, Walker J, Bandle J, Nieses K, Galarneau M. Venous thromboembolism after traumatic amputation: an analysis of 366 combat casualties. Am J Surg. 2016;212(2):230–4. https://doi.org/10.1016/j.amjsurg.2016.01.031.
32. Adams K. What is microgravity? (Grades 5–8) NASA STEM Team. 2024. https://www.nasa.gov/learning-resources/for-kids-and-students/what-is-microgravity-grades-5-8/#:~:text=Microgravity%20is%20the%20condition%20in,%2C%E2%80%9D%20but%20this%20is%20misleading.
33. Hargens AR, Richardson S. Cardiovascular adaptations, fluid shifts, and countermeasures related to space flight. Respir Physiol Neurobiol. 2009;169(Suppl 1):S30–3. https://doi.org/10.1016/j.resp.2009.07.005.

34. Fortrat JO, de Holanda A, Zuj K, Gauquelin-Koch G, Gharib C. Altered venous function during long-duration spaceflights. Front Physiol. 2017;8:694. https://doi.org/10.3389/fphys.2017.00694.
35. Marshall-Goebel K, Laurie SS, Alferova IV, Arbeille P, Auñón-Chancellor SM, Ebert DJ, Lee SMC, Macias BR, Martin DS, Pattarini JM, Ploutz-Snyder R, Ribeiro LC, Tarver WJ, Dulchavsky SA, Hargens AR, Stenger MB. Assessment of jugular venous blood flow stasis and thrombosis during spaceflight. JAMA Netw Open. 2019;2(11):e1915011. https://doi.org/10.1001/jamanetworkopen.2019.15011.
36. Kuzichkin DS, Morukov BV, Markin AA, Juravlyova OA, Zabolotskaya IV, Vostrikova LV. Hemostasis system indices after short-term space flights and during 7-day "dry" immersion experiment. Hum Physiol. 2010;36:478–82.
37. Larina IM, Percy AJ, Yang J, Borchers CH, Nosovsky AM, Grigoriev AI, Nikolaev EN. Protein expression changes caused by spaceflight as measured for 18 Russian cosmonauts. Sci Rep. 2017;7(1):8142. https://doi.org/10.1038/s41598-017-08432-w.
38. Chiu JJ, Chien S. Effects of disturbed flow on vascular endothelium: pathophysiological basis and clinical perspectives. Physiol Rev. 2011;91(1):327–87.
39. Berk BC, Juni ABE, Min W, Surapisitchat J, Yan C. Endothelial atheroprotective and anti-inflammatory mechanisms. Ann N Y Acad Sci. 2001;947(1):93–111.
40. Passerini AG, Milsted A, Rittgers SE. Shear stress magnitude and directionality modulate growth factor gene expression in preconditioned vascular endothelial cells. J Vasc Surg. 2003;37(1):182–90.
41. Traub O, Berk BC. Laminar shear stress: mechanisms by which endothelial cells transduce an atheroprotective force. Arterioscler Thromb Vasc Biol. 1998;18(5):677–85.
42. Navasiolava NM, Dignat-George F, Sabatier F, Larina IM, Demiot C, Fortrat JO, Gauquelin-Koch G, Kozlovskaya IB, Custaud MA. Enforced physical inactivity increases endothelial microparticle levels in healthy volunteers. Am J Physiol Heart Circ Physiol. 2010;299(2):H248–56. https://doi.org/10.1152/ajpheart.00152.2010.

High-Altitude-Induced Dysbiosis in Gut and Oral Microbiome: Strategies for Restoration of Microbiota

6

Manisha Kumari, Anjali Yadav, Anu Kumari, Chhavi Rai, Rajeev Varshney, and Ramesh C. Meena

Abstract

At high altitudes, the host microbiome undergoes significant changes due to reduced pressure and low oxygen levels. Both acute and chronic exposure to hypobaric hypoxia lead to alterations in the diversity, abundance, and composition of the gut and oral microbiota. These variations help the body to adapt to extreme environmental conditions, which pose serious health risks. The altered gut and oral microbiome can weaken the mucosal lining, leading to many gastrointestinal and oral problems and increased susceptibility to various symptoms and diseases. Results revealed that short-term exposure to high altitudes can harm the gut barrier, and prolonged exposure to low oxygen levels may further deteriorate the barrier by disrupting immune functions, mucous secretion and ultimately leading to dysbiosis. Preliminary studies have shown that hypobaric hypoxia may exacerbate intestinal damage by reducing blood flow to the intestine and increasing oxygen deficiency, thereby leading to additional stress on the body. In this chapter, we first summarize the R&D work done so far globally to demonstrate the effect of high altitude on gut and oral microbiota. Microbiome-targeted therapies are being investigated for their potential to modulate the microbiome and improve gut and oral health. As research continues to uncover the multifaceted relationships between the host and microbiome, we may see the emergence of therapeutic strategies to promote overall well-being and avert systemic diseases.

Keywords

Gut Microbiota · Oral Microbiota · High Altitude · Hypobaric Hypoxia · Probiotics

M. Kumari · A. Yadav · A. Kumari · C. Rai · R. Varshney · R. C. Meena (✉)
Defence Institute of Physiology and Allied Sciences (DIPAS), Defence Research and Development Organization (DRDO), Timarpur, Delhi, India
e-mail: rcmeena.dipas@gov.in

S. Srivastava et al. (eds.), *Health and Sustenance in Extreme Environment*,
https://doi.org/10.1007/978-981-96-8642-1_6

6.1 Introduction

High altitude (HA) presents strong environmental stressors for people residing at or visiting HA. Nearly, 140 million people worldwide live at HA [1]. Every year, approximately 100 million people are traveling to different high-altitude locations [2]. With the advent of modern travel facilities, people travel to high altitudes or mountains for adventure, to visit holy pilgrimage, and to compete in various sports events. HA regions are also frequently visited for military operations [3] and scientific research and development work [4]. Although HA has no precise definition, physiologically, it is considered to be an elevation of 1500–3500 m above sea level [5]. The amount of O_2 in the atmosphere is 20.93% which is constant at different altitudes, however, the partial pressure of O_2 (PaO_2) and barometric pressure fall with an ascent from sea level to high altitude (>2500 m elevation) in a logarithmic fashion. This condition is characterized as hypobaric hypoxia. Hypoxia-related physiologic changes that help to maintain oxygen delivery to host tissue show a marked increase in alveolar ventilation and cardiac output, erythrocyte concentration, and affinity. The genetic basis of HA adaptation includes single-nucleotide polymorphism located in genes involved in the HIF pathway, which is vital in response to hypoxia-induced changes [6]. Various other existing factors at high altitudes including environmental stressors such as low temperature, UV radiation, low relative humidity, high wind velocity, inability to keep proper individual hygiene, cramped living arrangements, low availability of fresh fruits, vegetables, and isolation from medical help is associated with increased inflammation [7], and illness and infection [8]. Travel to high altitudes has potentially significant pathophysiological consequences for the human body. Exposure to HA causes damage to many organs, especially the gastrointestinal tract and oral cavity. It has already been established that differences in barometric pressure result in insufficient oxygen in the air, which is known to cause neurological and cardiovascular disorders [9].

A very limited work has been done on studying the impact of HA hypoxia on the human gut and oral-related diseases. In fact, several gastrointestinal disturbances such as anorexia, epigastric discomfort, flatus expulsion, dyspepsia, nausea, severe acidity, vomiting, infectious diarrhea, and hematemesis [10, 11] gastrointestinal bleeding [12] and disorders associated with oral cavity like gingivitis, dental fluorosis [13], and periodontal inflammation [14] are frequently reported in mountaineers and altitude sojourners. It is evident that microbial diversity in the oral cavity and gastrointestinal tract performs multifarious functions to maintain host physiology such as metabolism of the nutrients, maintaining the integrity of the gut barrier, and development of the intestinal mucosa and immune system of the host [15]. Studies have reported that short exposure to HA [16] in addition to external factors (such as antibiotic usage, diet, psychological, and physical stress) and host factors [17] can alter the composition of the human microbiota. Several studies investigated and compared human microbiota at different altitudes and revealed a distinct taxonomical niche [17, 18]. Although the major altitude-related illnesses are well studied, more studies demonstrating the effect of hypobaric hypoxia on the microbial community of the intestinal tract and oral cavity are warranted. We have entered an era where therapeutics made up of microbes and/or their products are becoming a reality. Microbiota intervention strategy is an exciting area of research that may confer a plethora of beneficial effects by precision microbiome manipulation by using probiotics, prebiotics, and synbiotics in a variety of high altitude-induced gastrointestinal and oral disorders.

6.2 Gut Microbiome and High-Altitude Environments

The microbiome refers to the collective genomes of a remarkably diverse array of microorganisms. The microbiota lives symbiotically on and within four major colonization habitats of the human body [19]. Out of which, the majority of bacteria are found within the GIT, with the popular primarily anaerobic bacteria housed in the colon. The gut microbiota is comprised predominantly with *Firmicutes (>70%), Bacteriodetes (>30%), Proteobacteria (<5%), and Actinobacteria (<2%)* [20]. The gut microbial community is diverse and has adapted to colonize in the distinct microenvironment of the stomach, small intestine, and large intestine.

Utilization of available nutrients, cell surface receptors to adhere, and the ability to evade bacteriophages are some of the specific characteristics pivotal for the spatial colonization of gut microbes. The gut microbiota is also highly variable among individuals and numerous intrinsic as well as extrinsic factors can alter its diversity and development. One such environmental stressor is hypobaric hypoxia at high altitude, which decreases the integrity of the intestinal mucosal barrier function and leads to acute gastrointestinal injury [21, 22]. The cross-talk between the gut microbiota and the host is regulated and stabilized by microbial-synthesized proteins, peptides, and metabolites which exhibit pleiotropic effects. This intense cross-talk maintains the host-microbial homeostasis by triggering cell signaling and pathways that could physiologically link gut with other important systems and organs. However, altered microbial composition, known as dysbiosis can potentially disrupt these mechanisms and cause a diverse range of diseases. Considering the importance of the gut microbiome in homeostasis maintenance and the lack of sufficient literature demonstrating the mechanism responsible for alterations in gut microbiota due to hypoxia, more studies are warranted to explore the implication of rapid ascent to HA on host health.

6.2.1 Dysbiosis in Human Gut Microbiota at High-Altitude Conditions

Previous literature demonstrated that gut microbiota composition varies greatly with living conditions, lifestyle, genetics, and dietary habits [22–24] (Fig. 6.1). This section presented a detailed understanding of the results from all the studies conducted so far. A study of seven climbers at altitudes over 5000 m for 47 days found a significant increase in potentially pathogenic bacteria, i.e., *Escherichia coli*, and a decrease in commensal bacteria such as *Bifidobacterium*, *Corynebacterium*,

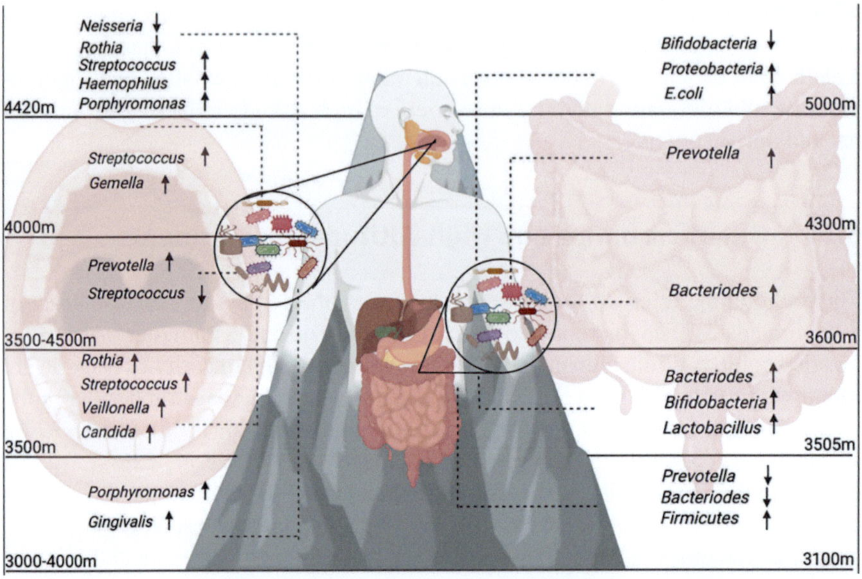

Fig. 6.1 Alterations in gut and oral microbiota at different heights of high altitude. (Made by BioRender)

Pseudomonas, and *Eggerthella* [16]. Another study also found increased facultative anaerobic bacteria and decreased aerobic bacteria in 15 soldiers during the adaptation at 3505 m for 15 days. These studies suggest that the low oxygen content at HA clearly alters the gut microbiota composition [11]. Another important study also found variations in gut microbiota between populations residing at HA and sea level (Papua New Guinea). These alterations are marked by differential phyla such as *Bacteroidetes, Firmicutes, Enterobacteriaceae, Actinobacteria,* and *Streptococcus* along with total bacterial count [25].

A study conducted in China analyzed the intestinal composition in three groups, i.e., Tibetans who were residing permanently in Tibet, Han Chinese who were in Tibet for more than 4 years, and people from plain areas. At the phylum level, the Tibetan and Han Chinese had a higher abundance of *Firmicutes*, while the plain areas population had a higher abundance of *Bacteroidetes*. These variations at the phylum level subsequently changed the genera which were marked by a notable decrease in *Prevotella* in Han Chinese, and a significant reduction and increase in *Bacteroides* and *Prevotella and Lactobacillus*, respectively, in the Tibetan population [26]. In contrast, a recent Indian study found that *Bacteroidetes* were higher in the HA population while *Firmicutes* and *Proteobacteria* were higher in plain populations [29]. A recent study found that dietary structures mainly composed of high animal protein and fats and low fiber played a vital role in shaping up the gut microbiota [17] (Fig. 6.1). Another study found that a lower *Bacteroides*-to-*Prevotella* ratio in participants who developed AMS at 4300 m could be a useful biomarker for identifying individuals susceptible to HA illnesses [27, 28].

6.2.2 Dysbiosis in Intestinal Microbiota in Animals Exposed to Hypobaric Hypoxia

Mammals living in HA regions showed significantly different gut microbiomes as compared to those in plain areas (Fig. 6.1). The gut microbiome of most of the mammals from high or low altitudes is characterized by two major phyla, i.e., *Firmicutes* and *Bacteroidetes* [26, 29–32]. Interestingly, the mammals from the HA region had a higher F/B ratio. Digestion of food and nutrient absorption are some of the vital roles played by *Firmicutes* and *Bacteroidetes*. A higher *Firmicutes/ Bacteroidetes* (F/B) ratio enhances energy extraction from the diet, helping to maintain energy balance and core body temperature in harsh environmental conditions [31, 33]. Several in vivo experiments have shown the impact of HA on the composition and diversity of gut microbiota. For instance, a study revealed that HA (5000 m) exposure for 30 days decreases the relative abundance of bacteria such as *Epsilonproteobacteria*, *Actinobacteria*, *Clostridia*, and *Spirochetes* while increasing the relative abundance of *Verrucomicrobiota*. These findings suggest that low oxygen levels, high-altitude exposure, and altered intestinal microbiota are directly associated [34] (Fig. 6.1).

In another study, acute hypobaric hypoxia (4100 m) exposure was given to Wistar rats for 7 days. Genera *Bacteroides* showed a notable increase while the genera *Prevotella* decreased as compared to the control group suggesting that atmospheric pressure is an important external regulating the intestinal microbiota [35]. A combined model of an AMS rat and hypobaric hypoxia exposure significantly decreased *Desulfovibrio* and increased the *Sutterella* genus [36]. With similar results from human expeditions, the relative abundance of aerobic bacteria was decreased and anaerobic bacteria was increased. All these studies concluded that hypobaric hypoxia greatly influences the host intestinal environment by altering its composition and diversity [28].

6.3 Oral Microbiome and High-Altitude Environmental Conditions

The oral microbiome is found in the human oral cavity, making it the second largest and complex microbiota after the gut [37, 38]. The teeth, the throat, the gingival sulcus, the floor of the mouth, the tongue, hard and soft palates, and tonsils are the different habitats found in the oral cavity [39]. Inside the oral cavity, the microbiome usually presents in the form of a biofilm consisting of microbes, extracellular matrix, gingival crevicular fluid (GCF), host cell components, and salivary glycoproteins. It is also very interesting to note that biofilms produced on hard surfaces are different from shedding surfaces due to the presence of an excellent binding surface for the adhesion of microbes [40] (Fig. 6.1). In addition, without significant fluctuations, an average normal temperature of the oral cavity is 37 °C, providing bacteria a stable environment to survive. Saliva, as the primary nutrient source provided a stable pH of 6.5–7.25, which is the preferred pH by oral microbes for flourishing [41, 42]. Approximately 1000 species of prokaryotes belong to 185 genera, mainly including the phyla *Actinobacteria, Bacteroidetes, Chlamydia, Euryarchaeota, Fusobacteria, Firmicutes, Proteobacteria, Spirochaetes*, and *Tenericutes* [39] have been identified. Out of which approximately 54% of species are officially named, 32% remain uncultivated, and 14% are cultivated but still nomenclature is pending [40].

The commensal oral microbiota is directly involved in maintaining oral homeostasis, protecting the oral cavity, and preventing disease development [43]. Several intrinsic and extrinsic host factors exert a direct or indirect effect on modulation and oral microbiota composition. Intrinsic factors such as genetics, immune response, circadian rhythm, saliva components (nutritional source), salivary pH, and temperature while factors such as lifestyle and behavior, diet, and oral hygiene are extrinsic factors [44]. The abovementioned factors underlie the transition of a symbiotic microbial community to a dysbiosis state and increase the potential for pathogenicity and promote microbiota-associated oral diseases [42] such as dental caries [40, 45] and periodontal diseases [46]. Further, several pieces of evidence pointed out that many diseases [43] such as diabetes [47], cardiovascular diseases [48], cystic fibrosis [49], esophageal and colorectal cancer [50], and rheumatoid arthritis [51] are also associated with disturbances in the oral ecosystem.

6.3.1 Alterations in Oral Microbiota After Exposure to High-Altitude Environments

It is important to mention that very little work has been done on high altitude and oral microbiota. Still, combined evidence from the studies show differences in oral microbiota composition between HA and sea-level residents. Chronic HH exposure in animals has been shown to increase the incidence of periodontitis due to the formation of ROS and inflammation in the submandibular glands [52, 53]. A study done at 3550 m reported that 23.2% of trekkers experienced dental issues [54]. Additionally, prolonged stays at high altitudes significantly decrease salivary flow, raising the risk of cavities [55, 56]. Other investigations also found that species diversity is also correlated with high altitude. For instance, data collected from studies conducted at 4000 m had decreased diversity along with an increased abundance of *Porphyromonas gingivalis* and *Prevotella* [57–60] (Fig. 6.1). A recent study found subtle alterations in oral microbiota at the genera level at HA. These genera are greatly linked to human oral and overall health status at HA (Fig. 6.1) [61].

6.4 Oral-Gut Microbiome Axis

If the human gut is considered a river for microbial flora, then the mouth is is the source of that river (Fig. 6.2). It means, in the swallowing process, the thousands of bacteria, that can be harmful or beneficial. However, oral and gut microbes are well separated due to the presence of physical as well as chemical hurdles, such as

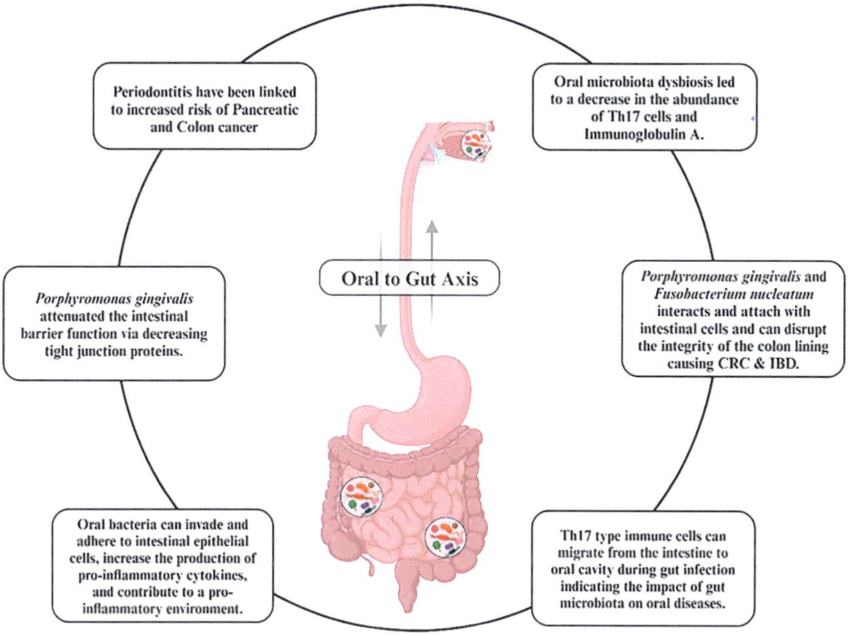

Fig. 6.2 The oral-gut axis: The axis involves the translocation of oral residents' microbiota to the gut due to compromised gut barrier integrity. It represents a dynamic and bidirectional relationship to alterations in gut and oral microbiome leading to dysbiosis

bile and gastric acids. But under conditions such as oral-intestine axis dysfunction, the oral microbiota could translocate to the gut via intestinal mucosa. This transmission of microbes via the oral cavity to the gut can change the microbial composition, eventually modulating the pathogenesis of diseases [62]. For instance, a common resident of the oral cavity, *Fusobacterium nucleatum*, has been found in the stool samples of IBD patients [63], suggesting an oral and gut axis [64, 65]. In addition, a prevalent intestine colonizer, i.e., *Klebsiella*, is known to promote Th1 cell induction and intestinal inflammation, which are key events in IBD pathogenesis. Another keystone pathogen *Porphyromonas gingivalis* causes periodontitis, a chronic inflammatory oral disease. This microbe when inoculated to mice attenuated the intestinal barrier function via decreasing tight junction proteins, altering gut microbiome (decreasing the enrichment of the family Clostridiaceae), and reduced systemic inflammation mediated by LPS [66, 67]. *P. gingivalis* not only disrupts the gut barrier but also stimulates the expression of Th1 and Th17-derived IFN-gamma and IL-17 in the colon [68] and mesenteric lymph nodes [69]. Similarly, *Klebsiella* and *P. gingivalis* induce Th1- and Th17-mediated inflammation after gut colonization, respectively [70, 71]. Oral-gut microbiome can also be seen in colorectal cancer (CRC), where several oral taxa have been detected in the gut of CRC patients [72] (Fig. 6.2).

All of the abovementioned studies and their results indicate a more direct association between gut microbiota and periodontal diseases. It was found that fecal microbiota transplantation (FMT) from mice fed a high-fat diet to mice with periodontitis resulted in severe alveolar bone destruction. *Enterococcus*, *Akkermansia*, and *Turicibacter* are some of the bacteria that increased in high-fat-fed mice. Metabolomic analysis showed an increase in purine nucleosides, which may be associated with elevated uric acid levels [73] (Fig. 6.2).

6.5 Microbial Therapeutic Strategies

6.5.1 Probiotics Intervention

Strains of the beneficial microbes *Lactobacillus*, *Bifidobacterium,* and *Saccharomyces* are considered as probiotics. Strains within these genera have the capability to enhance immunity, secrete compounds that assist digestion, and inhibit colonization by pathogens [74]. Probiotics improve health by promoting the growth of favorable gut microbes, competing with harmful species for adhesion sites, enhancing the intestinal immune response, or producing antimicrobial compounds that can kill harmful microbes [75].

Results from a study demonstrated that *Bacillus subtilis* and *Bacillus velezensis* isolated strains from the intestine of yaks when administered to rats improved the growth performance, the villi length, activity of the digestive enzyme, higher serum IgG, IgM and IgA, IL-10 and downregulated, major proinflammatory factor TNF-α, IL-6, and IL-8 [76]. These strains increased the population of *Lactobacillus* and *Ruminococcus* in the duodenum and *Candidatus arthromitus* in the jejunum [77]. Another study measured the efficacy of four strains: *Lactobacillus fermentum*, *Pediococcus pentosaceus,* and *Pediococcus acidilactici*, which were isolated healthy yaks, and found that *Pediococcus pentosaceus* had a positive effect on the growth of mice [78] (Fig. 6.3).

Another study supplemented three strains, namely *Enterococcus faecalis* EF1-mh, *Bacillus subtilis* BS1-ql Shu-qin, and *Lactobacillus sakei* LS-ql with probiotic potential isolated from Tibetan sheep in mice against *Clostridium perfringens* type C infection. These probiotic strains when administered single or as a complex to animals, no reduction in weight was seen, along with reduced expression of proinflammatory cytokines and enhanced expression of secretory IgA, suggesting that these probiotic ameliorated the inflammatory MyD88/NF-κB/MAPK pathway [79]. Another study isolated *Lactobacillus plantarum*, from the intestine of a healthy Tibet chicken and significantly decreased chicken mortality. The strain significantly increased the levels of IgA and IL-10, and decreased IL-1β, LITAF [80] (Fig. 6.3).

6.5.2 Prebiotics Supplementation

Prebiotics are nondigestible and fermentable food ingredients that positively alter the composition of gastrointestinal commensal bacteria [81]. FOSs, GOSs, isolmato-oligosaccharides (IMOs), resistant starch, and oligosaccharides are the main

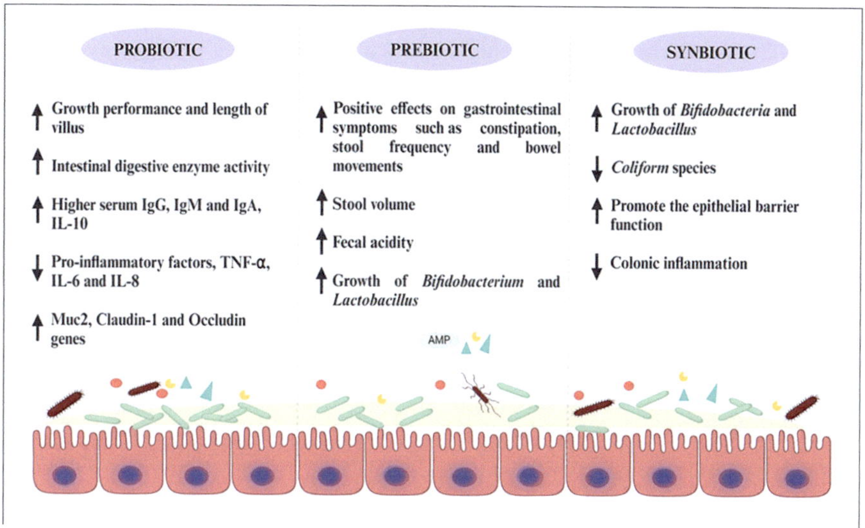

Fig. 6.3 A schematic diagram depicting the effect of probiotics, prebiotics, and synbiotics

groups of prebiotics. For instance, FOS (inulin and neosugar) help to increase stool volume, constipation relief, increase fecal acidity, and stimulate the growth of *Bifidobacterium* [82, 83]. Several clinical trials showed that the consumption of prebiotics had promising effects on gastrointestinal symptoms such as constipation, stool frequency, and bowel movements [84–87].

It was found that chicory inulin (15 g) when consumed for 28 days by the elderly population who had symptoms of constipation led to a shifting in gut microbiota toward enhanced levels of *Bifidobacterium* along with an improvement in constipation [84]. Consumption of inulin by constipated adults led to significantly increased stool frequency [85, 86]. Another trial showed that administration of Deshipu Stachyose granules has shown efficiency in relieving constipation by increased levels of commensals *Bifidobacteria* and *Lactobacilli* and decreased levels of *Clostridium perfringens* [87]. In summary, the existing evidence suggests that prebiotics have shown some promising results in improving gastrointestinal problems, possibly by modulating the gut microbiome (Fig. 6.3). However, their effects are yet to be explored in studies conducted at high-altitude environmental conditions.

6.5.3 Synbiotic Supplementation

Probiotics and prebiotics may exert a synergistic effect when administered together as synbiotics. Here, the prebiotic part improves the viability of the probiotic component, by providing resistance to acids, proteases, and bile salts [88]. A recent study has shown that synbiotics increase the number of *Bifidobacteria* and *Lactobacillus* with concomitant reduction in coliform species [89]. Ulcerative colitis shows similar symptoms as gastrointestinal injury induced by high altitude. An exploratory study compared two different doses (7.5 g/d or 15 g/d) of inulin-type fructans in active UC patients and found that high doses were associated with compositional shifts, where they supported the growth of β-fructans-fermenting bacteria namely *Lachnospiraceae* and *Bifidobacteriaceae* [90]. Butyrate induces the tight junction genes which promote the epithelial barrier function [91] associated with improvements in colonic inflammation (Fig. 6.3).

6.6 Conclusions and Future Perspective

The hypobaric hypoxia in the HA environment presents significant challenges for human adaptation, particularly due to reduced oxygen levels and pressure, which is only 50–60% of that at sea level. This hypoxic condition leads to severe stress responses in various organs, prominently affecting the gastrointestinal system. Acute mountain sickness often manifests as gastrointestinal discomfort, including diarrhea, nausea, vomiting, and anorexia, resulting from disrupted digestive functions due to hypoxia. The stress response induced by HA hypoxia, such as increased sympathetic nervous system activity and reduced blood flow to the gastrointestinal mucosa, exacerbates mucosal damage and increases permeability. This can lead to a vicious cycle of mucosal damage, bacterial translocation, and endotoxemia. Hypoxic stress at HA can disrupt this microbiota, leading to the overgrowth of opportunistic bacteria and bacterial translocation, where bacteria penetrate the mucosal barrier and enter systemic circulation, potentially causing systemic infections. Research indicates that the composition of intestinal microbiota varies with altitude, affecting immunity, metabolism, and susceptibility to diseases. The HA environments can alter the microbiota, potentially contributing to altitude sicknesses. The human intestine hosts a complex microbiota, consisting mostly of obligate anaerobes, which play a crucial role in maintaining the intestinal barrier and overall health status in extreme environmental conditions.

To study the dysbiosis in human microbiome that are crucial for understanding the health status at HA and for developing strategies to prevent microbial imbalances. Future studies should use advanced sequencing technologies to analyze microbiota from diverse populations at different altitudes, explore the roles of microbiota in HA illnesses, and development of interventions/supplementation such as diet modifications and probiotics to maintain microbiota balance. These insights could lead to new nutritional and health products that are beneficial for both HA residents and visitors.

Acknowledgments We are thankful to Directors of Defence Institute of Physiology and Allied Sciences (DIPAS), Delhi for their guidance and financial support to microbiome-related studies. The authors are also thankful to Ms. Harshita Gupta, Technician "B," Shri Dhruv Varshney and Ms. Guriya Sahi, Project Trainees, Department of Disruptive and Deterrence Technologies, DIPAS-DRDO for their constant technical and administrative support during the course of this study. MK and AY are supported by a fellowship from UGC-CSIR, Ministry of Human Resources, Govt. of India, Delhi, India.

Conflict of Interests The authors declare no conflict of interest.

Funding This work is supported by the Defence Institute of Physiology and Allied Sciences, Defence Research and Development Organization, Delhi-110054.

References

1. Singh LCG. High altitude dermatology. Indian J Dermatol. 2017;62(1):59–65.
2. Frisancho AR. Developmental functional adaptation to high altitude. Am J Hum Biol. 2013;25(2):151–68.
3. Purkayastha SS, Selvamurthy W. Soldier at high altitude: problem & preventive measures. Def Sci J. 2000;50(2):183.
4. West JB. Lessons from extreme altitudes. Front Physiol. 2019;10:465104.
5. Mathew TM, Sharma S. High Altitude Oxygenation. In: StatPearls. StatPearls Publishing, Treasure Island (FL); 2025. PMID: 30969523.
6. Bigham AW, Wilson MJ, Julian CG, Kiyamu M, Vargas E, Leon-Velarde F, Rivera-Chira M, Rodriquez C, Browne VA, Parra E, Brutsaert TD. Andean and Tibetan patterns of adaptation to high altitude. Am J Hum Biol. 2013;25(2):190–7.
7. Hartmann G, Tschöp M, Fischer R, Bidlingmaier C, Riepl R, Tschöp K, Hautmann H, Endres S, Toepfer M. High altitude increases circulating interleukin-6, interleukin-1 receptor antagonist and C-reactive protein. Cytokine. 2000;12(3):246–52.
8. Khanna K, Mishra KP, Ganju L, Kumar B, Singh SB. High-altitude-induced alterations in gut-immune axis: a review. Int Rev Immunol. 2018;37(2):119–26.
9. Luks AM, Swenson ER, Bärtsch P. Acute high-altitude sickness. Eur Respir Rev. 2017;26(143):160096.
10. Adak A, Maity C, Ghosh K, Mondal KC. Alteration of predominant gastrointestinal flora and oxidative damage of large intestine under simulated hypobaric hypoxia. Z Gastroenterol. 2014;52(02):180–6.
11. Adak A, Maity C, Ghosh K, Pati BR, Mondal KC. Dynamics of predominant microbiota in the human gastrointestinal tract and change in luminal enzymes and immunoglobulin profile during high-altitude adaptation. Folia Microbiol. 2013;58:523–8.
12. Wu TY. Take note of altitude gastrointestinal bleeding. Newsletter Int Soc Mountain Med. 2001;10:9–11.
13. Khandare AL, Gourineni SR, Validandi V. Dental fluorosis, nutritional status, kidney damage, and thyroid function along with bone metabolic indicators in school-going children living in fluoride-affected hilly areas of Doda district, Jammu and Kashmir, India. Environ Monit Assess. 2017;189:1–8.
14. Ashour AA. High altitude and its effects on oral health: a review of literature. J Adv Oral Res. 2020;11(2):143–7.
15. Macpherson AJ, Harris NL. Interactions between commensal intestinal bacteria and the immune system. Nat Rev Immunol. 2004;4(6):478–85.
16. Kleessen BRIGITTA, Schroedl WIELAND, Stueck MARCUS, Richter A, Rieck O, Krueger MONIKA. Microbial and immunological responses relative to high-altitude exposure in mountaineers. Med Sci Sports Exerc. 2005;37(8):1313–8.
17. Li K, Dan Z, Gesang L, Wang H, Zhou Y, Du Y, Ren YI, Shi Y, Nie Y. Comparative analysis of gut microbiota of native Tibetan and Han populations living at different altitudes. PLoS One. 2016;11(5):e0155863.
18. Das B, Ghosh TS, Kedia S, Rampal R, Saxena S, Bag S, Mitra R, Dayal M, Mehta O, Surendranath A, Travis SP. Analysis of the gut microbiome of rural and urban healthy Indians living in sea level and high altitude areas. Sci Rep. 2018;8(1):10104.

19. Lloyd-Price J, Abu-Ali G, Huttenhower C. The healthy human microbiome. Genome Med. 2016;8:1–11.
20. Belizário JE, Napolitano M. Human microbiomes and their roles in dysbiosis, common diseases, and novel therapeutic approaches. Front Microbiol. 2015;6:151578.
21. Alonso VR, Guarner F. Linking the gut microbiota to human health. Br J Nutr. 2013;109(S2):S21–6.
22. Yadav D, Ghosh TS, Mande SS. Global investigation of composition and interaction networks in gut microbiomes of individuals belonging to diverse geographies and age-groups. Gut Pathog. 2016;8:1–21.
23. Zhang F, Wu W, Deng Z, Zheng X, Zhang J, Deng S, Chen J, Ma Q, Wang Y, Yu X, Kang S. High altitude increases the expression of hypoxia-inducible factor 1α and inducible nitric oxide synthase with intest-inal mucosal barrier failure in rats. Int J Clin Exp Pathol. 2015;8(5):5189.
24. Hu B, Wang J, Li Y, Ge J, Pan J, Li G, He Y, Zhong H, Wang B, Huang Y, Han S. Gut microbiota facilitates adaptation of the plateau zokor (Myospalax baileyi) to the plateau living environment. Front Microbiol. 2023;14:1136845.
25. Greenhill AR, Tsuji H, Ogata K, Natsuhara K, Morita A, Soli K, Larkins JA, Tadokoro K, Odani S, Baba J, Naito Y. Characterization of the gut microbiota of Papua New Guineans using reverse transcription quantitative PCR. PLoS One. 2015;10(2):e0117427.
26. Li L, Zhao X. Comparative analyses of fecal microbiota in Tibetan and Chinese Han living at low or high altitude by barcoded 454 pyrosequencing. Sci Rep. 2015;5(1):14682.
27. Karl JP, Berryman CE, Young AJ, Radcliffe PN, Branck TA, Pantoja-Feliciano IG, Rood JC, Pasiakos SM. Associations between the gut microbiota and host responses to high altitude. Am J Physiol-Gastrointest Liver Physiol. 2018;315(6):G1003–15.
28. Zhang J, Chen Y, Sun Y, Wang R, Zhang J, Jia Z. Plateau hypoxia attenuates the metabolic activity of intestinal flora to enhance the bioavailability of nifedipine. Drug Deliv. 2018;25(1):1175–81.
29. Zhang Z, Xu D, Wang LI, Hao J, Wang J, Zhou X, Wang W, Qiu Q, Huang X, Zhou J, Long R. Convergent evolution of rumen microbiomes in high-altitude mammals. Curr Biol. 2016;26(14):1873–9.
30. Li JT, Gao YD, Xie L, Deng C, Shi P, Guan ML, Huang S, Ren JL, Wu DD, Ding L, Huang ZY. Comparative genomic investigation of high-elevation adaptation in ectothermic snakes. Proc Natl Acad Sci. 2018;115(33):8406–11.
31. Zhao J, Yao Y, Li D, Xu H, Wu J, Wen A, Xie M, Ni Q, Zhang M, Peng G, Xu H. Characterization of the gut microbiota in six geographical populations of Chinese rhesus macaques (Macaca mulatta), implying an adaptation to high-altitude environment. Microb Ecol. 2018;76:565–77.
32. Liu K, Yang J, Yuan H. Recent progress in research on the gut microbiota and highland adaptation on the Qinghai-Tibet Plateau. J Evol Biol. 2021;34(10):1514–30.
33. Zeng B, Zhang S, Xu H, Kong F, Yu X, Wang P, Yang M, Li D, Zhang M, Ni Q, Li Y. Gut microbiota of Tibetans and Tibetan pigs varies between high and low altitude environments. Microbiol Res. 2020;235:126447.
34. Zhang W, Jiao L, Liu R, Zhang Y, Ji Q, Zhang H, Gao X, Ma Y, Shi HN. The effect of exposure to high altitude and low oxygen on intestinal microbial communities in mice. PLoS One. 2018;13(9):e0203701.
35. Bai X, Liu G, Yang J, Zhu J, Wang Q, Zhou Y, Gu W, La L, Li X. Changes in the gut microbiota of rats in high-altitude hypoxic environments. Microbiol Spectr. 2022;10(6):e01626–2.
36. Zhao Z, Cui D, Wu G, Ren H, Zhu X, Xie W, Zhang Y, Yang L, Peng W, Lai C, Huang Y. Disrupted gut microbiota aggravates working memory dysfunction induced by high-altitude exposure in mice. Front Microbiol. 2022;13:1054504.
37. Grice EA, Segre JA. The human microbiome: our second genome. Annu Rev Genomics Hum Genet. 2012;13:151–70.
38. Kilian M, Chapple ILC, Hannig M, Marsh PD, Meuric V, Pedersen AML, Tonetti MS, Wade WG, Zaura E. The oral microbiome–an update for oral healthcare professionals. Br Dent J. 2016;221(10):657–66.

39. Dewhirst FE, Chen T, Izard J, Paster BJ, Tanner AC, Yu WH, Lakshmanan A, Wade WG. The human oral microbiome. J Bacteriol. 2010;192(19):5002–17.
40. Zhao H, Chu M, Huang Z, Yang X, Ran S, Hu B, Zhang C, Liang J. Variations in oral microbiota associated with oral cancer. Sci Rep. 2017;7(1):11773.
41. Aas JA, Paster BJ, Stokes LN, Olsen I, Dewhirst FE. Defining the normal bacterial flora of the oral cavity. J Clin Microbiol. 2005;43(11):5721–32.
42. Takahashi N. Microbial ecosystem in the oral cavity: metabolic diversity in an ecological niche and its relationship with oral diseases. In: International Congress Series, vol. 1284. Elsevier; 2005. p. 103–12.
43. Jia G, Zhi A, Lai PFH, Wang G, Xia Y, Xiong Z, Zhang H, Che N, Ai L. The oral microbiota–a mechanistic role for systemic diseases. Br Dent J. 2018;224(6):447–55.
44. Cornejo Ulloa P, van der Veen MH, Krom BP. Modulation of the oral microbiome by the host to promote ecological balance. Odontology. 2019;107:437–48.
45. Koo H, Bowen WH. Candida albicans and Streptococcus mutans: a potential synergistic alliance to cause virulent tooth decay in children. Future Microbiol. 2014;9(12):1295–7.
46. Philip N, Suneja B, Walsh LJ. Ecological approaches to dental caries prevention: paradigm shift or shibboleth? Caries Res. 2018;52(1–2):153–65.
47. Casarin RCV, Barbagallo A, Meulman T, Santos VR, Sallum EA, Nociti FH, Duarte PM, Casati MZ, Gonçalves RB. Subgingival biodiversity in subjects with uncontrolled type-2 diabetes and chronic periodontitis. J Periodontal Res. 2013;48(1):30–6.
48. Chhibber-Goel J, Singhal V, Bhowmik D, Vivek R, Parakh N, Bhargava B, Sharma A. Linkages between oral commensal bacteria and atherosclerotic plaques in coronary artery disease patients. NPJ Biofilms Microbiomes. 2016;2(1):7.
49. Whiley RA, Fleming EV, Makhija R, Waite RD. Environment and colonisation sequence are key parameters driving cooperation and competition between Pseudomonas aeruginosa cystic fibrosis strains and oral commensal streptococci. PLoS One. 2015;10(2):e0115513.
50. Kato I, Vasquez AA, Moyerbrailean G, Land S, Sun J, Lin HS, Ram JL. Oral microbiome and history of smoking and colorectal cancer. J Epidemiol Res. 2016;2(2):92.
51. Scher JU, Ubeda C, Equinda M, Khanin R, Buischi Y, Viale A, Lipuma L, Attur M, Pillinger MH, Weissmann G, Littman DR. Periodontal disease and the oral microbiota in new-onset rheumatoid arthritis. Arthritis Rheum. 2012;64(10):3083–94.
52. Terrizzi AR, Fernandez-Solari J, Lee CM, Conti MI, Martínez MP. Deleterious effect of chronic continuous hypoxia on oral health. Arch Oral Biol. 2016;72:1–7.
53. Terrizzi AR, Conti MI, Martínez MP, Fernández-Solari J. The process of acclimation to chronic hypoxia leads to submandibular gland and periodontal alterations: an insight on the role of inflammatory mediators. Mediat Inflamm. 2018;2018(1):6794508.
54. Küpper T, Hettlich M, Horz HP, Lechner K, Scharfenberg C, Conrads G, Yekta SS, Lampert F, Gore C. Dental problems and emergencies of trekkers—epidemiology and prevention. Results of the ADEMED Expedition 2008. High Alt Med Biol. 2014;15(1):39–45.
55. Kumar N, Dhiman RK, Arora V, Baranwal VK, Krishnaprasad KR, Pearl B. Changes in salivary output after induction at high-altitude areas and its effects on dental caries among Indian Army troops. Med J Armed Forces India. 2019;75(3):288–92.
56. Stookey GK. The effect of saliva on dental caries. J Am Dent Assoc. 2008;139:11S–7S.
57. Dong K, Wu K, Zheng T, Yue J, Wang W, Luo R, You L, He X, Li J, Hong Z, Zuo H. Comparative study of oral bacteria and fungi microbiota in Tibetan and Chinese han living at different altitude. Tohoku J Exp Med. 2021;254(2):129–39.
58. Hao Y, Tang C, Du Q, Zhou X, Peng X, Cheng L. Comparative analysis of oral microbiome from Zang and Han populations living at different altitudes. Arch Oral Biol. 2021;121:104986.
59. Damgaard C, Danielsen AK, Enevold C, Massarenti L, Nielsen CH, Holmstrup P, Belstrøm D. Porphyromonas gingivalis in saliva associates with chronic and aggressive periodontitis. J Oral Microbiol. 2019;11(1):1653123.
60. Liu F, Liang T, Zhang Z, Liu L, Li J, Dong W, Zhang H, Bai S, Ma L, Kang L. Effects of altitude on human oral microbes. AMB Express. 2021;11:1–13.

61. Kumari M, Bhushan B, Eslavath MR, Srivastava AK, Meena RC, Varshney R, Ganju L. Impact of high altitude on composition and functional profiling of oral microbiome in Indian male population. Sci Rep. 2023;13(1):4038.
62. Park SY, Hwang BO, Lim M, Ok SH, Lee SK, Chun KS, Park KK, Hu Y, Chung WY, Song NY. Oral–gut microbiome axis in gastrointestinal disease and cancer. Cancers. 2021;13(9):2124.
63. Brennan CA, Garrett WS. Fusobacterium nucleatum—symbiont, opportunist and oncobacterium. Nat Rev Microbiol. 2019;17(3):156–66.
64. Huh JW, Roh TY. Opportunistic detection of Fusobacterium nucleatum as a marker for the early gut microbial dysbiosis. BMC Microbiol. 2020;20:1–17.
65. Strauss J, Kaplan GG, Beck PL, Rioux K, Panaccione R, DeVinney R, Lynch T, Allen-Vercoe E. Invasive potential of gut mucosa-derived Fusobacterium nucleatum positively correlates with IBD status of the host. Inflamm Bowel Dis. 2011;17(9):1971–8.
66. Nakajima M, Arimatsu K, Kato T, Matsuda Y, Minagawa T, Takahashi N, Ohno H, Yamazaki K. Oral administration of P. gingivalis induces dysbiosis of gut microbiota and impaired barrier function leading to dissemination of enterobacteria to the liver. PLoS One. 2015;10(7):e0134234.
67. Arimatsu K, Yamada H, Miyazawa H, Minagawa T, Nakajima M, Ryder MI, Gotoh K, Motooka D, Nakamura S, Iida T, Yamazaki K. Oral pathobiont induces systemic inflammation and metabolic changes associated with alteration of gut microbiota. Sci Rep. 2014;4(1):4828.
68. Liu Y, Huang W, Dai K, Liu N, Wang J, Lu X, Ma J, Zhang M, Xu M, Long X, Liu J. Inflammatory response of gut, spleen, and liver in mice induced by orally administered Porphyromonas gingivalis. J Oral Microbiol. 2022;14(1):2088936.
69. Sato K, Takahashi N, Kato T, Matsuda Y, Yokoji M, Yamada M, Nakajima T, Kondo N, Endo N, Yamamoto R, Noiri Y. Aggravation of collagen-induced arthritis by orally administered Porphyromonas gingivalis through modulation of the gut microbiota and gut immune system. Sci Rep. 2017;7(1):6955.
70. Atarashi K, Suda W, Luo C, Kawaguchi T, Motoo I, Narushima S, Kiguchi Y, Yasuma K, Watanabe E, Tanoue T, Thaiss CA. Ectopic colonization of oral bacteria in the intestine drives TH1 cell induction and inflammation. Science. 2017;358(6361):359–65.
71. Nagao JI, Kishikawa S, Tanaka H, Toyonaga K, Narita Y, Negoro-Yasumatsu K, Tasaki S, Arita-Morioka KI, Nakayama J, Tanaka Y. Pathobiont-responsive Th17 cells in gut-mouth axis provoke inflammatory oral disease and are modulated by intestinal microbiome. Cell Rep. 2022;40(10):111314.
72. Nakatsu G, Li X, Zhou H, Sheng J, Wong SH, Wu WKK, Ng SC, Tsoi H, Dong Y, Zhang N, He Y. Gut mucosal microbiome across stages of colorectal carcinogenesis. Nat Commun. 2015;6(1):8727.
73. Rock KL, Kataoka H, Lai JJ. Uric acid as a danger signal in gout and its comorbidities. Nat Rev Rheumatol. 2013;9(1):13–23.
74. Hill C, Guarner F, Reid G, Gibson GR, Merenstein DJ, Pot B, Morelli L, Canani RB, Flint HJ, Salminen S, Calder PC. The International Scientific Association for Probiotics and Prebiotics consensus statement on the scope and appropriate use of the term probiotic. Nat Rev Gastroenterol Hepatol. 2014;11(8):506–14.
75. Plaza-Diaz J, Ruiz-Ojeda FJ, Gil-Campos M, Gil A. Mechanisms of action of probiotics. Adv Nutr. 2019;10:S49–66.
76. Li A, Wang Y, Li Z, Qamar H, Mehmood K, Zhang L, Liu J, Zhang H, Li J. Probiotics isolated from yaks improves the growth performance, antioxidant activity, and cytokines related to immunity and inflammation in mice. Microb Cell Factories. 2019;18:1–12.
77. Li A, Wang Y, Pei L, Mehmood K, Li K, Qamar H, Iqbal M, Waqas M, Liu J, Li J. Influence of dietary supplementation with Bacillus velezensis on intestinal microbial diversity of mice. Microb Pathog. 2019;136:103671.
78. He Y, Li F, Xu M, Jin C, Zhang Y, Nawaz S, Kulyar MFEA, Iqbal M, Qin Z, Li J. Probiotic potential of lactic acid bacteria isolated from yaks. Anim Dis. 2024;4(1):17.

79. He X, Ye G, Xu S, Chen X, He X, Gong Z. Effects of three different probiotics of Tibetan sheep origin and their complex probiotics on intestinal damage, immunity, and immune signaling pathways of mice infected with Clostridium perfringens type C. Front Microbiol. 2023;14:1177232.
80. Wang L, Fu G, Liu S, Li L, Zhao X. Effects of oxygen levels and a Lactobacillus plantarum strain on mortality and immune response of chickens at high altitude. Sci Rep. 2019;9(1):16037.
81. Gibson GR, Scott KP, Rastall RA, Tuohy KM, Hotchkiss A, Dubert-Ferrandon A, Gareau M, Murphy EF, Saulnier D, Loh G, Macfarlane S. Dietary prebiotics: current status and new definition. Food Sci Technol Bull Funct Foods. 2010;7(1):1–19.
82. Coussement, P. Pre-and synbiotics with inulin and oligofructose: promising developments in functional foods. Food Tech Europe 1996;2(4):102–104.
83. Gibson GR, Beatty ER, Wang XIN, Cummings JH. Selective stimulation of bifidobacteria in the human colon by oligofructose and inulin. Gastroenterology. 1995;108(4):975–82.
84. Odamaki T, Kato K, Sugahara H, Hashikura N, Takahashi S, Xiao JZ, Abe F, Osawa R. Age-related changes in gut microbiota composition from newborn to centenarian: a cross-sectional study. BMC Microbiol. 2016;16:1–12.
85. Li B, Ge Y, Cheng L, Zeng B, Yu J, Peng X, Zhao J, Li W, Ren B, Li M, Wei H. Oral bacteria colonize and compete with gut microbiota in gnotobiotic mice. Int J Oral Sci. 2019;11(1):10.
86. Del Castillo E, Meier R, Chung M, Koestler DC, Chen T, Paster BJ, Charpentier KP, Kelsey KT, Izard J, Michaud DS. The microbiomes of pancreatic and duodenum tissue overlap and are highly subject specific but differ between pancreatic cancer and noncancer subjects. Cancer Epidemiol Biomarkers Prev. 2019;28(2):370–83.
87. Gevers D, Kugathasan S, Denson LA, Vázquez-Baeza Y, Van Treuren W, Ren B, Schwager E, Knights D, Song SJ, Yassour M, Morgan XC. The treatment-naive microbiome in new-onset Crohn's disease. Cell Host Microbe. 2014;15(3):382–92.
88. Markowiak P, Śliżewska K. Effects of probiotics, prebiotics, and synbiotics on human health. Nutrients. 2017;9(9):1021.
89. Yang SC, Chen JY, Shang HF, Cheng TY, Tsou SC, Chen JR. Effect of synbiotics on intestinal microflora and digestive enzyme activities in rats. World J Gastroenterol. 2005;11(47):7413.
90. Valcheva R, Koleva P, Martínez I, Walter J, Gänzle MG, Dieleman LA. Inulin-type fructans improve active ulcerative colitis associated with microbiota changes and increased short-chain fatty acids levels. Gut Microbes. 2019;10(3):334–57.
91. Zheng L, Kelly CJ, Battista KD, Schaefer R, Lanis JM, Alexeev EE, Wang RX, Onyiah JC, Kominsky DJ, Colgan SP. Microbial-derived butyrate promotes epithelial barrier function through IL-10 receptor–dependent repression of claudin-2. J Immunol. 2017;199(8):2976–84.

Multi-omics Approaches for Understanding High-Altitude Pathophysiology

7

Anurag Kumar Upadhyay and Niroj Kumar Sethy

Abstract

The ever-increasing use of multi-omics investigations has become a pivotal and major technique for decoding complex biological systems. These high-throughput technologies produce a mammoth load of data in comparison to analysing just one layer of data and are frequently constrained by the type and variety of data being analysed. However, the use of multi-omics data integration strategies allows researchers to correlate cellular function levels, including genomes, epigenomes, transcriptomes, proteomes, modified proteomes, metabolomes, lipidomics, microbiomes and phenomes, this helps in data analysis and interpretation to elucidate the underlying mechanism of biological processes. This chapter provides detailed information on state-of-the-art analytical tools used in modern research of omics, particularly on computational techniques for amalgamating various omics data layers. We have covered algorithms, analytical pipelines, and web-based multi-omics integration tools available for data analysis and visualization. In addition, we have also conducted a concise review of multi-omics studies available to date for high-altitude acclimatization and maladaptation.

Keywords

Multi-omics · Transcriptomics · Proteomics · Metabolomics · High altitude

A. K. Upadhyay · N. K. Sethy (✉)
Defence Institute of Physiology and Allied Sciences (DIPAS), Defence Research and Development Organisation (DRDO), Timarpur, Delhi, India
e-mail: niroj.dipas@gov.in

S. Srivastava et al. (eds.), *Health and Sustenance in Extreme Environment*,
https://doi.org/10.1007/978-981-96-8642-1_7

7.1 Introduction

With advances in contemporary biology and high-throughput technologies like DNA sequencing and mass spectrometry, there has been a discernible surge in the number of omics studies (e.g. genomics, proteomics, metagenomics, metabolomics, etc.). All these studies have helped researchers gain greater insight into human disease [1], agriculture [2], plant science [3], microbiology [4], and the environment, which has been largely successful over the past 10 years. This omics revolution has created an enormous wealth of data that, when integrated into an appropriate statistical and mathematical framework, can aid in answering wider questions in fundamental and applied biology. A broader and deeper cognizance of biology may be obtained by combining different omes, each of which is crucial for a profound knowledge of the biological functions of living systems. However, researchers often find it difficult to integrate and analyse this huge data across multiple biological levels, diverse nature of omics measurements and difficulty of fitting the data types to the best model of statistical distributions.

'Multi-omics integration' is an umbrella term used to describe many distinct methods developed for studying the information flow through central dogma, with suitable experimental designs and study contexts. Over time, the utilization of these technologies has accumulated a massive amount of data prompting for understanding causes like gene regulation, protein expression dynamics, metabolite abundance, epigenetic variation, and host defence mechanisms, which have revolutionized the biological understanding of health and disease. In this chapter, we will highlight how multi-omics technologies have evolved over the past decade, along with their biological and clinical effects and the bioinformatics platforms that have become integral to system-based approaches.

7.2 The Omics Techniques

Single Omics data cannot uncover all the biological complexities and alterations of cellular micro- and macromolecules during the test conditions. Because each omics technique only yields a specific kind of data, it can offer only a limited understanding of the biological pathways and mechanisms that differ between the control and test groups. Nevertheless, understanding of individual Omics data types is critical for understanding multi-omics data integration.

7.2.1 Genomics

Genomics is the untargeted identification and characterization quantification of all the genes in the genome of an organism. The genome is the central information storage core and it essentially codes for all the basic information for all the other Omes. This encompasses revealing the interrelationship, interactions, and the overall impact of the environment on the organism. The Human Genome Project is an example of the oldest and most popular use of genomics technology [5]. Genomics routinely covers high-throughput DNA sequencing, sequence assembly, and genome annotation. A typical genomics application includes a Genome-wide association study (GWAS), a genotyping array, exome sequencing, micro-RNA and related small RNA sequencing.

7.2.2 Transcriptomics

Transcriptomics is the study of all the RNA species from a cell or tissue which offers a global molecular dynamic and temporal changes induced by experimental conditions, disease states, or pathogenic challenges. The transcriptome covers protein-coding messenger RNAs (mRNAs), long noncoding RNAs (lncRNAs), circular RNAs (cRNAs), short noncoding RNAs including microRNAs (miRNAs), small-interfering RNAs (siRNAs), small nuclear RNAs (snRNAs), piwi-interacting RNAs (piwiRNAs), and enhancer RNAs (eRNAS). It is important to note that mRNAs account for only 1–4% of the overall transcript, warranting the study of noncoding RNAs in health and disease. With the evolution of sequencing technology, single-cell RNA sequencing (scRNA-seq) has become an integral part of transcriptomics.

7.2.3 Epigenomic Analysis

Epigenomics focuses on the condition-specific post-transcriptional reversible modifications of DNA or histones that affect gene expression. The major alterations of DNA and histones are DNA methylation and histone modification, respectively, which provide functional context to the genome. The chromatin content and modifications regulate which genes are accessible to transcriptional machinery and which are insulated from active translation. These alterations are often affected by environmental or genetic causes and may occasionally be enduring or heritable [3]. Epigenetic alterations are tissue-specific and affect gene expression across organs based on their epigenetic state. NGS is the common technology for the comprehensive mapping of epigenetic patterns. The recent development in epigenetic analysis is the assay for transposase-accessible chromatin sequencing (ATAC-seq). This technique provides advantages like (1) direct evidence of genomic positions of nucleosome-depleted chromatin, which are permissible for transcriptional machinery binding, and (2) minimal sample input for the assay (requiring only 10,000–50,000 cells).

7.2.4 Single-Cell RNA Sequencing

Single-cell RNA sequencing (scRNA-seq) technology is becoming a state-of-the-art approach to understanding the heterogeneity and complexity of RNA transcripts within individual cells. Considering its first discovery in 2009, single-cell RNA sequencing (scRNA-seq) technologies have revolutionized molecular biology by enabling the measurement of transcriptome profiles at an unprecedented scale and resolution. scRNA-seq aims to decipher the transcriptome status and alterations of heterogeneous cell populations. Single-cell sequencing can elucidate the heterogeneity of the genome, transcriptome, and epigenome in cellular populations as well as corresponding changes in the levels. This rapid advancement in single-cell sequencing technology and the increasing availability of single-cell data for diverse species have paved paths for the development of novel computational methods to help in the understanding of biological complexity at a single-cell level.

7.2.5 Spatial Transcriptomics

Spatial transcriptome (ST) is a technology that combines the imaging-based spatial location of tissues and simultaneous next-generation sequencing (NGS)-based measurement of expression levels of genes systematically throughout the tissue space. This omics approach is routinely used to analyse and characterize the expression patterns of certain cell types in spatial dimensions in order to better understand variations in transcription in organs, tissues, and disease states. In 2021, the 'spatial transcriptomics' technique was designated by Nature Methods as 'Method of the year 2020'. Spatial transcriptomics has important applications like hypothesis testing using multiple experimental designs that compare time points or conditions, including genetic or environmental perturbations.

7.2.6 Metagenomic Analysis

The term metagenome was coined by Handelsman et al. (1988), which refers to the collective genomes of microbial populations found in any biological samples, whereas 'metagenomics' allude to the analysis of whole genomes of these microbial populations. This approach helps in generating information for microbial diversity, functional activity, cooperative and evolutionary relationship, composition of community and relationship with environment [6]. It can be distinguished as functional metagenomics and sequencing metagenomics. Functional metagenomics covers the discovery of new microbial genes or the identification of new microbial-derived bioactive compounds, whereas sequencing metagenomics refers to the identification of microbial species, diversity, and abundance in the sample of interest.

7.2.7 Proteomics

Proteomics is the study of protein composition, structure, expression, function, modification status, interactions, signalling networks and their cellular activities. In 1995, Marc Wilkins introduced the term 'proteome'. It provides a deeper insight into the structure and function of organisms and offers more practical information than genomics. Proteomics approaches have evolved over the last decade from traditional methods to high-throughput methods including tissue microarray (TMA), protein pathway arrays, and high-resolution mass spectrometry (HRMS). Proteomics holds enormous potential to identify new biomarkers by revealing the essential proteins and associated molecular pathways that underlie disease states. These biomarkers have the potential to be used as specific disease markers, diagnostic assays, prognostic predictors, and therapeutic targets to facilitate personalized medicine. Proteomics can be further subdivided into three main types based on applications: expression proteomics, functional proteomics, and structural proteomics, based on the experimental approaches, techniques used, and data generated. The investigation of protein production in specific circumstances, focusing on both quantity and quality, is referred to as Expression proteomics. The goal is to detect and measure variations in protein levels between test and control conditions through methods like protein electrophoresis, protein chips, DNA chips, microsequencing, mass spectrometry, and proximity ligation assay-based quantitation techniques. Functional proteomics allows the characterization of molecular-protein networks and dynamics involved in a living cell to any particular stimuli. It also aims to compare organisms at a global level to extract phylogenetic information. Techniques like yeast two-hybrid systems, deletion analysis, motif analysis, ligand chips, and high-throughput functional assays are a part of functional analysis. The 3D configurations and structural intricacies resulting in functional proteins are identified through structural proteomics. This proteomics approach also investigates protein roles, binding associates, protein–protein interactions. It employs high-throughput x-ray crystallography and NMR spectroscopy to elucidate the structural information of specific proteins.

7.2.8 Metabolomics

Metabolomics focuses on the profiling of metabolites in biological specimens like biofluids, cells, and tissues. These metabolites include small molecules, substrates, metabolic intermediates, and end products of biochemical processes such as carbohydrates, fatty acids, lipids, and amino acids (<1700 Da). The main objective of metabolomics is the detailed characterization of metabolites, including quantitative and qualitative analyses, for diagnostic, prognostic, and therapeutic purposes because the change in metabolites closely reflect an organism phenome at any given time as metabolites are closely linked to, and placed downstream of the transcriptomics, genomics, and proteomics. In general, metabolome analysis of a biological sample at a given point in time provides a functional summary of all the dynamic metabolic activities. Metabolomes have two benefits over the genome: (i) their capacity to reflect environmental impacts and (ii) they offer a summary of an individual's pathophysiological status at a given point of time. The widely used untargeted and targeted metabolomics techniques are meant for the measurement of a specific set of metabolites and thorough study of all detectable metabolites in the target sample.

7.2.9 Phenomics

Phenomics refers to the study of phenome, i.e. acquisition of high-dimensional phenotypic data on an organism-wide scale. In general, phenomics refers to the systematic study and analysis of qualitative and quantitative traits including physiological, haematological, biochemical, genetic, and imaging methodologies for the characterization of a phenotype. The phenotype is the ultimate expression of biological organization at the organismal level and phenotypic variation is produced through a complex interaction between genotype and environment, and this 'genotype-phenotype' map is unavailable without the precise phenotypic data that is required to study these relationships. Phenomics requires comprehensive recording of a wide variety of phenotypes or traits across biological scales and over time (referred to as deep phenotyping) and evaluation of phenotypes and their relationship with genotype, biological stage, and environment and they can be influenced (referred to as phenomics analysis).

7.3 Objective of Multi-omics

The primary goal of analysing multi-omics data is to uncover valuable information for identifying molecular patterns, biological pathways, molecular signatures, biomarkers, and drug targets to help in drug discovery. One important application of multi-omics analysis is to identify and validate potential therapeutic targets for different human diseases by uncovering specific connections between biomolecules and disease characteristics. The uses of multi-omics analysis in drug discovery include: (i) Identification and validation of novel therapeutic targets for an array of maladies is one of the primary uses of multi-omics [7]. Drug targets are the molecules whose alteration can change illness states or phenotypes. It includes metabolites, proteins, genes and epigenetic markers that have a role in disease aetiology or progression. (ii) Building of molecular pathways and networks of malady and environmental/drug response by interpretating different omics data and establishing a interrelation between different biological molecules involved in any drug or malady mechanisms of action [8]. (iii) Validation of drug and its mode of action by experimental and computational methods on various aspects affected by disease [9]. Multi-omics data reveal the effect of alterations on drug targets using various machine learning models and help in designing experiments for drug targets. (iv) Finding the drug response on different individuals by identifying genetic [10] variants like copy number variations (CNV), gene expressions levels, metabolite levels, and epigenetic alterations give an insight into how a drug affects different individuals, leading to new avenues for personalized medicines. (v) Predicting optimal drug responses and prioritizing drug targets by ranking genes, proteins, and metabolites affecting disease and their responses to drug, thereby lowering toxicity and adverse effects of drugs and levering safety and efficacy [11].

7.4 Challenges in Multi-omic Analysis

Multi-omics analysis inherits challenges from single omics datasets like data cleaning, normalization, biomolecule identification, data dimensionality reduction, biological contextualization, and statistical validation. In addition, heterogeneity in nomenclature among data types (such as genomics, transcriptomics, proteomics, and metabolomics) with confounding new challenges of missing value handling, integration, clustering, visualization, and functional characterization makes multi-omics data analysis challenging.

An integration method for identifiers (IDs) and mapping genes with its associated metabolites, proteins, and transcriptomics become a congested traffic jam for data analysis. Storage and annotation of data is a big challenge for the integration of multi-omics huge data owing to different dimensionality (e.g. a large metagenomic data with thousands of samples as compared to phosphoproteome data from a few samples) of the datasets. Integrated annotation of multi-omics data often requires commercially available software packages and is generally not available in public. Despite the growing abundance of multi-omics data analysis tools, storing and retrieving multi-omics data for specific analysis of biological aspects still continues to exist an opposition. The numbers of identified and annotated molecules across various omics techniques vary significantly, ranging from very good coverage of the genome, selective coverage of the metabolome, and partially annotated phosphoproteome. Hence, elucidation of biological interrelations, abstracting information from complex biological pathways and co-mapping still remains elusive.

Reproducing results in the multi-omics domain remains challenging because of the use of diverse data analysis methods, tools, and statistical methods for data processing. More recently, there has been a consensus exploration effort for findability, accessibility, interoperability, and reusability (FAIR) standards. Thus, the latest advancements in ammunition sharing and replication can address this issue.

7.5 Prerequisites for Multi-omics Integration

7.5.1 Dimensionality Reduction

Dimensionality reduction is the first and foremost step in an omics study. This step reduces data to fewer dimensions which will represent most of the variation in the data. These methods can discover the batch effects and inform additional correction steps required for multi-omics analysis. This technique is capable of identifying instances that did not meet the criteria, which would be classified as anomalies within the dataset. After achieving data quality, these techniques can identify any patterns in the data that could be linked to biologically significant variations.

7.5.2 Data Normalization

After structuring the data, omics measurements are grouped together for integration. Data generated from different omics platforms are transformed (usually log, log2, or other suitable transformations) to facilitate downstream statistics. In routine practice, the log transformation is normally used, and this transformation makes highly skewed data to approximately approach a normal distribution. These transformations facilitate the data to meet the assumption of statistical models and for making data patterns more interpretable. Another method for normalization is Z-score that calculates the probability in a normal distribution. This approach reduces the issue of significantly varied expression levels in input data and makes grouping of numerous separate omes easier for integrations. Data can be normalized to baseline measurements, allowing one to quantify deltas, limiting the impact of inter-individual variability in longitudinal research.

7.5.3 Correlation Networks Between Data Sets

After the normalizations and transformations of the data, correlation metrics give information on the most basic connection between the analytes. Pearson correlation coefficients (PCC) and Spearman correlation coefficients (SCC) are the most used correlation metrics. PCC is the preferred and more accurate method for correlation analysis having parametric metrics, whereas SCC is more vigorous if outlier samples are there. For longitudinal data, correlation networks effectively lessen inter-individual variability. During multi-omics analysis, data should be rectified for multiple hypotheses to reduce false positives using the Benjamini–Hochberg amendment. Few studies have used the Bonferroni correction, but the use of such method results in the loss of true positives. This correlation analysis often plotted a visual network diagram, giving informative overviews of interactions as well as differences across different conditions or disease states. A network analysis also provides information on the total number of nodes (molecules) that are associated in the network, the total number of edges between the networks (significant correlations), and the nature of correlations (positive and negative). The summary statistics also provide an excellent overview of correlations among different analytes under analysis completing the visual representation. This type of analysis helps identify network hubs and centres which are key regulators for overlapping pathways during multi-omics analysis.

7.5.4 Statistical Analyses

In cross-sectional studies, the differences of a single analyte comparing two separate datasets should be performed with student's t-test or Wilcoxon rank sum test. But in studies where two categories are being tested against, it is typically advised to utilize logistic regression for regression analysis within these categories. This regression model determines a curve for binary data, producing an odds ratio and p-value. When comparing data across multiple categories, it is important to utilize a non-parametric analysis of variance (ANOVA), such as Kruskal-Wallis. This method can be used to test the trend of data across different categories. Correction of multi-omics data can also be performed by using ANNOVA, but it should be kept in mind that ANNOVA assumes normal distribution only. Permutational multivariate analysis of variance (PERMANOVA) can be performed where the distribution is not normal.

7.5.5 Clustering Algorithms

Clustering algorithms group together similar samples or analysts. There are two primary clustering approaches: (i) hierarchical and (ii) K-means clustering. This clustering is very helpful in determining if samples cluster by person, batch, time point or any biological parameters for dimensionality reduction. This is used for finding extremes of the data which can be find of particular interest. Finding the ideal number of clusters is the most confounding difficulty in clustering analysis as selecting a high number of clusters might lead to redundancy of cluster detection while choosing less number of clusters might end in the loss of detection of certain relevant molecular trends and clusters. Furrow c-means clustering is an effective method for identifying patterns in longitudinal data.

7.5.6 Machine Learning

Machine learning is a branch of artificial intelligence that is frequently used to solve classification and prediction challenges. Utilizing machine learning for profiling omics data has now been widely used, which helps in the characterization and interpretation of this huge stats from different omics [12]. Categorization and forecasting, the primary divisions of machine learning, rely on the specific tasks or issues that machine learning aims to address, and can be either categorical (categorization) or continuous (forecasting). Three primary types of machine learning algorithms include supervised learning, unsupervised learning, and reinforcement learning [13].

7.6 Integration and Analysis Methods of Omics Data

The integration of various omics data can be brought together through two main factors.

1. Bottom-up method.
2. Top-down method.

In bottom-up approach, all the information is collected from different omics techniques and compiled to enable the manual clustering of the corresponding partitions. In contrast, in top-down approach, all information is concurrently aggregated with reduced dimensions facilitating simultaneous data annotation. Integration strategies can be unsupervised, semi-supervised, supervised, fusion-based, similar, Bayesian, and other multivariants.

7.6.1 Multivariate Methods for Data Integration

Principal Component Analysis is built on an unsupervised approach to identify features that will serve as indicators of the response data where Partial least square is a supervised alternative for PCA. It identifies the linear combination of variables that make a new set of characteristics. The variable with the strongest correlation receives the most weight.

Joint and individual variation explained (JIVE) is an unsupervised method that aims to separate the data into independent and combined, allowing the identification of similar and distinctive alterations in various omics data, which is very beneficial in sequentially analysing datasets knowing specific modifications and fundamental biological signals. It is capable of breaking down complicated information into small, understandable parts. JIVE makes easy interpretation of the relevance of biological findings while focusing on important variables leading to more reliable and robust results [14].

Joint non-negative matrix factorization (NMF) is the simplest unsupervised data integration which works by dividing a positive matrix into positive loading and factors. It requires correct and positive input matrices which can be more laborious and demanding of memory [15].

7.6.2 Bayesian Approach

A Bayesian framework establishes an empirical approach to handle the variability and unpredictability that are inherited in systems of biology via integrating previous knowledge alongside with current facts. Bayesian calculus cups together a comprehensive statistical array of omics variables. The distinct and discrete analysis of each kind of data has an absence of recognition and size inception of some interconnection. On the other hand, cumulative analysis of datasets from different omics approaches lacks heterogeneity of each dataset and may have underrepresentation of foremost features that are unique to each source of data [16].

An integrative clustering (icluster) approach of proteomic data, DNA methylation, mRNA expression, microRNA expression, and biological component shows a high interrelation and reliant connection. A biological integrative cluster approach is important for combining the weighted power of each source of data. icluster is an unsupervised method based on a Gaussian latent variable that minimizes the complexity of the datasets by creating a single cluster based on contemporaneous interpretation from several different clusters of data. It integrates multiple genetic and biological characteristics without lacking negative restrictions. Its objective is to collectively cluster samples and identify cluster-relevant elements throughout the datasets [17].

iCluster+ has been developed to cope with binary, orderly, categorized, and continuous variables in conjunction that include different analytical model constraints arising from genomic, transcriptomic, and epigenomic data. iCluster+ is configured to cut down on all kinds of omics data into compositional variables and transferring factors. This produces the most precise identification for modelled datasets juxtaposed with all other methods [18].

7.6.3 Network-Based Integration

The supervised network-based tool 'iOmicsPASS' is used for integration of transcriptomics, and proteomics data for prediction of subnetworks for molecular interactions. Based on user provided network data and omics data sets, this tool compute a score for each molecular interaction and applies a shrunken gene-centroid algorithm, which examines all network particulars in identify association pair of pattern guise anticipating best phenotype class. This method effectively addresses data variability and detects unique molecular markers corresponding to the phenotypic class. Its key benefit is picking predictive signatures from strongly interrelated subnetworks, which narrows the search pool of prognostic factors and is also acceptable with small sample sizes [19].

7.6.4 Fusion-Based Integration

Pattern fusion analysis identifies patterns, interconnections, and comprehensions that may not be evident when examining each dataset individually [20]. It can determine consequential patterns from different omics profiles through unbiased and automated data distribution. It can identify distinct subgroups having substantial aggregation compared to SNF and icluster besides increased proactivity, although it neither provides insights into fundamental processes nor reveals additional biomarkers.

7.6.5 Similarity-Based Integration

Similarity network fusion (SNF) identifies disease subtypes and forecasts their phenotypes by focusing on patient resemblance matrices extracted from each omics independently. In contrast to other networks, SNF uses sample–sample nets that are formed by correlation analysis. Using the local k-nearest neighbours (KNN) method in conjunction with the overall matrices of the other layers, the global patient corelation matrices of every layer has been updated by this method constantly and concurrently [21]. Fusion is completed by averaging the similar matrix when upgrade is finished.

A different machine learning biomedical data method, Multiple Kernel learning, combines data from different sources using a preset set of kernels [22]. rMKL-LPP (Regularized multiple kernels learning locality preserving projections) is an unsupervised method developed by Speicher and Pfeifer that derives kernel learning by the graph embedding framework algorithm, which substantially reduces clustering of sample and its repeated data analysis. It allows us to manage numerical and sequence matrices as inputs and maintain stability.

Neighbourhood-based multi-omics clustering (NEMO) is used when there is non-availability of data from different omics of each sample. It is quicker and sample reduction is not required. Nemo operates on three stages: firstly, the construction of a similarity matrix of inter-patient. Secondly, the integration of a single matrix, and lastly, a grouping of resultant networks. It is not a fit method to find biomarkers [23].

7.6.6 Correlation-Based Integration

Canonical correlation analysis (CCA) is a standard procedure for evaluating the relationship among methylation, CNV, and gene expression data. CCA evaluates feature correlation analysis by integrating group effects as well as selected individual features. It is not very useful for the evaluation of molecular subtypes and the identification of biomarkers.

7.7 Tools for Multi-omics Analysis

Multi-omics data analysis approaches are commonly classified as: (a) Unsupervised-exploratory interrelation-based clustering and analysis (e.g. disease subtype discovery, correlation/covariance and network models). (b) Supervised-categorizing quest for concluding a distinct upshot like control/disease status or forecast task for continuous result (e.g. pain score, survival, latency). In the last decade, a sizeable number of distinguished forms of multi-omics platforms have been developed; quite a few of them gained traction in recent years, apparently due to rise in users, well-founded user supports, and citations. Some of the examples of these platforms are mixOmics [24], SNF [25], Paintomics [26], 3Omics [27], miodin [28], and MOFA [29, 30] and additional specialized tools identified for the analysis of Omics data are outlined in Table 7.1.

Table 7.1 Examples of multi-omics tools

Tool/software	Category	Approach	Based language	Data type supported	References
iCluster	Unsupervised	Joint latent variable model-based clustering method	R package	Copy number; DNA methylation; gene expression; (numerical)	McLendon et al. [31]
PARADIGM	Unsupervised	Directed graphs factor giving probabilistic graphical models	Python	Multi-omics (numerical)	McLendon et al. [32]
iCluster+	Unsupervised	Matrix factorization and general linear regression	R package	Multi-omics (numerical) and categorical	Mo et al. [33]
iClusterBayes	Unsupervised	Matrix factorization	R package	Multi-omics (numerical) and categorical	Mo et al. [34]
rMKL-LPP	Unsupervised	Multiple kernel learning	Similarity based	Multi-omics (numerical)	Speicher and Pfeifer [22]
NetICS	Unsupervised	Network based	MATLAB	Multi-omics (numerical)	Dimitrakopoulos et al. [35]
BCC	Unsupervised	Bayesian consensus clustering	R	Multi-omics (numerical)	Lock and Dunson [36]
JIVE	Unsupervised	Decomposition method with rank approximation	MATLAB	Multi-omics (numerical)	Lock et al. [14]
SNF	Unsupervised	Local K-nearest neighbours (KNN)	R/MATLAB	Multi-omics (numerical) and categorical	Wang et al. [21]
iOmicsPASS	Supervised	Network based	R package	Multi-omics (numerical) and categorical	Koh et al. [19]
NEMO	Unsupervised	Similarity based clustering	R package	Multi-omics (numerical) and categorical	Rappoport and Shamir [23]
PFA	Unsupervised	Fusion method using PCA k-means clustering	MATLAB	DNA methylation, miRNA expression, gene expression, protein expression (numerical)	Shi et al. [37]
CCA	Unsupervised	Correlation based	R	CNV, methylation, gene expression	ElManzalawy et al. [38]
mixOmics	Supervised and unsupervised	Multivariate methods such as PLS, SPLS, sPLSDA,	R package	mixOmics (categorical numerical)	TCGA [39]

7.8 Multi-omics Studies for High Altitude

A handful number of studies have used multi-omics approach to study high altitude acclimatization and adaptation. Han et al. investigated the microbiome and plasma metabolome profile of 472 healthy young Chinese men at high altitudes (>4300 m, 1 week, 6 months and de-induction from high altitudes) and 138 indigenous Tibetan men at same altitude. During prolonged stay at high altitude, the lowlander Chinese men gut microbial composition was moderately similar to the Tibetans gut microbiome. The study further reveals that volunteers at high altitude have a representative enterotype irrespective of their sea level or high altitude ancestry. The intestinal microbial structure of Han volunteers gradually changed towards that of Tibetans after chronic high altitude exposure, proposing 'assimilation' of intestine microbiota towards Tibetan may help Han people in acclimatizing to high altitude. The gut microbiota significantly influenced plasma clinical parameters, accounting for 48.4% of variation. The study identified notable metabolic shift of lowlander Han volunteers closely aligning with Tibetan metabolic signatures after prolonged stay at high altitude. The plasma metabolome profiling of Han volunteers identified elevated levels of 41 metabolites key plasma metabolites, including lactic acid, sphingosine-1-phosphate, taurine, and inositol, reflecting a metabolic adaptation pattern comparable to Tibetans. These collective results emphasize that altitude-induced shifts in the gut microbial composition of lowlanders are closely associated with dynamic alterations in the host metabolome and hypoxia-related responses. This underscores the gut microbiota as a key mediator in high-altitude acclimatization and suggests that targeted modulation of the intestinal microbiome may serve as a promising strategy to enhance altitude adaptation in lowland populations [40].

Yin et al. explored high altitude acclimatization of mountaineers by using single-cell RNA sequencing (scRNA-seq) to closely examine immune cells, along with plasma metabolomics and lipidomics. The authors collected blood samples of 11 healthy high mountain climbers at 5 different time points (both at sea level and high altitude) and performed single-cell RNA sequencing (scRNA-seq) analysis along with plasma metabolomics and lipidomics. In this study, a total of 375722 single cells with 1083 gene per cell were identified. Using an array of multi omics methods, the authors reported enhanced immune activity and reduced inflammation, supported by AMPK-driven metabolic adjustments during acclimatization. In contrast, extreme-altitude exposure resulted in immune suppression, increased inflammation, oxidative stress, and a metabolic shift toward glycolysis and fatty acid utilization.

This study demonstrate that the dynamic and altitude-dependent nature of immune adaptation and underscore the critical role of metabolic plasticity and antioxidant defences in preserving immune competence under sustained hypoxic stress [41].

7.9 Sharing of Results by Means of Several Public Databases

The pool of open accessible biological databases that house omics data is enormous and perpetually developing. Omics technologies gained popularity in the 2000s, with more than 1500 'molecular biology' databases published and with an increase of 100 new databases per year. These values describe only the visible portion of the current databases; there are many databases that have not been published on the World Wide Web (WWW), the internet platform that enables the establishment of online databases. Omics data are often stored in their raw form in publicly available data repositories. These databases preserve raw data files from the published research articles in structured and accessible manner, facilating reproducibility and transparent reuse. This practice not only ensures that researchers receive proper recognition for their contributions but also actively supports the FAIR principles—making data Findable, Accessible, Interoperable, and Reusable—thereby fostering open science and collaborative advancement.

A database is a host of information drawn out from raw archive material extracted like an NCBI database for Genome and EMBL-EBI database known as ENSEMBL. The database facilitates sequence comparison and visual exploration through specialized analytical tools for genomic sequences accompanied by detailed functional annotations for enhanced interpretation. Transcriptomics data can be stored and found in multiple databases like GENE Expression Omnibus (GEO) [42]. For mass-spectroscopy proteomics data, a reference database PRIDE [43] with a raw file with spectra peak is accessible along with its accompanying quantification and identification data. Metabolight is an archive repository for metabolic data together with reference raw spectra, their location, role and structure of metabolite [44]. Apart from this general database, more specialized databases like aging atlas have been developed, which contains multi-omics data of aging-related scenario only [45]. Table 7.2 provides a curated compilation of repositories for multi-omics data, highlighting several notable examples.

Table 7.2 Examples of multi-omics repository

Data Repository	Web link	Disease	Type of multi-omics data	References
The Cancer Genome Atlas (TCGA)	https://cancergenome.nih.gov/	Cancer	Cancer RNA-Seq, DNA-Seq, miRNA-Seq, SNV, CNV, DNA methylation, and RPPA	Weinster et al. [46]
Clinical Proteomic Tumor Analysis Consortium (CPTAC)	https://cptac-data-portal.georgetown.edu/cptacPublic/	Cancer	Proteomics data corresponding to TCGA Cohorts	Lindgren et al. [47]
International Cancer Genomics Consortium (ICGC)	https://icgc.org/	Cancer	Whole genome sequencing, genomic variations data (somatic and germline mutation)	Xing et al. [48]
Cancer Cell Line Encyclopaedia (CCLE)	https://portals.broadinstitute.org/ccle	Cancer cell line	Gene expression, copy number, and sequencing data; pharmacological profiles of 24 anticancer drugs	Ghandi et al. [49]
Molecular Taxonomy of Breast Cancer International Consortium (METABRIC)	http://molonc.bccrc.ca/aparicio-lab/research/metabric/	Breast cancer	Clinical traits, gene expression, SNP, and CNV	Richard et al. [50]
TARGET	https://ocg.cancer.gov/programs/target	Paediatric cancers	Gene expression, miRNA expression, copy number, and sequencing data	Mullighan et al. [51]
Omics Discovery Index	https://www.omicsdi.org	Consolidated datasets from 17 repositories in a uniform framework	Genomics, transcriptomics, proteomics, and metabolomics	Drouard et al. [52]
China National Gene Bank Database	https://db.cngb.org/scientific_database/	Ten types of data are interconnected and indexed	Genomics, transcriptomics, proteomics, and metabolomics	(CNGBdb) [53]
GeneLab Data System (GLDS)	https://genelab.nasa.gov/publications/omics_database	Spaceflight and spaceflight-relevant data from experiments using model organisms	Epigenomics, genomics, transcriptomics, proteomics, metabolomics	Ray et al. [54]

(continued)

Paired Omics Data Platform	https://pairedomicsdata bioinformatics.nl/	Natural products discovery	Genomic and metabolomics	Schorn et al. [55]
ColPortal	https://colportal.imib.es/	Cancer	Transcriptomic, microtranscriptomic, methylomic	Esteban-Gil et al. [56]
Fibromine	http://www.fibromine.com	Pulmonary fibrosis	Transcriptomic and proteomic pulmonary fibrosis datasets	Fanidis et al. [57]
RNAactDrug	http://bio-bigdata.hrbmu.edu.cn/RNAactDrug	Drug sensitivity	mRNAs, long noncoding RNAs (lncRNAs) and microRNAs (miRNAs)	Dong et al. [58]
Aging Cell Atlas	https://ngdc.cncb.ac.cn/aging/index	Aging genes	Genomics	Kedlian et al. [59]
C/VDdb	www.padb.org/cvd	Cardiovascular disease	microRNA, genomics, proteomics, and metabolomics	Fernandes et al. [60]
PeptiCKDdb	www.peptickddb.com	Kidney disease	Peptide and RNA	Krochmal et al. [61]
iMETHYL	http://imethyl.iwate-megabank.org/	CD4+ T-lymphocytes, monocytes, and neutrophils	Whole-DNA methylation, whole-genome, and whole-transcriptome	Komaki et al. [62]

7.10 Conclusions

There are various shortcomings in multi-omics data integration and analysis, which include potential data mathematical overfitting, different analytical distributions, and throughput restrictions on certain techniques. Despite these, multi-omics is an ensemble of technologies that enable scientists to address issues in an unprecedented level of detail. These measures are not ideal in themselves, but consistency across omics measurements will make sure that the discoveries are precise to fundamental biological reality. While there is no perfect way to analyse the data, a researcher should trust their conclusion when the same results are found in various ways such as different datasets or statistical techniques. The analysis methods we have covered are not all-encompassing, but it is crucial to follow standardized and effective procedures for generating, analysing, interpreting, reporting, review, validating and verifying multi-omics data in any research. It should be shared with FAIR principles respecting the liberty and involvement of various stakeholders in the production and use of multi-omics data. To addresses the inherent imbalance and complexities of multi-omics input data, an essential robust data management system combined with artificial intelligence, data mining, advanced deep learning methods complemented by sophisticated analytical and interpretative tools is need of the hour. Different biomarker, therapeutic targets, precision medicine and more intricate clinical diagnosis and prognostic obtainable by analysis of different Omics levels giving a valuable penetration into molecular basis of disease and drug action. Multi-omics analysis is a powerful analytical tool which enables faster biomarker identification, therapeutic targets, advancing precision medicine, facilitating clinical diagnostic and prognostic assessments. This comprehensive approach provides profound and actionable insight in fundamental mechanism of diseases progression and drug response to it, catering personalized healthcare and therapeutic strategies.

References

1. Karczewski KJ, Snyder MP. Integrative omics for health and disease. Nat Rev Genet. 2018;19(5):299–310. https://doi.org/10.1038/nrg.2018.4.
2. Ichihashi Y, et al. Multi-omics analysis on an agroecosystem reveals the significant role of organic nitrogen to increase agricultural crop yield. Proc Natl Acad Sci. 2020;117(25):14552–60. https://doi.org/10.1073/pnas.1917259117.
3. (PDF) Liu et al 2016. Accessed July 23, 2024. https://www.researchgate.net/publication/303408121_Liu_et_al_2016.
4. Unique Features of Long Non-Coding RNA Biogenesis and Function | Nature Reviews Genetics. Accessed July 23, 2024. https://www.nature.com/articles/nrg.2015.10.
5. Collins FS, Morgan M, Patrinos A. The human genome project: lessons from large-scale biology. Science (New York, NY). 2003;300(5617):286–90. https://doi.org/10.1126/science.1084564.
6. Clinical Metagenomics | Nature Reviews Genetics. Accessed July 16, 2024. https://www.nature.com/articles/s41576-019-0113-7.

7. Kreitmaier P, Katsoula G, Zeggini E. Insights from multi-omics integration in complex disease primary tissues. Trends Genet. 2023;39(1):46–58. https://doi.org/10.1016/j.tig.2022.08.005.
8. Doran S, et al. Multi-omics approaches for revealing the complexity of cardiovascular disease. Brief Bioinform. 2021;22(5):bbab061. https://doi.org/10.1093/bib/bbab061.
9. Li K, et al. Bioinformatics approaches for anti-cancer drug discovery. Curr Drug Targets. 2020;21(1):3–17. https://doi.org/10.2174/1389450120666190923162203.
10. Allesøe RL, et al. Discovery of drug-omics associations in type 2 diabetes with generative deep-learning models. Nat Biotechnol. 2023;41(3):399–408. https://doi.org/10.1038/s41587-022-01520-x.
11. Doherty LM, et al. Integrating multi-omics data reveals function and therapeutic potential of deubiquitinating enzymes. eLife. 2022;11:e72879. https://doi.org/10.7554/eLife.72879.
12. Wei L, et al. Artificial intelligence (AI) and machine learning (ML) in precision oncology: a review on enhancing discoverability through multiomics integration. Br J Radiol. 2023;96(1150):20230211. https://doi.org/10.1259/bjr.20230211.
13. Reel PS, et al. Using machine learning approaches for multi-omics data analysis: a review. Biotechnol Adv. 2021;49:107739. https://doi.org/10.1016/j.biotechadv.2021.107739.
14. Lock EF, et al. Joint and individual variation explained (jive) for integrated analysis of multiple data types. Ann Appl Stat. 2013;7(1):523–42. https://doi.org/10.1214/12-AOAS597.
15. Meng C, et al. A multivariate approach to the integration of multi-omics datasets. BMC Bioinform. 2014;15(1):162. https://doi.org/10.1186/1471-2105-15-162.
16. Pačínková A, Popovici V. IntOMICS: a Bayesian framework for reconstructing regulatory networks using multi-omics data. J Comput Biol. 2023;30(5):569–74. https://doi.org/10.1089/cmb.2022.0149.
17. Chalise P, et al. Statistical methods for integrative clustering of multi-omics data. Methods Mol Biol (Clifton, NJ). 2629(2023):73–93. https://doi.org/10.1007/978-1-0716-2986-4_5.
18. Chauvel C, et al. Evaluation of integrative clustering methods for the analysis of multi-omics data. Brief Bioinform. 2020;21(2):541–52. https://doi.org/10.1093/bib/bbz015.
19. iOmicsPASS: Network-Based Integration of Multiomics Data for Predictive Subnetwork Discovery | Npj Systems Biology and Applications. Accessed July 16, 2024. https://www.nature.com/articles/s41540-019-0099-y.
20. Ding Z, et al. Integrative network fusion-based multi-omics study for biomarker identification and patient classification of rheumatoid arthritis. Chin Med. 2023;18(1):48. https://doi.org/10.1186/s13020-023-00750-8.
21. Similarity Network Fusion for Aggregating Data Types on a Genomic Scale | Nature Methods. Accessed July 16, 2024. https://www.nature.com/articles/nmeth.2810.
22. Speicher NK, Pfeifer N. Integrating different data types by regularized unsupervised multiple kernel learning with application to cancer subtype discovery. Bioinformatics. 2015;31(12):i268–75. https://doi.org/10.1093/bioinformatics/btv244.
23. Rappoport N, Shamir R. NEMO: cancer subtyping by integration of partial multi-omic data. Bioinformatics. 2019;35(18):3348–56. https://doi.org/10.1093/bioinformatics/btz058.
24. Rohart F, et al. mixOmics: an R package for 'omics feature selection and multiple data integration. PLoS Comput Biol. 2017;13(11):e1005752. https://doi.org/10.1371/journal.pcbi.1005752.
25. Wang M, et al. Sharing and community curation of mass spectrometry data with global natural products social molecular networking. Nat Biotechnol. 2016;34:828–37. https://doi.org/10.1038/nbt.3597.
26. Volume 46 Issue W1 | Nucleic Acids Research | Oxford Academic. Accessed July 16, 2024. https://academic.oup.com/nar/issue/46/W1.
27. 3Omics: A Web-Based Systems Biology Tool for Analysis, Integration and Visualization of Human Transcriptomic, Proteomic and Metabolomic Data | BMC Systems Biology | Full Text. Accessed July 16, 2024. https://bmcsystbiol.biomedcentral.com/articles/10.1186/1752-0509-7-64.

28. Vertical and Horizontal Integration of Multi-Omics Data with Miodin | BMC Bioinformatics | Full Text. Accessed July 16, 2024. https://bmcbioinformatics.biomedcentral.com/articles/10.1186/s12859-019-3224-4.
29. Multi-Omics Factor Analysis—a Framework for Unsupervised Integration of Multi-omics Data Sets | Molecular Systems Biology. Accessed July 16, 2024. https://www.embopress.org/doi/full/10.15252/msb.20178124.
30. Multi-Omics Data Integration, Interpretation, and Its Application – PubMed. Accessed July 16, 2024. https://pubmed.ncbi.nlm.nih.gov/32076369/.
31. Comprehensive genomic characterization defines human glioblastoma genes and core pathways. Nature. 2008;455(7216):1061–8. https://doi.org/10.1038/nature07385.
32. Comprehensive Genomic Characterization Defines Human Glioblastoma Genes and Core Pathways.
33. Pattern Discovery and Cancer Gene Identification in Integrated Cancer Genomic Data | PNAS. Accessed July 16, 2024. https://www.pnas.org/doi/abs/10.1073/pnas.1208949110.
34. Pattern Discovery and Cancer Gene Identification in Integrated Cancer Genomic Data | PNAS.
35. Dimitrakopoulos C, et al. Network-based integration of multi-omics data for prioritizing cancer genes. Bioinformatics (Oxford, England). 2018;34(14):2441–8. https://doi.org/10.1093/bioinformatics/bty148.
36. Fang et al. Bayesian integrative model for multi-omics data with missingness.
37. Shi et al. Pattern fusion analysis by adaptive alignment of multiple heterogeneous omics data.
38. El-Manzalawy Y. CCA based multi-view feature selection for multi-omics data integration; 2018. https://doi.org/10.1101/243733.
39. The Cancer Genome Atlas Program (TCGA) – NCI. cgvMiniLanding, May 13, 2022, nciglobal, ncienterprise. https://www.cancer.gov/ccg/research/genome-sequencing/tcga.
40. Longitudinal Multi-Omics Analysis Uncovers the Altered Landscape of Gut Microbiota and Plasma Metabolome in Response to High Altitude | Microbiome | Full Text. Accessed July 23, 2024. https://microbiomejournal.biomedcentral.com/articles/10.1186/s40168-024-01781-5.
41. Yin J, Lv J, Yang S, et al. Multi-omics reveals immune response and metabolic profiles during high-altitude mountaineering. Cell Rep. 2025;44(1). https://doi.org/10.1016/j.celrep.2024.115134
42. Clough E, Barrett T. The gene expression omnibus database. In: Mathé E, Davis S, editors. Statistical genomics: methods and protocols. New York, NY: Springer; 2016. p. 93–110. https://doi.org/10.1007/978-1-4939-3578-9_5.
43. PRIDE Database and Related Tools and Resources in 2019: Improving Support for Quantification Data | Nucleic Acids Research | Oxford Academic. Accessed July 16, 2024. https://academic.oup.com/nar/article/47/D1/D442/5160986.
44. Yurekten O, et al. MetaboLights: open data repository for metabolomics. Nucleic Acids Res. 2024;52(D1):D640–6. https://doi.org/10.1093/nar/gkad1045.
45. Basisty N, et al. A proteomic atlas of senescence-associated secretomes for aging biomarker development. PLoS Biol. 2020;18(1):e3000599. https://doi.org/10.1371/journal.pbio.3000599.
46. Cancer Genome Atlas Research Network, et al. The cancer genome atlas pan-cancer analysis project. Nat Genet. 2013;45(10):1113–20. https://doi.org/10.1038/ng.2764.
47. Simplified and Unified Access to Cancer Proteogenomic Data | Journal of Proteome Research. Accessed July 16, 2024. https://pubs.acs.org/doi/10.1021/acs.jproteome.0c00919.
48. Diagnostic and Prognostic Values of Pyroptosis-Related Genes for the Hepatocellular Carcinoma | BMC Bioinformatics | Full Text. Accessed July 16, 2024. https://bmcbioinformatics.biomedcentral.com/articles/10.1186/s12859-022-04726-7.
49. Ghandi M, et al. Next-generation characterization of the cancer cell line encyclopedia. Nature. 2019;569(7757):503–8. https://doi.org/10.1038/s41586-019-1186-3.
50. Richard V, et al. MicroRNAs in molecular classification and pathogenesis of breast tumors. Cancers. 2021;13(21):5332. https://doi.org/10.3390/cancers13215332.
51. Mullighan CG, et al. JAK mutations in high-risk childhood acute lymphoblastic leukemia. Proc Natl Acad Sci USA. 2009;106(23):9414–8. https://doi.org/10.1073/pnas.0811761106.

52. Drouard G, et al. Multi-omics integration in a twin cohort and predictive modeling of blood pressure values. OMICS: J Integr Biol. 2022;26(3):130–41. https://doi.org/10.1089/omi.2021.0201.
53. CNGBdb – China National GeneBank DataBase. Accessed July 16, 2024. https://db.cngb.org/.
54. GeneLab: Omics Database for Spaceflight Experiments | Bioinformatics | Oxford Academic. Accessed July 16, 2024. https://academic.oup.com/bioinformatics/article/35/10/1753/5134065.
55. A Community Resource for Paired Genomic and Metabolomic Data Mining | Nature Chemical Biology. Accessed July 16, 2024. https://www.nature.com/articles/s41589-020-00724-z.
56. ColPortal, an Integrative Multiomic Platform for Analysing Epigenetic Interactions in Colorectal Cancer | Scientific Data. Accessed July 16, 2024. https://www.nature.com/articles/s41597-019-0198-z.
57. Fibromine Is a Multi-Omics Database and Mining Tool for Target Discovery in Pulmonary Fibrosis | Scientific Reports. Accessed July 15, 2024. https://www.nature.com/articles/s41598-021-01069-w.
58. Dong Q, et al. RNAactDrug: a comprehensive database of RNAs associated with drug sensitivity from multi-omics data. Brief Bioinform. 2020;21(6):2167–74. https://doi.org/10.1093/bib/bbz142.
59. Human Skeletal Muscle Aging Atlas | Nature Aging. Accessed July 16, 2024. https://www.nature.com/articles/s43587-024-00613-3.
60. C/VDdb: A Multi-Omics Expression Profiling Database for a Knowledge-Driven Approach in Cardiovascular Disease (CVD) | PLOS ONE. Accessed July 16, 2024. https://journals.plos.org/plosone/article?id=10.1371/journal.pone.0207371.
61. Krochmal M, et al. PeptiCKDdb—peptide- and protein-centric database for the investigation of genesis and progression of chronic kidney disease. Database. 2016;2016:baw128. https://doi.org/10.1093/database/baw128.
62. iMETHYL: An Integrative Database of Human DNA Methylation, Gene Expression, and Genomic Variation | Human Genome Variation. Accessed July 16, 2024. https://www.nature.com/articles/hgv20188.

High-Altitude Maladies: Unravelling Diagnostic and Therapeutic Potentials of microRNAs

8

Sunanda Arya and Swati Srivastava

Abstract

Exposure to high altitudes presents considerable challenges due to the reduced availability of oxygen, which may lead to conditions such as acute mountain sickness (AMS), high-altitude pulmonary edema (HAPE), and high-altitude cerebral edema (HACE). This chapter investigates the diagnostic and therapeutic potential of microRNAs (miRNAs) in mitigating high-altitude-related maladies. It underscores the critical role of miRNAs, particularly miR-210 and miR-21, in hypoxia-induced mechanisms such as angiogenesis and inflammation, which are vital for physiological adaptation to elevated altitudes. The discussion encompasses the mechanisms underlying miRNA regulation, their application as non-invasive biomarkers, and the development of novel therapeutic approaches such as miRNA mimics and inhibitors. By integrating recent empirical findings, this chapter illustrates the ability of microRNAs (miRNAs) to support the advancement of personalized medicine and improve health outcomes for individuals subjected to high-altitude conditions.

Keywords

High-altitude physiology · AMS · HAPE · HACE · MicroRNAs (miRNAs) · Diagnostic biomarkers · Personalized medicine

S. Arya · S. Srivastava (✉)
Defence Institute of Physiology and Allied Sciences (DIPAS), Defence Research and Development Organization (DRDO), Timarpur, Delhi, India
e-mail: swati-sri.dipas@gov.in

S. Srivastava et al. (eds.), *Health and Sustenance in Extreme Environment*, https://doi.org/10.1007/978-981-96-8642-1_8

8.1 Introduction

A significant number of individuals are venturing to elevated altitudes for recreational, occupational, and sports-related purposes. Approximately a quarter of travelers who venture to altitudes exceeding 8500 feet experience symptoms of high-altitude illness [1]. High-altitude illness, whether in the manifestation of high-altitude pulmonary edema (HAPE) or high-altitude cerebral edema (HACE), carries the potential for fatality. The emergence of both these conditions, alongside a milder yet prevalent type of high-altitude illness known as acute mountain sickness (AMS), is commonly observed in individuals lacking acclimatization following rapid ascent to high-altitude environments. The underlying mechanisms of these conditions are complex and multifaceted, involving hypoxia-induced alterations in physiological and molecular pathways.

Recent advancements in molecular biology have uncovered a critical role for microRNAs (miRNAs) in the adaptive and maladaptive responses to hypoxia. MicroRNAs (miRNAs) represent a category of diminutive noncoding RNAs that hold significance in overseeing the posttranscriptional control of genetic information in eukaryotic organisms. Typically spanning 18–25 nucleotides in length, these molecules participate in processes such as gene suppression, inhibition of protein synthesis, and degradation of messenger RNA [2]. Impaired regulation of specific groups of miRNAs is associated with various diseases, such as cancer, heart disease, and neurological disorders [3]. In the context of high altitude, miRNAs have emerged as key regulators of the cellular response to hypoxia, influencing processes such as angiogenesis, erythropoiesis, and inflammation [4].

This chapter delves into the diagnostic and therapeutic potential of miRNAs in high-altitude maladies. The study will investigate the role of specific miRNAs in the pathophysiology of high-altitude maladies and discuss their potential as biomarkers for early detection and risk stratification. Furthermore, an analysis will be carried out on the potential treatment options that result from miRNA manipulation, providing a fresh perspective on the prevention and management of these conditions. Through the integration of recent research discoveries, this section seeks to offer a comprehensive examination of the role of miRNAs in high-altitude physiology and pathology, emphasizing their potential in enhancing outcomes for individuals exposed to high altitudes.

8.2 Pathophysiology of High-Altitude Maladies

High altitude, characterized by elevations exceeding 1500 m, requires acclimatization in order to mitigate hypoxemia and maintain cellular function in spite of decreased oxygen levels [5]. The process of acclimatization involves complex physiological adaptations that are not completely understood, impacting virtually all bodily systems [6]. The primary environmental alteration experienced at high altitudes is a decline in barometric pressure, resulting in a reduction of oxygen partial pressure from the atmosphere to the mitochondria of cells. This hypobaric hypoxia initiates a cascade of physiological reactions that generally facilitate adaptation to diminished oxygen conditions [7]. The most prevalent condition is acute mountain sickness (AMS), which typically initiates shortly after ascending and is commonly characterized by headache, often accompanied by reduced appetite, queasiness, throwing up, disrupted sleep, weariness, and dizziness [8]. These symptoms demonstrate rapid relief if further progression is stopped; conversely, if symptoms persist despite the progression, there is a potential for developing HAPE or HACE, where HAPE and HACE refer to acute pulmonary edema and acute cerebral edema, which are the two main severe expressions of altitude sickness [9]. Comprehension of the genetic elements that underlie these conditions offers valuable insights into their pathophysiology and potential targets for therapy. The pathophysiology of altitude-related illnesses involves genetic components such as variations in hypoxia-inducible factors, erythropoietin, nitric oxide synthase, and cytokine genes. These genetic alterations impact the transportation of oxygen, vascular reactivity, and inflammation, thereby affecting an individual's vulnerability and the intensity of conditions like AMS, HAPE, and HACE [10].

8.3 MicroRNAs: Biogenesis and Its Function

MicroRNA (miRNA), initially unveiled by Victor Ambros and his colleagues in the year 1993 and dubbed lin-4 within Caenorhabditis elegans, [11, 12] is ubiquitously present in the majority of eukaryotic organisms, encompassing humans, thereby instigating a profound transformation within the realm of molecular biology. miRNAs, which are nonprotein-coding regulatory genes, are evolutionarily conserved and exhibit partial complementarity to one or more messenger RNA (mRNA) molecules, as described by Farazi and coworkers [13]. The mature form of miRNAs consists of short, single-stranded RNA molecules approximately 21–23 nucleotides long, a length that allows for specific interactions with target genes while also enabling flexibility for adaptation to new gene targets. These molecules modulate the expression of homologous target-gene transcripts at both transcriptional and posttranscriptional levels through the process of RNA interference (RNAi). Their regulatory effects can lead to both up- and downregulation of gene expression through distinct mechanisms, taking place in both the cytoplasm and the nucleus, where miRNAs are localized. miRNA biogenesis commences with the processing of RNA polymerase II/III transcripts either post- or cotranscriptionally [14]. The process of miRNA biogenesis is categorized into canonical and noncanonical pathways. Within the canonical miRNA biogenesis pathway, pri-miRNAs are transcribed by RNA polymerase II, characterized by a 5′-7 methyl-guanosine cap and a 3′ polyadenylated tail. These pri-miRNAs undergo cleavage by the microprocessor complex, comprising Drosha and its cofactor DGCR8, leading to the formation of pre-miRNAs with a hairpin structure of approximately 60–70 nucleotides [15–17]. This cleavage event takes place within the nucleus. Subsequently, the pre-miRNAs are transported to the cytoplasm by exportin-5 (EXP 5) and the Ras-related nuclear protein guanosine triphosphate (RAN-GTP) [18]. In the cytoplasm, Dicer along

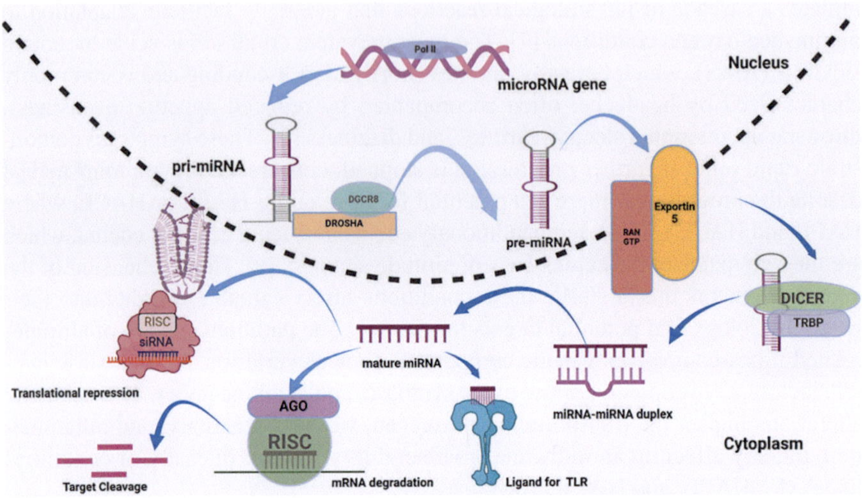

Fig. 8.1 MicroRNAs: biogenesis and its function. Created with BioRender.com

with its cofactor transactivation-responsive RNA-binding protein (TRBP) eliminates the terminal loop of the pre-miRNA, generating a miRNA duplex of around 20–22 nucleotides [19]. This duplex comprises a mature miRNA guide strand and a complementary passenger strand [20]. The duplex is then integrated into an Argonaute protein (AGO), resulting in the formation of the RNA-induced silencing complex (RISC). The passenger strand is degraded, while the guide strand becomes part of the miRNA-induced silencing complex (miRISC). Through binding of the miRNA's "seed" region (positions 2–7) to complementary sequences in the 3′-UTRs of target mRNAs, this complex regulates gene expression, leading to mRNA degradation or inhibition of translation. Furthermore, noncanonical pathways, such as mirtrons (derived from mRNA introns), Dicer-independent miRNAs, and tRNA-derived miRNAs, can also generate miRNAs. These alternative pathways involve different combinations of proteins from the canonical pathway, including Dicer, exportin-5, and AGO2 (Fig. 8.1). In contrast to canonical miRNA biogenesis, noncanonical pathways may not always necessitate Drosha and DGCR8.

8.4 Role of MicroRNAs in Complex Diseases

Gene regulation through microRNAs (miRNAs) is of significant importance in various biological contexts such as development, physiology, and disease. These miRNAs specifically interact with matching sequences within the 3′ untranslated regions (3′ UTR) of target messenger RNAs, resulting in either degradation of the mRNA or inhibition of translation. The pivotal role of miRNAs is evident in a wide array of biological processes like development, cellular differentiation, cell proliferation, and programmed cell death. Aberrant expression of miRNAs has been linked to the onset and progression of diverse complex diseases, including but not limited to cancer, cardiovascular ailments, neurodegenerative conditions, diabetes, and autoimmune disorders.

8.4.1 MicroRNAs in Cancer

In the realm of cancer, miRNAs can operate as either oncogenes or tumor suppressors [21]. Specifically, miR-21 exhibits frequent upregulation in a variety of cancers, such as breast, lung, and colon cancers, where it enhances tumor growth and survival by targeting tumor suppressor genes like PTEN and PDCD4 [22]. Conversely, let-7 family miRNAs function as tumor suppressors by inhibiting oncogenes such as RAS and MYC; [23]; the diminished expression of these miRNAs is correlated with unfavorable outcomes in cancer patients.

8.4.2 MicroRNAs in Cardiovascular Diseases

Within the domain of cardiovascular diseases, miRNAs play a crucial role. Muscle-specific miRNAs like miR-1 and miR-133 are essential for cardiac development and function. Changes in the levels of these miRNAs are associated with conditions like myocardial infarction and heart failure. For instance, reduced levels of miR-1 are linked to cardiac arrhythmias, while decreased miR-133 levels contribute to cardiac hypertrophy by regulating genes involved in muscle contractility [24].

8.4.3 MicroRNAs in Neurodegenerative Disorders

In the context of neurodegenerative disorders such as Alzheimer's and Parkinson's diseases, miRNAs influence neuronal function and survival. miR-29a/b, miR-107, and miR-124 are implicated in neuronal differentiation and synaptic plasticity. Disruption in the expression of these miRNAs is evident in Alzheimer's disease, impacting genes associated with amyloid precursor protein processing and tau phosphorylation [25]. Likewise, a notable reduction in miR-133b is observed in Parkinson's disease, affecting the function of dopaminergic neurons [26].

8.4.4 MicroRNAs in Diabetes

MicroRNAs are pivotal in the pathophysiology of diabetes as they control the processes of insulin production, secretion, and beta-cell function [27]. Specifically, miR-375 regulates insulin secretion, whereas miR-29a/b and miR-124a impact insulin sensitivity and beta-cell apoptosis, respectively [28].

8.4.5 MicroRNAs in Autoimmune Diseases

MicroRNAs play a significant role in the development of autoimmune diseases by affecting the differentiation and function of immune cells. In systemic lupus erythematosus (SLE), the downregulation of miR-146a results in the heightened expression of its target genes like TRAF6 and IRAK1, which participate in the inflammatory response [29]. Conversely, in rheumatoid arthritis (RA), miR-155 is upregulated in synovial fibroblasts and macrophages, exacerbating inflammation by targeting SHIP1 and suppressor of cytokine signaling 1 (SOCS1) [30].

8.5 Role of MicroRNAs in High-Altitude Maladies

MicroRNAs (miRNAs) are known to play a significant role in the pathogenesis of high-altitude maladies, such as acute mountain sickness (AMS), high-altitude pulmonary edema (HAPE), high-altitude cerebral edema (HACE), and other related conditions. Research studies have demonstrated that miRNAs participate in the modulation of gene expression, the precise adjustment of downstream effectors like hypoxia-inducible factor (HIF)-1α and VEGF, and the maintenance of endothelial barrier function to prevent vascular permeability [31, 32]. The High-Altitude Human miRNA Database (HAHmiR DB) serves as a comprehensive repository of experimentally validated miRNAs linked to high-altitude stress, providing valuable insights into the intricate regulatory processes involved in acclimatization. This database is anticipated to facilitate the identification of the intricate interplay among miRNAs, transcription factors (TF), and genes that regulate the fine-tuning of gene expression during high-altitude acclimatization [33]. Consequently, it is expected to aid in the identification of novel therapeutic targets and critical regulatory modules essential for adapting to high-altitude environments.

8.6 Hypoxia Response and MicroRNAs

The Pathway of Hypoxia-inducible Factor (HIF): An essential aspect of the cellular response to hypoxia is the HIF pathway. Stabilization of HIF-1α under conditions of low oxygen levels results in the transcription of various genes related to angiogenesis, metabolism, and erythropoiesis. MiR-210, known as the "hypoxamir," [56] presents a significant upsurge under hypoxic circumstances, aiding in angiogenesis and metabolic adjustments crucial for cellular adaptation to decreased oxygen levels. MiR-210 influences mitochondrial metabolism through the targeting of iron-sulfur cluster assembly proteins and impacts angiogenesis by modulating the expression of ephrin-A3 (EFNA3) (Fig. 8.2).

Adaptation of the Cardiovascular System: Modifications in the cardiovascular system at high altitudes involve heightened heart rate, blood pressure, and vascular restructuring. Studies have indicated that miR-210 plays a role in regulating VEGF expression [44], thus promoting angiogenesis and vascular adaptation to hypoxia.

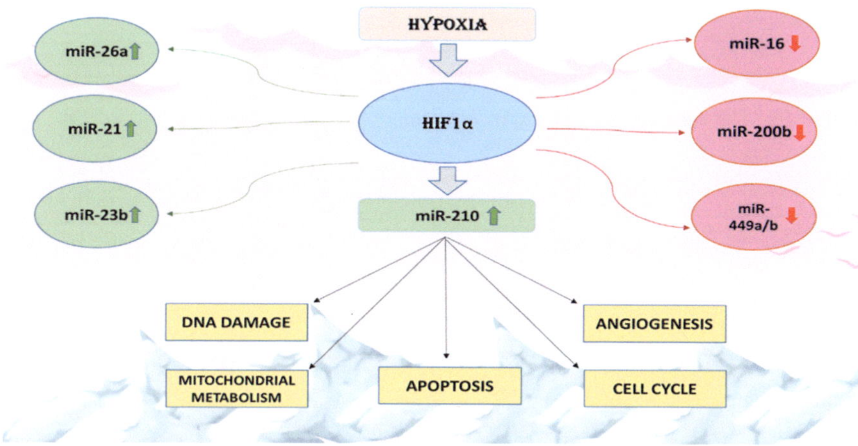

Fig. 8.2 MicroRNAs and HIF1α cross talk and miR-210 role in biological system

Moreover, miR-126 participates in endothelial cell functionality and vascular maintenance, contributing to the appropriate vascular responses under hypoxic conditions [34].

Erythropoiesis process: Erythropoiesis is critical for augmenting the capacity to carry oxygen in response to hypoxia. Key regulators of erythropoietin (EPO) signaling are miR-144 and miR-451. These microRNAs adjust the expression of genes involved in the production of red blood cells, thereby assisting in the body's adaptive responses at elevated altitudes [35].

8.7 MicroRNAs in High-Altitude Maladies

Acute Mountain Sickness (AMS): AMS presents symptoms such as headache, nausea, dizziness, and fatigue. The involvement of MiR-155 in the inflammatory response related to AMS is notable as it targets the suppressor of cytokine signaling 1 (SOCS1), thus influencing cytokine signaling and inflammation [45]. The differential expression of miRNAs in AMS patients indicates their potential contribution to the inflammatory and oxidative stress responses induced by hypoxia.

High-Altitude Pulmonary Edema (HAPE): HAPE is a severe condition characterized by the accumulation of fluid in the lungs. MiR-21 and miR-126 play significant roles in endothelial barrier function and vascular permeability. MiR-21's interaction with PTEN and PDCD4 promotes cell survival and proliferation, potentially worsening pulmonary edema in hypoxic environments (Table 8.1). Moreover, miR-126 aids in enhancing vascular integrity and endothelial repair, offering a possible protective mechanism against hypoxia-induced pulmonary hypertension [37].

High-Altitude Cerebral Edema (HACE): HACE involves dangerous brain swelling resulting from hypoxia. MiR-125 and miR-146 are linked to the maintenance of blood-brain barrier (BBB) integrity and regulation of neuroinflammation. MiR-125

Table 8.1 Hypoxic response and biological significance of microRNAs

MicroRNAs	Expression in hypoxia	Physiological Function	References
miR-210	Upregulated	Negative regulator of HIF signaling	Mutharasan et al. [46], Kulshreshtha et al. [47]
miR-26a	Upregulated	Airways smooth muscle hypertrophy, cell growth, and tumerogenesis	Kulshreshtha et al. [47], Mohamed et al. [48], Lu et al. [49]
miR-21	Upregulated	Pulmonary arterial hypertension	Drake et al. [50], Liu et al. [52]
miR-23b	Upregulated	Endothelial cell growth, TGFB signaling	Rogler et al. [51], Wang et al. [53]
miR-16	Downregulated	VEGF Expression	Dejean et al. [54]
miR-200b	Downregulated	Hypoxia-induced angiogenic response, cell cycle progression	Xia et al. [55], Chan et al. [56]
miR-449a/b	Downregulated	Cell proliferation	Marcet et al. [57], Muth et al. [58]

affects various components of the BBB, while miR-146 influences inflammatory pathways by targeting TRAF6 and IRAK1 [36]. Any disruption in the regulation of these miRNAs may lead to an increase in BBB permeability and the development of cerebral edema.

8.8 Therapeutic Strategies

1. *MiRNA Mimics and Inhibitors:* Synthetic RNA molecules known as miRNA mimics endogenous miRNAs, heightening their cellular activity. Conversely, miRNA inhibitors, also referred to as antagomirs, are formulated to impede the functions of particular miRNAs. These mechanisms are instrumental in adjusting miRNA levels and reinstating normal gene expression under high-altitude circumstances.

   MiR-210 Mimics: MiR-210, an miRNA inducible by hypoxia, governs mitochondrial metabolism and angiogenesis. Employing MiR-210 mimics could stabilize HIF signaling, thereby fostering adaptive reactions to hypoxia [38].

   MiR-21 Inhibitors: Suppression of miR-21 has the potential to diminish pulmonary hypertension and endothelial permeability in cases of HAPE, thereby ameliorating patient prognoses [39].
2. *Targeted Delivery Systems:* The development of targeted delivery systems for miRNA-centered therapies is imperative to heighten specificity and curtail off-target repercussions [40]. Various delivery techniques such as nanoparticles, liposomes, and viral vectors are under scrutiny to ensure the effective and precise conveyance of miRNA mimics or inhibitors to the afflicted tissues.
3. *Combination Therapies:* The amalgamation of miRNA-based therapies with prevailing pharmacological remedies could augment therapeutic efficiency. For instance, the concurrent application of miR-126 mimics with vasodilators [41]

might elicit synergistic outcomes in managing HAPE by enhancing vascular integrity and reducing pulmonary pressure.

8.9 Future Directions

1. *Clinical Trials:* The execution of clinical trials to scrutinize the safety and effectiveness of miRNA-based therapies for high-altitude afflictions is imperative [42]. These trials must evaluate the optimal dosages, delivery modalities, and potential adverse effects associated with miRNA mimics and inhibitors.
2. *Biomarker Establishment:* The identification of distinct miRNA expression patterns as biomarkers [43] for early detection and risk classification of high-altitude maladies can facilitate timely interventions and management of these conditions.
3. *Personalized Health Care*: Profound comprehension of individual variances in miRNA expression and reactions to hypoxia could pave the way for personalized medical approaches. Customizing miRNA-centered therapies based on individual genetic and epigenetic profiles has the potential to amplify treatment efficiency and diminish negative outcomes.

8.10 SWOT Analysis of MicroRNAs in High-Altitude Maladies

8.10.1 Strengths

1. Regulatory precision—microRNAs exhibit precise regulatory functions in gene expression, aiding in the comprehension of the molecular mechanisms involved in high-altitude adaptation and maladaptation.
2. Biomarker potential—microRNAs have the potential to function as biomarkers for the diagnosis and monitoring of high-altitude ailments like acute mountain sickness (AMS), high-altitude pulmonary edema (HAPE), and high-altitude cerebral edema (HACE).
3. Noninvasive testing—The detection of microRNAs in diverse body fluids (e.g., blood, saliva) allows for noninvasive diagnostic procedures.
4. Therapeutic targets—Targeting microRNAs for therapeutic purposes could pave the way for innovative treatments for high-altitude maladies (Fig. 8.3).

8.10.2 Weaknesses

1. Complexity of miRNA networks—The intricate regulatory networks of microRNAs pose challenges in identifying specific microRNAs involved in high-altitude maladies.
2. Variability and standardization—Fluctuations in microRNA expression due to environmental influences and the absence of standardized protocols for microRNA analysis can impact result reproducibility.

Fig. 8.3 SWOT analysis of microRNAs in high-altitude maladies

3. Limited understanding—The functions of microRNAs in high-altitude physiology and pathophysiology remain inadequately understood, necessitating extensive research efforts for elucidation.
4. Ethical and regulatory issues—Ethical and regulatory obstacles may arise in research involving microRNAs, particularly concerning clinical applications and genetic alterations (Fig. 8.3).

8.10.3 Opportunities

1. Advanced research—Continued research endeavors hold the promise of uncovering novel microRNAs linked to high-altitude adaptation and illnesses, enriching our knowledge and therapeutic options.
2. Personalized medicine—Utilizing microRNA profiles can facilitate the development of personalized medical strategies for individuals vulnerable to high-altitude maladies.
3. Technological advancements—Progress in sequencing technologies and bioinformatics has the potential to enhance microRNA detection, profiling, and functional analysis, fostering more precise, and comprehensive studies.

4. Cross-disciplinary collaboration—Collaborative efforts among molecular biologists, clinicians, and high-altitude researchers can expedite the translation of microRNA research into clinical settings (Fig. 8.3).

8.10.4 Threats

1. Technical challenges—Technical obstacles in microRNA extraction, quantification, and analysis may impede research progress and clinical implementation.
2. Funding limitations—Scarce funding for high-altitude research could restrict the exploration of microRNAs in this domain.
3. Genetic and environmental interactions—The intricate interplay between genetic predisposition and environmental factors at high altitudes can complicate the interpretation of microRNA data.
4. Ethical concerns—Concerns regarding genetic research and manipulation may present hurdles in the advancement of microRNA-based therapies (Fig. 8.3).

8.11 Conclusion and Future Directions

This chapter highlights the pivotal role of microRNAs (miRNAs) in conditions related to high altitudes, elucidating their substantial potential as both diagnostic biomarkers and targets for therapeutic intervention. Noteworthy miRNAs such as miR-210 and miR-21 have been recognized as pivotal regulators in the cellular response to hypoxia, influencing fundamental processes including angiogenesis, erythropoiesis, and inflammation. Gaining insight into the specific mechanisms through which miRNAs operate in the context of high-altitude adaptation and maladaptation offers a valuable understanding of the pathophysiology underlying conditions like acute mountain sickness (AMS), high-altitude pulmonary edema (HAPE), and high-altitude cerebral edema (HACE).

Subsequent research endeavors should emphasize comprehensive profiling of miRNAs in populations residing at high altitudes to pinpoint new miRNAs linked to both adaptation to high altitudes and susceptibility to altitude-related maladies. Progress in sequencing technologies and bioinformatics holds promise for enhancing the precision and breadth of such investigations. Furthermore, investigating the therapeutic possibilities of miRNA mimics and inhibitors has the potential to pave the way for innovative treatments addressing high-altitude afflictions. Synergistic collaborations spanning molecular biology, clinical investigation, and high-altitude physiology are imperative for translating these discoveries into clinical applications. Through integrating miRNA studies with personalized medicine strategies, tailored interventions can be formulated to enhance outcomes for individuals exposed to high altitudes, ultimately mitigating the frequency and intensity of altitude-induced illnesses.

References

1. Harris MD, Terrio J, Miser WF, Yetter JF 3rd. High-altitude medicine. Am Fam Physician. 1998;57(8):1907–14, 1924–6. Erratum in: Am Fam Physician 1998;58(4):866.
2. Wu M, Wang G, Tian W, Deng Y, Xu Y. MiRNA-based therapeutics for lung cancer. Curr Pharm Des. 2018;23(39):5989–96. https://doi.org/10.2174/1381612823666170714151715.
3. Agarwal V, Bell GW, Nam JW, Bartel DP. Predicting effective microRNA target sites in mammalian mRNAs. elife. 2015;4:e05005. https://doi.org/10.7554/eLife.05005.
4. Serocki M, Bartoszewska S, Janaszak-Jasiecka A, Ochocka RJ, Collawn JF, Bartoszewski R. miRNAs regulate the HIF switch during hypoxia: a novel therapeutic target. Angiogenesis. 2018;21(2):183–202. https://doi.org/10.1007/s10456-018-9600-2.
5. Baniya S, Holden C, Basnyat B. Reentry high altitude pulmonary edema in the Himalayas. High Alt Med Biol. 2017;18(4):425–7. https://doi.org/10.1089/ham.2017.0088.
6. Gallagher SA, Hackett PH. High-altitude illness. Emerg Med Clin North Am. 2004;22(2):329–55, viii. https://doi.org/10.1016/j.emc.2004.02.001.
7. Luks AM, Swenson ER, Bärtsch P. Acute high-altitude sickness. Eur Respir Rev. 2017;26(143):160096. https://doi.org/10.1183/16000617.0096-2016.
8. Taylor AT. High-altitude illnesses: physiology, risk factors, prevention, and treatment. Rambam Maimonides Med J. 2011;2(1):e0022. https://doi.org/10.5041/RMMJ.10022.
9. Savioli G, Ceresa IF, Gori G, Fumoso F, Gri N, Floris V, Varesi A, Martuscelli E, Marchisio S, Longhitano Y, Ricevuti G, Esposito C, Caironi G, Giardini G, Zanza C. Pathophysiology and therapy of high-altitude sickness: practical approach in emergency and critical care. J Clin Med. 2022;11(14):3937. https://doi.org/10.3390/jcm11143937.
10. Mallet RT, Burtscher J, Pialoux V, Pasha Q, Ahmad Y, Millet GP, Burtscher M. Molecular mechanisms of high-altitude acclimatization. Int J Mol Sci. 2023;24(2):1698. https://doi.org/10.3390/ijms24021698.
11. Lee RC, Feinbaum RL, Ambros V. The C. elegans heterochronic gene lin-4 encodes small RNAs with antisense complementarity to lin-14. Cell. 1993;75(5):843–54. https://doi.org/10.1016/0092-8674(93)90529-y.
12. Wightman B, Ha I, Ruvkun G. Posttranscriptional regulation of the heterochronic gene lin-14 by lin-4 mediates temporal pattern formation in C. elegans. Cell. 1993;75(5):855–62. https://doi.org/10.1016/0092-8674(93)90530-4.
13. Farazi TA, Juranek SA, Tuschl T. The growing catalog of small RNAs and their association with distinct Argonaute/Piwi family members. Development. 2008;135(7):1201–14. https://doi.org/10.1242/dev.005629.
14. Ha M, Kim VN. Regulation of microRNA biogenesis. Nat Rev Mol Cell Biol. 2014;15(8):509–24. https://doi.org/10.1038/nrm3838.
15. O'Brien J, Hayder H, Zayed Y, Peng C. Overview of microRNA biogenesis, mechanisms of actions, and circulation. Front Endocrinol (Lausanne). 2018 Aug;3(9):402. https://doi.org/10.3389/fendo.2018.00402.
16. Felekkis K, Touvana E, Stefanou C, Deltas C. microRNAs: a newly described class of encoded molecules that play a role in health and disease. Hippokratia. 2010;14(4):236–40.
17. Peng Y, Croce CM. The role of MicroRNAs in human cancer. Signal Transduct Target Ther. 2016;1:15004. https://doi.org/10.1038/sigtrans.2015.4.
18. Wong LL, Wang J, Liew OW, Richards AM, Chen YT. MicroRNA and heart failure. Int J Mol Sci. 2016;17(4):502. https://doi.org/10.3390/ijms17040502.
19. Kim VN, Han J, Siomi MC. Biogenesis of small RNAs in animals. Nat Rev Mol Cell Biol. 2009;10(2):126–39. https://doi.org/10.1038/nrm2632.
20. Inui M, Martello G, Piccolo S. MicroRNA control of signal transduction. Nat Rev Mol Cell Biol. 2010;11:252–63.
21. Iorio MV, Croce CM. microRNA involvement in human cancer. Carcinogenesis. 2012;33(6):1126–33. https://doi.org/10.1093/carcin/bgs140.

22. Chan JA, Krichevsky AM, Kosik KS. MicroRNA-21 is an antiapoptotic factor in human glioblastoma cells. Cancer Res. 2005;65(14):6029–33. https://doi.org/10.1158/0008-5472.CAN-05-0137.
23. Johnson SM, Grosshans H, Shingara J, Byrom M, Jarvis R, Cheng A, Labourier E, Reinert KL, Brown D, Slack FJ. RAS is regulated by the let-7 microRNA family. Cell. 2005;120(5):635–47. https://doi.org/10.1016/j.cell.2005.01.014.
24. Carè A, Catalucci D, Felicetti F, Bonci D, Addario A, Gallo P, Bang ML, Segnalini P, Gu Y, Dalton ND, Elia L, Latronico MV, Høydal M, Autore C, Russo MA, Dorn GW 2nd, Ellingsen O, Ruiz-Lozano P, Peterson KL, Croce CM, Peschle C, Condorelli G. MicroRNA-133 controls cardiac hypertrophy. Nat Med. 2007;13(5):613–8. https://doi.org/10.1038/nm1582.
25. Hébert SS, Horré K, Nicolaï L, Bergmans B, Papadopoulou AS, Delacourte A, De Strooper B. MicroRNA regulation of Alzheimer's Amyloid precursor protein expression. Neurobiol Dis. 2009;33(3):422–8. https://doi.org/10.1016/j.nbd.2008.11.009.
26. Kim J, Inoue K, Ishii J, Vanti WB, Voronov SV, Murchison E, Hannon G, Abeliovich A. A MicroRNA feedback circuit in midbrain dopamine neurons. Science. 2007;317(5842):1220–4. https://doi.org/10.1126/science.1140481.
27. Poy MN, Eliasson L, Krutzfeldt J, Kuwajima S, Ma X, Macdonald PE, Pfeffer S, Tuschl T, Rajewsky N, Rorsman P, Stoffel M. A pancreatic islet-specific microRNA regulates insulin secretion. Nature. 2004;432(7014):226–30. https://doi.org/10.1038/nature03076.
28. LaPierre MP, Stoffel M. MicroRNAs as stress regulators in pancreatic beta cells and diabetes. Mol Metab. 2017;6(9):1010–23. https://doi.org/10.1016/j.molmet.2017.06.020.
29. Pauley KM, Satoh M, Chan AL, Bubb MR, Reeves WH, Chan EK. Upregulated miR-146a expression in peripheral blood mononuclear cells from rheumatoid arthritis patients. Arthritis Res Ther. 2008;10(4):R101. https://doi.org/10.1186/ar2493.
30. Stanczyk J, Pedrioli DM, Brentano F, Sanchez-Pernaute O, Kolling C, Gay RE, Detmar M, Gay S, Kyburz D. Altered expression of MicroRNA in synovial fibroblasts and synovial tissue in rheumatoid arthritis. Arthritis Rheum. 2008;58(4):1001–9. https://doi.org/10.1002/art.23386.
31. Feng S, Ma J, Long K, Zhang J, Qiu W, Li Y, Jin L, Wang X, Jiang A, Liu L, Xiao W, Li X, Tang Q, Li M. Comparative microRNA transcriptomes in domestic goats reveal acclimatization to high altitude. Front Genet. 2020;11:809. https://doi.org/10.3389/fgene.2020.00809.
32. Su T, Gu C, Draga D, Zhou C, Lhamo T, Zheng Z, Qiu Q. Integrative analysis of miRNA-mRNA network in high altitude retinopathy by bioinformatics analysis. Biosci Rep. 2021;41(1):BSR20200776. https://doi.org/10.1042/BSR20200776.
33. Khurana P, Gupta A, Sugadev R, Sharma YK, Kumar B. HAHmiR.DB: a server platform for high-altitude human miRNA-gene coregulatory networks and associated regulatory circuits. Database (Oxford). 2020;2020:baaa101. https://doi.org/10.1093/database/baaa101.
34. Wang S, Aurora AB, Johnson BA, Qi X, McAnally J, Hill JA, Richardson JA, Bassel-Duby R, Olson EN. The endothelial-specific microRNA miR-126 governs vascular integrity and angiogenesis. Dev Cell. 2008;15(2):261–71. https://doi.org/10.1016/j.devcel.2008.07.002.
35. Dore LC, Amigo JD, Dos Santos CO, Zhang Z, Gai X, Tobias JW, Yu D, Klein AM, Dorman C, Wu W, Hardison RC, Paw BH, Weiss MJ. A GATA-1-regulated microRNA locus essential for erythropoiesis. Proc Natl Acad Sci USA. 2008;105(9):3333–8. https://doi.org/10.1073/pnas.0712312105.
36. Iyer A, Zurolo E, Prabowo A, Fluiter K, Spliet WG, van Rijen PC, Gorter JA, Aronica E. MicroRNA-146a: a key regulator of astrocyte-mediated inflammatory response. PLoS One. 2012;7(9):e44789. https://doi.org/10.1371/journal.pone.0044789.
37. Fish JE, Santoro MM, Morton SU, Yu S, Yeh RF, Wythe JD, Ivey KN, Bruneau BG, Stainier DY, Srivastava D. miR-126 regulates angiogenic signaling and vascular integrity. Dev Cell. 2008;15(2):272–84. https://doi.org/10.1016/j.devcel.2008.07.008.
38. Devlin C, Greco S, Martelli F, Ivan M. miR-210: more than a silent player in hypoxia. IUBMB Life. 2011;63(2):94–100. https://doi.org/10.1002/iub.427.

39. Yang S, Banerjee S, Freitas A, Cui H, Xie N, Abraham E, Liu G. miR-21 regulates chronic hypoxia-induced pulmonary vascular remodeling. Am J Physiol Lung Cell Mol Physiol. 2012;302(6):L521–9. https://doi.org/10.1152/ajplung.00316.2011.
40. Dasgupta I, Chatterjee A. Recent advances in miRNA delivery systems. Methods Protoc. 2021;4(1):10. https://doi.org/10.3390/mps4010010.
41. Yu B, Jiang Y, Wang X, Wang S. An integrated hypothesis for miR-126 in vascular disease. Med Res Arch. 2020;8(5):2133. https://doi.org/10.18103/mra.v8i5.2133.
42. Chakraborty C, Sharma AR, Sharma G, Lee SS. Therapeutic advances of miRNAs: a preclinical and clinical update. J Adv Res. 2020;28:127–38. https://doi.org/10.1016/j.jare.2020.08.012.
43. Wang J, Chen J, Sen S. MicroRNA as biomarkers and diagnostics. J Cell Physiol. 2016;231(1):25–30. https://doi.org/10.1002/jcp.25056.
44. Pizzicannella J, Cavalcanti M, Trubiani O, Diomede F. MicroRNA 210 mediates VEGF upregulation in human periodontal ligament stem cells cultured on 3DHydroxyapatite ceramic scaffold. Int J Mol Sci. 2018;19(12):3916. https://doi.org/10.3390/ijms19123916.
45. Hu J, Huang S, Liu X, Zhang Y, Wei S, Hu X. miR-155: an important role in inflammation response. J Immunol Res. 2022;2022:7437281. https://doi.org/10.1155/2022/7437281.
46. Mutharasan RK, Nagpal V, Ichikawa Y, Ardehali H. microRNA-210 is upregulated in hypoxic cardiomyocytes through Akt- and p53-dependent pathways and exerts cytoprotective effects. Am J Physiol Heart Circ Physiol. 2011;301(4):H1519–30. https://doi.org/10.1152/ajpheart.01080.2010.
47. Kulshreshtha R, Ferracin M, Wojcik SE, Garzon R, Alder H, Agosto-Perez FJ, Davuluri R, Liu CG, Croce CM, Negrini M, Calin GA, Ivan M. A microRNA signature of hypoxia. Mol Cell Biol. 2007;27(5):1859–67. https://doi.org/10.1128/MCB.01395-06.
48. Mohamed JS, Hajira A, Li Z, Paulin D, Boriek AM. Desmin regulates airway smooth muscle hypertrophy through early growth-responsive protein-1 and microRNA-26a. J Biol Chem. 2011;286(50):43394–404. https://doi.org/10.1074/jbc.M111.235127.
49. Lu J, He ML, Wang L, Chen Y, Liu X, Dong Q, Chen YC, Peng Y, Yao KT, Kung HF, Li XP. MiR-26a inhibits cell growth and tumorigenesis of nasopharyngeal carcinoma through repression of EZH2. Cancer Res. 2011;71(1):225–33. https://doi.org/10.1158/0008-5472.CAN-10-1850.
50. Drake KM, Zygmunt D, Mavrakis L, Harbor P, Wang L, Comhair SA, Erzurum SC, Aldred MA. Altered MicroRNA processing in heritable pulmonary arterial hypertension: an important role for Smad-8. Am J Respir Crit Care Med. 2011;184(12):1400–8. https://doi.org/10.1164/rccm.201106-1130OC.
51. Rogler CE, Levoci L, Ader T, Massimi A, Tchaikovskaya T, Norel R, Rogler LE. MicroRNA-23b cluster microRNAs regulate transforming growth factor-beta/bone morphogenetic protein signaling and liver stem cell differentiation by targeting Smads. Hepatology. 2009;50(2):575–84. https://doi.org/10.1002/hep.22982.
52. Liu Y, Nie H, Zhang K, Ma D, Yang G, Zheng Z, Liu K, Yu B, Zhai C, Yang S. A feedback regulatory loop between HIF-1α and miR-21 in response to hypoxia in cardiomyocytes. FEBS Lett. 2014;588(17):3137–46. https://doi.org/10.1016/j.febslet.2014.05.067.
53. Wang KC, Garmire LX, Young A, Nguyen P, Trinh A, Subramaniam S, Wang N, Shyy JY, Li YS, Chien S. Role of microRNA-23b in flow-regulation of Rb phosphorylation and endothelial cell growth. Proc Natl Acad Sci USA. 2010;107(7):3234–9. https://doi.org/10.1073/pnas.0914825107.
54. Dejean E, Renalier MH, Foisseau M, Agirre X, Joseph N, de Paiva GR, Al Saati T, Soulier J, Desjobert C, Lamant L, Prósper F, Felsher DW, Cavaillé J, Prats H, Delsol G, Giuriato S, Meggetto F. Hypoxia-microRNA-16 downregulation induces VEGF expression in anaplastic lymphoma kinase (ALK)-positive anaplastic large-cell lymphomas. Leukemia. 2011;25(12):1882–90. https://doi.org/10.1038/leu.2011.168.
55. Xia W, Li J, Chen L, Huang B, Li S, Yang G, Ding H, Wang F, Liu N, Zhao Q, Fang T, Song T, Wang T, Shao N. MicroRNA-200b regulates cyclin D1 expression and promotes S-phase entry by targeting RND3 in HeLa cells. Mol Cell Biochem. 2010;344(1–2):261–6. https://doi.org/10.1007/s11010-010-0550-2.

56. Chan YC, Khanna S, Roy S, Sen CK. miR-200b targets Ets-1 and is down-regulated by hypoxia to induce angiogenic response of endothelial cells. J Biol Chem. 2011;286(3):2047–56. https://doi.org/10.1074/jbc.M110.158790.
57. Marcet B, Chevalier B, Luxardi G, Coraux C, Zaragosi LE, Cibois M, Robbe-Sermesant K, Jolly T, Cardinaud B, Moreilhon C, Giovannini-Chami L, Nawrocki-Raby B, Birembaut P, Waldmann R, Kodjabachian L, Barbry P. Control of vertebrate multiciliogenesis by miR-449 through direct repression of the Delta/Notch pathway. Nat Cell Biol. 2011;13(6):693–9. https://doi.org/10.1038/ncb2241. Epub 2011. Erratum in: Nat Cell Biol. 2011;13(10):1280
58. Muth M, Theophile K, Hussein K, Jacobi C, Kreipe H, Bock O. Hypoxia-induced down-regulation of microRNA-449a/b impairs control over targeted SERPINE1 (PAI-1) mRNA – a mechanism involved in SERPINE1 (PAI-1) overexpression. J Transl Med. 2010;8:33. https://doi.org/10.1186/1479-5876-8-33. Erratum in: J Transl Med. 2011;9:24. doi: 10.1186/1479-5876-9-24

Cognitive Challenges at High Altitude: Human Responses and Adaptation Strategies

9

Harsh Pawar, Koushik Ray, and Krishna Kishore

Abstract

Humans have successfully adapted to various terrains across the planet due to their robust homeostatic mechanisms. However, high-altitude environments remain an extreme challenge. The acclimatization process at high altitudes involves physiological adaptations such as hyperventilation, accelerated erythropoiesis, increased heart rate, and enhanced cerebral blood flow. These changes, while essential for survival, can disrupt normal brain functions. Cognitive function, critical for effective survival, encompasses the processes of gathering, processing, and acting upon environmental information. Impairments in cognitive function at high altitudes can pose serious risks, particularly for mountaineers and soldiers operating in these environments. Currently, resources to enhance cognition under such conditions are limited. Emerging research has identified mechanisms such as oxidative stress (reactive oxygen species), blood–brain barrier disruptions, microvascular damage, impaired autoregulation, altered neurotransmitter dynamics, and inflammatory cytokines as key factors contributing to cognitive decline at high altitudes. While various tools are available for assessing cognitive function—including neuropsychological tests (e.g., Cambridge Neuropsychological Test Automated Battery), questionnaire-based scoring, electrophysiological markers (e.g., event-related potential P300, contingent negative variance), fMRI, and molecular assessments of neurotransmitters and cytokines—wearable technologies for real-time cognitive monitoring remain underdeveloped. Effective acclimatization remains the most reliable strategy to enhance cognitive performance at high altitudes. Pharmacological interventions

H. Pawar · K. Ray (✉) · K. Kishore
Defence Institute of Physiology and Allied Sciences (DIPAS), Defence Research and Development Organization (DRDO), Timarpur, Delhi, India
e-mail: kray.dipas@gov.in

S. Srivastava et al. (eds.), *Health and Sustenance in Extreme Environment*,
https://doi.org/10.1007/978-981-96-8642-1_9

such as acetazolamide and dexamethasone, nutraceuticals like quercetin and curcumin, and non-pharmacological measures, including supplemental oxygen, have proven to support acclimatization and cognitive function.

Keywords

High-altitude cerebral edema (HACE) · Cognitive function · Acute mountain sickness (AMS) · Contingent negative variance (CNV) · Transcranial doppler (TCD)

9.1 Introduction

Over time, humans have majorly evolved and occupied lowland regions as they are not able to provide ample oxygen and other necessary resources at high altitudes. Global data suggest that only about 1.1% of the entire population (81.6 M peoples) reside at high altitudes, which is 2500 m, and the majority of them belong to specialized highland populations such as Tibetan, Andean, and Sherpa populations [1]. The rest only encounter such high-altitude situations only occasionally, typically for sports, military operations, research, or recreational activities. High-altitude research indicates that over a quarter of unacclimatized individuals experience symptoms of acute mountain sickness (AMS) when ascending above 2500 m. These symptoms include headaches, light-headedness, nausea or vomiting, shortness of breath, fatigue, reduced appetite, dizziness, insomnia, chest pain or fullness, muscular weakness, and an inability to exercise [2]. Severe conditions like high-altitude cerebral edema (HACE) and high-altitude pulmonary edema (HAPE) can also occur at even higher elevations or in the absence of proper countermeasures, such as immediate descent to lower altitudes. Unlike un-acclimatized individuals, even acclimatized individuals experience high-altitude exposure-related cognitive dysfunctionscclimatized individuals, even acclimatized individuals, since the brain is among the highest oxygen-requiring organs (nearly about 20%) of the body, it is quite possible that an inefficient supply of oxygen to the brain may hamper its normal functioning [3].

9.1.1 Background

Cognitive function refers to the range of mental processes involved in acquiring knowledge, reasoning, memory, attention, and problem solving. In humans, these cognitive abilities are exceptionally advanced compared to other life forms, characterized by complex language, abstract thought, and sophisticated problem-solving capabilities. These higher-order cognitive functions are crucial for human survival, enabling individuals to adapt to a variety of environmental challenges, including the physiological stress of hypoxia encountered at high altitudes [4]. Hypoxia, a condition characterized by insufficient availability of oxygen to the body, poses a significant threat to cognitive function due to the brain's high demand for oxygen. When exposed to high-altitude conditions, individuals may experience cognitive impairments such as decreased memory, attention, and executive function [5], all of which are essential for making critical decisions and responding effectively to environmental stressors. Research has shown that both acute and chronic exposure to high-altitude hypoxia can lead to significant reductions in cognitive performance, with the degree of impairment influenced by factors such as altitude level, duration of exposure, and individual susceptibility [6]. For instance, numerous studies have documented that high-altitude hypoxia can significantly impair tasks that require sustained attention [7] and executive function [8], which are vital for coping with such challenging environments. Understanding the impact of hypoxia on cognitive function is therefore essential for developing strategies to mitigate its effects and enhance human performance and safety in high-altitude conditions.

9.2 Objective of the Chapter

This chapter aims to comprehensively discuss the effects of high-altitude environments on human neurophysiology, with a specific focus on the cognitive and physiological changes induced by hypoxia and other environmental stressors. Central questions addressed include How does altitude exposure modulate the neurophysiological processes and molecular mechanisms involved? What adaptive responses do humans develop to mitigate the neurocognitive impairments associated with high-altitude hypoxia? And furthermore, how do additional stressors, such as extreme cold and dehydration, compound neurocognitive dysfunction? Finally, what mitigation strategies are commonly employed to address hypoxia-induced cognitive deficits including both pharmacological and non-pharmacological?

9.3 High Altitude and Human Neurophysiology

Altitude is categorized based on elevation above sea level, with each range imposing unique physiological challenges as shown in Fig. 9.1. Low altitude (0–1500 m) causes minimal physiological disturbance, while moderate altitude (1500–2500 m) begins to induce mild hypoxia, slightly reducing arterial oxygen saturation. At this altitude, sleep-related changes, mood changes, and headaches have been reported among susceptible individuals; similarly, high altitude (2500–3500 m) is often accompanied by slower reaction times, dizziness, and headaches, and very high altitude (3500–5500 m) leads to more pronounced reductions in oxygen pressure, requiring greater physiological adaptation. At the neurological level, it can cause confusion, memory retrieval problems, spatial and working memory impairment, and learning impairment [6]. At extreme altitudes (above 5500 m), severe hypoxia occurs, and even acclimatized individuals may experience significant cognitive and physiological impairments [9, 10]. Cortical atrophy and loss of consciousness, along with microhemorrhages and cerebral edema, are life-threatening conditions at such altitudes [11, 12].

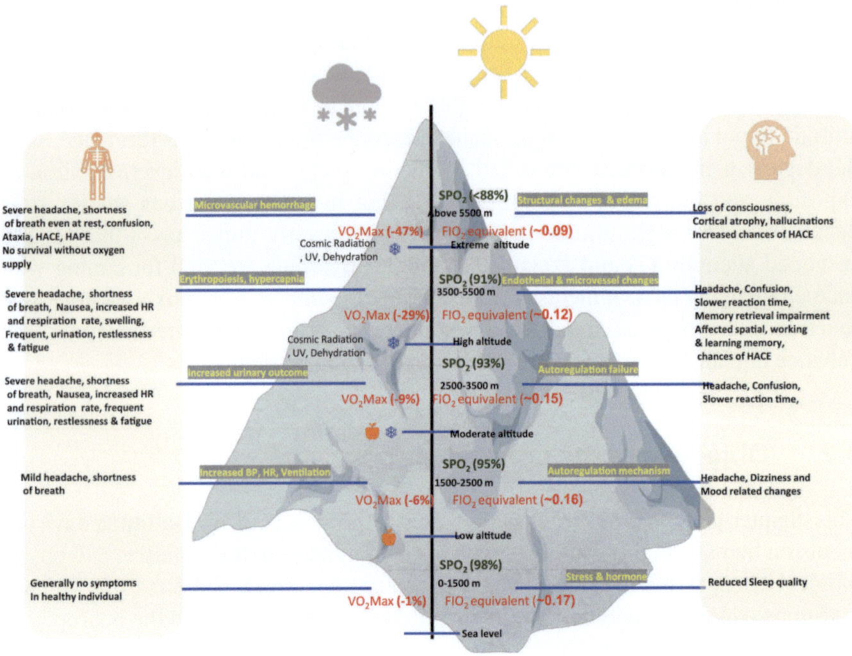

Fig. 9.1 This diagram represents the effects of high-altitude exposure on human physiology depending on the different levels of high altitude

9.3.1 Acclimatization Processes

Acclimatization is the physiological process by which the body adapts to the reduced oxygen availability at high altitudes. The key adaptive responses include increased respiratory rate and depth (the hypoxic ventilatory response), enhanced red blood cell production (erythropoiesis), and greater capillary density to optimize oxygen delivery to tissues [13]. Neurologically, acclimatization also involves adjustments in cerebral blood flow [14], metabolism [15], and neurotransmitter activity [16], which help to maintain cognitive function despite reduced oxygen levels [17].

9.4 Molecular Alterations Behind High-Altitude Cognitive Impairments

Exposure to high-altitude hypoxia triggers a series of neurobiological alterations that collectively impair cognitive function. Researchers have streamlined various interconnected mechanisms that include neurotransmitter release and clearance, microvascular changes, autoregulation, vasoconstriction, Blood–brain barrier integrity, edema, excitotoxicity and pH alterations, etc. (Fig. 9.2). Lower oxygen levels can significantly interfere with the brain's neurotransmitter systems, which are

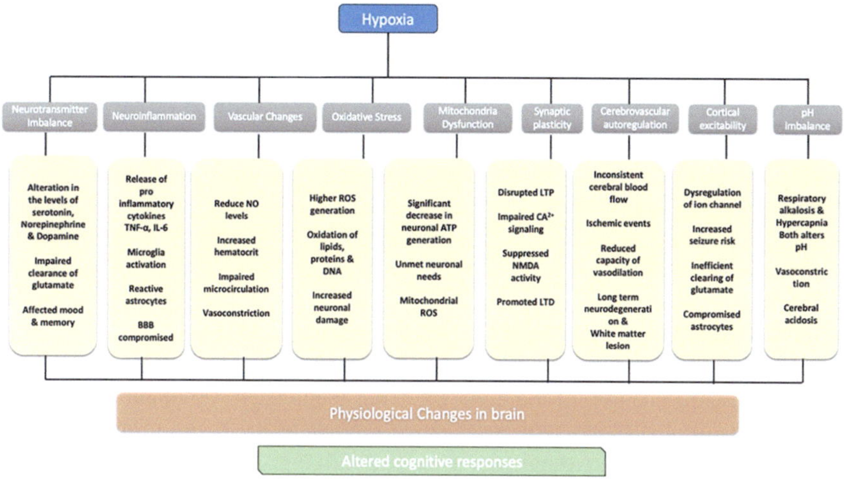

Fig. 9.2 Various potential molecular mechanisms that could be affected by hypoxia at high altitudes and, in turn, may impair cognitive functions

crucial for maintaining cognitive function. High altitude leads to impaired O_2 supply and the enzymes responsible for producing key neurotransmitters like dopamine, serotonin, glutamate, norepinephrine, and acetylcholine, leading to neurotransmitter imbalances that negatively affect thinking, mood, and memory [18–20]. Dopamine production, which is important for memory, decision-making, and motor control, is reduced due to the limited activity of tyrosine hydroxylase, which results in reduced cognitive flexibility and less effective decision-making under hypoxic conditions [21]. Affected mood regulation and anxiety and depression are quite common at high altitudes, which can be due to a reduction in serotonin production, which can further contribute toward cognitive difficulties [22]. Excitotoxicity-related hippocampal neuronal damage due to impaired glutamate clearance is reported to contribute to the reduction in learning and memory due to hypoxia [23]. Norepinephrine, which plays a role in attention and stress response, is also affected, resulting in difficulty in concentrating and maintaining attention. Similarly, the production of acetylcholine—important for focus and memory—declines, further impairing cognitive function [20]. Similarly, the lack of oxygen activates microglia and astrocytes—key immune cells in the brain—causing them to release pro-inflammatory cytokines such as interleukin-1 beta (IL-1β), tumor necrosis factor-alpha (TNF-α), and interleukin-6 (IL-6) [24]. This microglial transition from their resting state to an activated form, releases inflammatory mediators that not only promote neuroinflammation but also contribute to oxidative stress and neuronal damage, contributes to cognitive impairments [25]. Astrocytes, normally responsible for maintaining homeostasis and supporting neurons, also become reactive during hypoxia, exacerbating the situation [26]. They produce additional inflammatory cytokines and decrease their glutamate uptake capacity, which increases the risk of excitotoxicity—an overactivation of neurons leading to cell death—thus further impairing cognitive function. Hypoxia also weakens the blood-brain barrier (BBB) by elevating matrix metalloproteinase (MMP) activity, which increases BBB permeability [27].

Hypoxia induces profound changes in cerebral blood flow and vascular function, which critically limit the brain's oxygen supply. Reduction in the nitric oxide (NO) production [28] and impaired endothelial function [29] are two primary players behind the vascular changes, leading to the reduction in cerebral perfusion and compromised oxygen delivery to crucial brain areas, such as the prefrontal cortex, which is vital for executive functions and decision-making [30]. Increased hematocrit observed at high altitudes, a compensatory response to low oxygen levels, elevates blood viscosity. This thickening of the blood impairs microcirculation, which further restricts oxygen flow to neurons and other brain cells [31]. Vasoconstriction, which is an adaptive mechanism of the brain for better blood supply at altitude, paradoxically causes limited blood delivery to some regions of the brain, further straining cognitive function [32]. Overproduction of reactive oxygen species (ROS) in hypoxia conditions also leads to significant oxidative damage to neurons, primarily due to the brain's high oxygen demand and its vulnerability to oxidative stress. This excessive generation of ROS results in the oxidation of lipids, proteins, and DNA within neural cells [33], affecting synaptic plasticity, neuronal communication, and overall cognitive function [34]. The cellular damage impairs these regions' ability to support cognitive tasks, leading to declines in memory, decision-making, and complex problem-solving abilities [35]. Additionally, the dysfunction of mitochondria under hypoxia leads to the overproduction of reactive oxygen species (ROS), which exacerbates cellular damage by triggering apoptotic pathways [36]. Mitochondria are vulnerable to hypoxia, leading to widespread energy deficits in neurons that compromise cognitive function [37]. Disruption in oxidative phosphorylation results in a significant decrease in ATP production, the primary energy source for cells [38], and this reduction in the available ATP impairs various neuronal functions, contributing to cognitive deficits such as diminished memory capacity, slower information processing, and reduced attentional control [39]. Hypoxia also disrupts long-term potentiation (LTP), a crucial mechanism for memory formation, by impairing calcium signaling and diminishing NMDA receptor function within the hippocampus [40]; this, in turn, affects synaptic plasticity, which is brain's intrinsic capacity to increase synaptic strength in response to learning and experiences [41]. Hypoxia also promotes long-term depression (LTD), which further weakens synaptic connections [42]. Scientific studies have also demonstrated that decreased oxygen levels can lead to altered neurotransmitter release and receptor sensitivity, compounding the challenges of maintaining cognitive function in hypoxic environments [43].

Cortical hyperexcitability is also reported in hypoxia conditions, primarily due to dysregulation of ion channels, as it impairs the normal balance of sodium, potassium, and calcium ions, leading to overactive neuronal firing [44]. This impairment of ion channels can lead to an increased risk of seizures and cognitive alterations. Similarly, astrocytes also lose their ability to effectively clear excess glutamate, resulting in astrocyte-mediated excitotoxicity. The accumulation of glutamate in the synaptic cleft overstimulates neurons, causing further neuronal damage [45].

Hypoxia also triggers an increase in stress hormones such as cortisol and norepinephrine, which are essential for managing stress but can become detrimental when elevated for prolonged periods [46]. Chronic high cortisol levels impair the hippocampus, a critical area for memory formation and retrieval, while also promoting neuroinflammation, which exacerbates cognitive decline [47]. Additionally, the body's response to hypoxia leads to frequent sleep disturbances, particularly characterized by periodic breathing and fragmented sleep cycles [48]. These sleep disruptions, especially during the REM phase, interfere with the brain's ability to consolidate memories, further impairing cognitive functions such as learning, attention, and decision-making.

Cerebrovascular autoregulation is the brain's ability to maintain stable blood flow despite fluctuations in blood pressure, which becomes compromised in hypoxic conditions at high altitudes. Research indicates that this impairment leads to inconsistent cerebral blood flow (CBF), which negatively affects cognitive functions such as sustained attention, working memory, and decision-making [49]. Hypoxia also weakens the brain's capacity for vasodilation, reducing blood flow precisely when cognitive demand is high, which worsens performance on tasks requiring concentration and executive function [50]. In severe cases, high-altitude cerebral edema (HACE) can occur due to the breakdown of the blood-brain barrier, highlighting the serious consequences of impaired cerebrovascular regulation in hypoxic conditions [51]. Similarly, respiratory alkalosis and hypercapnia, both consequences of altered gas exchange at high altitudes, are significant contributors to cognitive dysfunction in such environments. Respiratory alkalosis that occurs due to a drop in CO_2 levels during hyperventilation constricts cerebral blood vessels, reducing cerebral blood flow and limiting oxygen delivery to brain tissues [52]. Research has shown that this vasoconstriction negatively impacts cognitive functions, particularly those related to memory, concentration, and problem solving, as the brain struggles to maintain adequate oxygen levels for optimal performance [53]. Hypercapnia can cause cerebral acidosis, where the brain's pH drops, disrupting the normal function of neurons [54]. Studies indicate that this acidosis interferes with synaptic activity and overall brain metabolism, worsening cognitive impairments, particularly in memory retention, decision-making, and cognitive flexibility [54].

9.5 Other Physiological Stressors

Beyond hypoxia, high-altitude environments impose additional stressors, such as cold exposure (frostbite, hypothermia, and even hallucinations), dehydration (low humidity level), increased ultraviolet radiation (sunburn, fatigue, and mild dizziness), increased exposure to cosmic radiations, and nutritional deficiency (deficiency of fresh food). Exposure to cold at high altitudes leads to peripheral vasoconstriction, reducing blood flow to extremities and impairing dexterity, which can impact critical motor tasks required for survival [55]. Dehydration, often a consequence of increased respiratory water loss and altitude-induced diuresis, further impairs cognitive performance and exacerbates the symptoms of acute mountain sickness [56]. These stressors add to the overall neurophysiological burden on individuals, complicating their ability to perform cognitively demanding tasks in extreme environments.

9.6 Impact of High Altitude on Cognitive Functions

High-altitude environments present reduced oxygen levels, leading to a condition known as hypoxia. This hypoxic state impairs brain function by decreasing ATP production and increasing oxidative stress and inflammatory responses. These biochemical changes disrupt neuronal function and contribute to cognitive impairments. Sharma et al. [26] have documented that chronic exposure to hypoxia can result in both adaptive neural changes and detrimental effects on cognitive function. Similarly, Heerlein et al. [38] and Hochachka et al. [15] elaborate on how cellular metabolism and neurophysiological processes are compromised under hypoxic conditions, further elucidating the mechanisms by which cognitive decline occurs.

9.6.1 Memory Impairments

At high elevations, cognitive functions related to memory, especially working memory, can be significantly impacted by lower oxygen levels. Working memory is essential for temporarily storing and manipulating information, which plays a crucial role in decision-making and problem solving. Reduced oxygen availability impairs the brain's ability to encode and retrieve information, making it harder to learn new tasks and remember important details [57]. Hypoxia adversely affects the hippocampus, a brain region essential for memory formation and consolidation. This impact leads to memory impairments due to decreased neurogenesis and disrupted synaptic plasticity within the hippocampus. Evidence from Hencz et al. [58] indicates that hypoxia at high altitudes impairs hippocampal function, resulting in significant memory deficits. Osborne et al. [47] provide a comprehensive review of how hippocampal changes related to hypoxia contribute to cognitive decline.

9.6.2 Attention and Concentration Issues

Attention, which involves focusing on specific tasks while ignoring irrelevant stimuli, is also affected at high altitudes. Decreased oxygen supply can lead to difficulties in maintaining sustained attention and selectively focusing on necessary tasks. This impairment can be particularly problematic for activities that require constant awareness and hazard detection, such as mountaineering or military operations. The ability to maintain attention and concentration is compromised at high altitudes due to hypoxia's effects on the prefrontal cortex, which is vital for these cognitive functions. Pun et al. [57] investigate the broader impacts of hypoxia on cognitive performance, including attention, emphasizing the challenges faced in high-altitude environments.

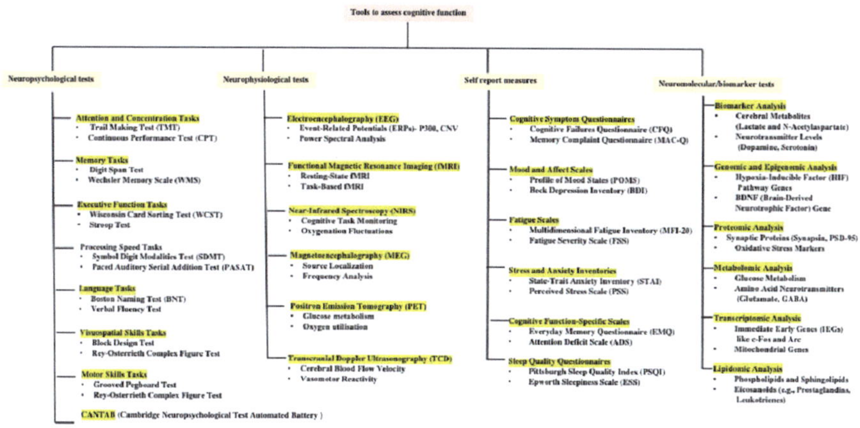

Fig. 9.3 Tools commonly employed to assess cognitive function

9.6.3 Executive Function Disruptions

Executive functions, including planning, reasoning, and adapting to new situations, are compromised due to hypoxia. These higher-order cognitive processes are essential for managing complex tasks and adjusting strategies. Hypoxia reduces cognitive flexibility, which affects decision-making under pressure and the ability to adapt to changing conditions, posing challenges in both research and operational environments. Executive functions, including planning, decision-making, and problem solving, are particularly vulnerable to hypoxic stress. The prefrontal cortex, critical for these higher-order cognitive tasks, suffers under hypoxic conditions. De Wachter et al. [59] explore how hypoxia impairs the prefrontal cortex function and executive cognitive abilities. Wei et al. [60] further examine the effects of high altitude on executive functions, providing evidence of significant disruptions.

9.6.4 Psychomotor Performance Decline

The synchronization of cognitive processes with physical movements, known as psychomotor performance, is diminished at high altitudes. Factors such as hypoxia and exposure to cold can slow reaction times and reduce motor coordination. This decline in psychomotor performance can impact tasks requiring precision and control, such as handling equipment or climbing. High-altitude hypoxia also leads to declines in psychomotor performance, affecting motor coordination and reaction times. This decline is attributed to disruptions in the basal ganglia and motor cortex. Li et al. [61] discuss how hypoxia impairs motor performance and coordination, while Davis et al. [62] review the psychomotor deficits associated with altitude exposure, providing a thorough understanding of these effects. Various tools are commonly used to assess cognitive function at high-altitude hypoxia (Fig. 9.3).

9.7 Factors Influencing Cognitive Response at High Altitude

There are multiple factors that play a significant role in deciding the actual influence of high-altitude hypoxia on an individual's cognitive performance. These can be categorized broadly based on individual, psychological, and environmental factors (Fig. 9.4). Genetic variations can influence an individual's cognitive response to high-altitude exposure. Specific genetic markers related to oxygen transport and cellular stress responses can affect susceptibility to cognitive impairments [63]. Zhang further added that variants in EPAS1 and VEGFA could be linked to the susceptibility to AMS in the Chinese Han population with potential use in screening [64]. Beall [65] explores genetic adaptations to high altitude and their impact on cognitive function. Simonson [66] offers additional insights into the genetic factors influencing high-altitude adaptation and cognitive resilience like the EGLN1 gene and PPARA gene highlighted to reduce hemoglobin at high altitudes in Tibetans. Similarly, an individual's physical fitness and history of acclimatization significantly affect cognitive resilience at high altitudes. Individuals with higher physical fitness levels and successful acclimatization experiences typically exhibit better cognitive performance. Su et al. [67] examined whether high-intensity aerobic exercises in high-altitude acclimatized individuals improve attentional function using attentional network task (ANT) and Stroop color and word task (SCWT); however, acute aerobic exercises were found to impair cognitive function at high altitude [68]. Furian et al. [69] explained that reexposure to high altitude reduces the hypoxia and breathing issues due to the potential role of previous acclimatization. Basnyat and Murdoch [70] discuss how an individual's acclimatization history plays a key role in determining cognitive performance at high altitudes. They emphasize that individuals who have undergone previous high-altitude exposure tend to fare better cognitively when facing new altitude challenges [70]. However, a few studies have also shown that prior acclimatization at high altitudes does not affect the cognitive function of subsequent altitude ascents. The duration of high-altitude exposure is a crucial factor in determining cognitive function. Extended exposure exacerbates

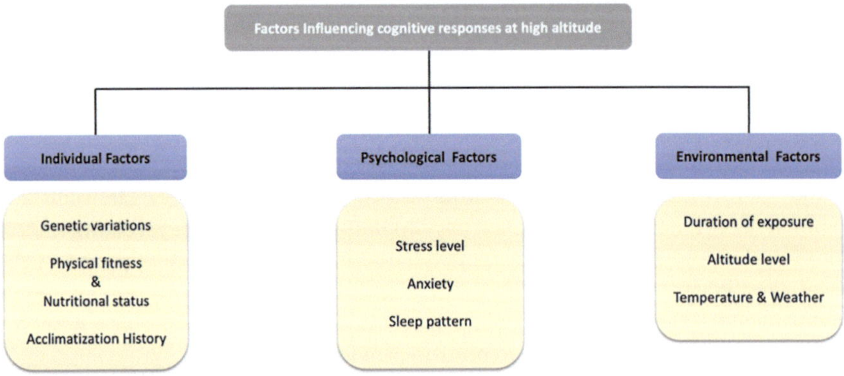

Fig. 9.4 Various factors that affect cognitive functions at high altitude exposure

cognitive impairments due to prolonged hypoxic stress. Su et al. [5] provide an overview of how prolonged exposure to high altitudes affects cognitive function, suggesting that extended periods of hypoxia can lead to more severe cognitive deficits. Pun et al. [57] review the impact of varying exposure durations on cognitive performance, emphasizing the cumulative effects of hypoxia. The cognitive impairments observed at high altitudes are directly related to the altitude level. Higher altitudes intensify hypoxic stress and lead to more pronounced cognitive deficits. Aboouf et al. [6] presented the relationship between altitude and cognitive function, highlighting the increased severity of cognitive decline at higher elevations. Lieberman et al. [71] provide empirical data demonstrating how cognitive performance varies with altitude, emphasizing the altitude-dependent nature of cognitive impairment. Environmental factors such as temperature and weather conditions can further impact cognitive function at high altitudes. Extreme temperatures and adverse weather conditions add additional stress, exacerbating the effects of hypoxia. Taylor [72] examines how environmental conditions influence cognitive performance at high altitudes. Barbara Shukitt-Hale and Harris R. Lieberman documented the combined effects of temperature and altitude on cognitive function, illustrating the interaction between environmental stressors and cognitive performance [73]. Psychological factors such as stress and anxiety can exacerbate cognitive impairments associated with high altitude. Stress affects neurotransmitter systems and overall brain function, leading to increased cognitive deficits. Scott et al. [74] discuss the impact of stress on cognitive function, while Bliemsrieder et al. [75] provide a broader review of the relationship between psychological changes and cognitive performance, highlighting how stress can worsen cognitive impairments at high altitudes. Disruptions in sleep quality and patterns due to high-altitude conditions contribute to cognitive impairments. Poor sleep adversely affects various cognitive domains, including memory and attention. San et al. [76] examine how sleep disturbances at high altitudes affect cognitive function. Frost et al. [77] provide insights into how sleep quality and patterns influence cognitive performance under hypoxic conditions, emphasizing the role of sleep in mitigating cognitive deficits.

9.8 Mitigation and Adaptation Strategies

Mitigation and adaptation strategies are essential for maintaining cognitive function at high altitudes, where hypoxia can severely impair brain performance. Hypoxia-induced cognitive decline, such as reduced memory, attention, and decision-making capabilities, can be particularly dangerous in high-altitude environments where individuals face greater physical hazards. Implementing strategies like gradual acclimatization, supplemental oxygen use, and pharmacological interventions help maintain adequate oxygen delivery to the brain, minimizing cognitive deficits. For those undertaking cognitively demanding activities at altitude—such as climbers, military personnel, or long-term residents—these strategies promote clearer thinking, enhance safety, and help preserve cognitive health. Additionally, proper hydration, antioxidant-rich nutrition, and physical and mental conditioning can bolster the brain's ability to resist the impacts of oxygen deprivation. Utilizing these approaches supports cognitive function, improves decision-making, and ensures long-term brain health in high-altitude settings.

9.8.1 Acclimatization Protocols

Acclimatization protocols for high-altitude environments are developed to mitigate the adverse effects of hypoxia and reduce the risk of altitude-related conditions such as acute mountain sickness (AMS), high-altitude pulmonary edema (HAPE), and high-altitude cerebral edema (HACE). These protocols, varying across the world, generally emphasize a gradual ascent, allowing the body time to adapt by incorporating intermediate stops at specific elevations and scheduled rest days. For instance, protocols used by mountaineers in the Himalayas often advocate for "climb high, sleep low" strategies, while military forces may employ structured acclimatization schedules with controlled altitude exposure over several phases. In India, the Defence Research and Development Organisation (DRDO) has developed a specialized acclimatization protocol tailored for military personnel operating in the high-altitude regions of the Himalayas. The DRDO protocol follows a phased approach, which includes a gradual increase in altitude with rest periods and the potential use of oxygen supplementation and medications such as acetazolamide to facilitate adaptation. The protocol also emphasizes the importance of physical conditioning, proper hydration, and adequate nutrition to optimize the body's ability to acclimatize while maintaining both cognitive and physical performance in high-altitude conditions.

9.8.2 Gradual Ascent Strategies

Gradual ascent strategies are instrumental in mitigating the cognitive decline often associated with high-altitude environments by allowing the brain adequate time to adjust to reduced oxygen availability. A slow and controlled ascent promotes physiological adaptations, such as increased cerebral blood flow and enhanced oxygen delivery, which help sustain cognitive functions like attention, memory, and decision-making. For instance, the "climb high, sleep low" method, where individuals ascend to higher elevations during the day but descend to lower altitudes to sleep, exposes the brain to altitude stress while providing recovery opportunities overnight. These intermittent exposures to hypoxia stimulate the brain's ability to adapt, reducing the risk of hypoxia-induced cognitive impairment. Similarly, scheduled rest days during ascent give the brain time to recuperate, which helps maintain cognitive performance throughout the exposure to higher elevations [78].

9.8.3 Pre-acclimatization Techniques

Pre-acclimatization techniques play a crucial role in protecting cognitive function by preparing the brain to cope with hypoxia before exposure to high-altitude environments. Methods such as intermittent hypoxic exposure (IHE) or the use of hypoxic tents simulate low-oxygen conditions, encouraging physiological adaptations like increased red blood cell production and improved oxygen utilization. These adaptations enhance the brain's resilience to hypoxia by improving oxygen transport and utilization, helping to prevent cognitive decline, such as impaired judgment, reduced attention, and memory lapses, which are common in oxygen-deprived environments. By priming the brain for reduced oxygen levels, pre-acclimatization techniques help individuals maintain cognitive sharpness and decision-making capacity, which are especially important in high-risk activities like military operations and high-altitude expeditions where cognitive performance is critical to safety and success [79].

9.8.4 Pharmacological Interventions

Pharmacological interventions have been explored as a means to preserve and improve cognitive functions at high altitudes, where hypoxia can lead to impairments in attention, memory, and executive functioning. These interventions target various physiological pathways to enhance oxygen delivery and utilization, reduce oxidative stress, and improve brain function under hypoxic conditions as tabulated in Table 9.1.

Table. 9.1 Different pharmacological and nutraceutical interventions to improve cognitive function

S No	Intervention	Potential mechanism	References
1	Acetazolamide	Favourable impact on enhancing the body's capacity to withstand hypoxia	Wang et al. [80]
2	Dexamethasone	Act by reducing vascular leakage and cerebral edema	Lafleur et al. [81]
3	Oxiracetam	Mild stimulant improves synaptic plasticity	Hu et al. [82]
4	Zolpidem	$GABA_A$ receptor modulator improves reaction time in climbers	Bouzat et al. [83]
5	Physostigmine, Galantamine	Acetylcholinesterase inhibitors	Muthuraju et al. [84]
6	Nimodipine, Iradipine	Ca^{2+} channel blocker reduces apoptosis	Uema et al. [85]
7	Memantine	NMDA antagonist reduces hypoxia induces excitotoxicity	Ji et al. [86]
8	Guanfacine	α_2 receptor agonist	Kauser et al. [87]

(continued)

Table 9.1 (continued)

S. No	Intervention	Potential mechanism	References
Antioxidants			
9	Quercetin	Antioxidant reduces expressions of caspase 3, reduces hippocampal neurodegeneration and improves memory function	Prasad et al. [88]
10	Salidroside	Antioxidant inhibits NFkB/NLRP3 pathway	Jiang et al. [89]
11	Epigallocatechin-3-gallate (EGCG)	Antioxidant, anti-inflammatory promotes nerve regeneration	Chen et al. [90]
12	5,6,7,8-Tetrahy droxyflavone (THF)	Antioxidant effect	Jing et al. [91]
13	Verbascoside	Antioxidant, anti-inflammatory	Zhou et al. [92]
14	Crocin	Antioxidant regulates SIRT1/ PGC-1alpha pathway	Zhang et al. [93]
15	Astragalus	Antioxidant promotes mTOR signaling	Du et al. [94]
16	Propolis	Antioxidant, anti-inflammatory	Zhu et al.[95]
17	Echinacea glycoside (ECH)	Antioxidant supports signaling pathways (Keap1-Nrf2-ARE) for neuronal adaptation	Zheng et al.[96]
Supplementations used			
18	Melatonin	Neurohormone, antioxidant effect, helpful in HH-induced cognitive alterations	Alonso-Alconada et al. [97]
19	Tyrosine	Act as precursor to catecholamines	Kishore et al.[98]
20	Vitamin B6/B12/ folate and choline	Reduces homocysteine and tau phosphorylation improves memory in HH-induced impairments	Yu et al.[99]

Acetazolamide: Acetazolamide is one of the most commonly used medications for altitude sickness prevention and has been shown to improve cognitive performance by accelerating acclimatization. It works by inducing metabolic acidosis, which stimulates ventilation and improves oxygenation. This increase in oxygen delivery to the brain helps mitigate hypoxia-induced cognitive impairments such as impaired concentration and decision-making [100]. However, a few studies also suggested the impairing effect of acetazolamide on neuropsychological performances when given prophylactically [80].

Dexamethasone: This corticosteroid is often used to prevent and treat high-altitude cerebral edema (HACE) but also has cognitive benefits. Dexamethasone reduces inflammation and cerebral edema, which can help protect brain function at high altitudes. Studies have shown that it can improve mental clarity and reduce the risk of cognitive decline associated with severe hypoxia [101].

Nootropics and cognitive enhancers: Nootropic drugs such as modafinil and piracetam are explored for their potential to enhance cognitive performance under high-altitude conditions. Modafinil, a wakefulness-promoting agent, has been shown to counteract fatigue and maintain alertness in hypoxic conditions, thus supporting sustained cognitive function. Piracetam, oxiracetam, and other cognitive enhancers may improve memory and learning by enhancing neuroplasticity and increasing blood flow to the brain [82].

Ginkgo biloba: Ginkgo biloba, a herbal supplement, has been studied for its neuroprotective properties at high altitudes. It may improve cognitive function by enhancing cerebral blood flow and providing antioxidant effects that protect brain cells from hypoxia-induced oxidative stress. Some studies suggest that Ginkgo biloba may reduce symptoms of acute mountain sickness (AMS) and preserve cognitive clarity [102].

Oxygen supplementation: Although not a drug, oxygen supplementation can be considered a pharmacological intervention. Supplemental oxygen is often used to combat hypoxia directly, preventing cognitive decline by restoring adequate oxygen levels in the brain. This intervention is particularly useful in preventing the immediate cognitive deficits caused by severe hypoxia at very high altitudes.

Antioxidants: High altitudes induce oxidative stress, which can impair cognitive function. Antioxidant supplementation with vitamins C and E or coenzyme Q10 may protect neurons from oxidative damage. These antioxidants help neutralize free radicals produced during hypoxia, potentially preserving memory, attention, and cognitive processing at high elevations.

9.8.5 Behavioral and Cognitive Strategies

Training and Preparation

Effective preparation for high-altitude environments involves both physical and cognitive training aimed at reducing the detrimental effects of hypoxia on cognitive performance. Physical conditioning primarily targets enhancing cardiovascular efficiency, which plays a critical role in optimizing oxygen utilization and improving overall resilience to altitude-induced stress [103]. Engaging in regular aerobic exercise prior to ascent can increase capillary density and improve oxygen transport mechanisms, thus enhancing the body's ability to cope with the physiological demands of high altitudes.

In addition to physical fitness, structured acclimatization protocols—such as gradual ascent strategies—are essential for improving cognitive performance at altitude. These protocols allow for progressive physiological adaptations, reducing the risk of acute mountain sickness and minimizing the impact of hypoxia on cognitive function [104]. By incrementally exposing the body to lower oxygen levels, acclimatization facilitates better maintenance of cognitive abilities, even at extreme elevations.

Cognitive Training Exercises

Cognitive training is an essential strategy for improving mental resilience in high-altitude environments, especially for tasks that require sustained attention, working memory, and executive control. Exercises designed to enhance cognitive flexibility, processing speed, and overall cognitive endurance are particularly beneficial in mitigating the effects of hypoxia on brain function. For instance, computer-based cognitive training programs focus on improving critical areas such as attention, memory, and problem-solving skills, all of which are susceptible to decline at high altitudes.

Training exercises that stimulate the prefrontal cortex, such as dual-task activities and working memory drills, can help build the mental endurance required to function effectively in low-oxygen conditions [105]. Additionally, mindfulness-based cognitive interventions have shown promise in reducing stress and enhancing focus, which are key factors in maintaining cognitive performance in challenging environments [106].

By consistently engaging in these cognitive training activities, individuals can develop greater cognitive resilience and adaptive strategies, such as improved task-switching abilities and better management of cognitive load under stress. These exercises reinforce neural circuits associated with critical thinking and sustained attention, providing a buffer against the cognitive decline often seen in hypoxic conditions.

9.9 Case Studies and Real-World Applications

9.9.1 Military and High-Altitude Missions

Military operations at high altitudes, such as in the Himalayas or the Andes, pose significant cognitive challenges due to the effects of hypoxia on brain function. Case studies from operations on the Siachen Glacier and the Tibetan Plateau reveal impairments in cognitive domains, including attention, memory, and decision-making, when soldiers are exposed to extreme altitudes. For example, soldiers stationed at elevations exceeding 5000 m have experienced reduced multitasking abilities and slower reaction times, as reported by both Indian and Chinese military personnel. These operational challenges have prompted the development of specialized training programs aimed at enhancing cognitive resilience for high-altitude deployments. Such programs include cognitive drills focused on memory retention, attention control exercises, and gradual acclimatization protocols. Additionally, military strategies now incorporate periodic rotations to lower altitudes to help sustain cognitive performance during extended operations in high-altitude environments. The use of supplemental oxygen and medications such as acetazolamide has further been integrated to help mitigate the cognitive deficits induced by hypoxia. These interventions highlight the critical need to address cognitive impairments to ensure that military personnel maintain high levels of operational effectiveness in demanding high-altitude settings.

9.9.2 Mountaineering and Adventure Sports

Mountain climbers ascending to extreme altitudes, particularly above 3500 m, encounter considerable cognitive challenges due to hypoxia's detrimental effects on brain function. Documented cases from expeditions on peaks like Mount Everest and K2 illustrate frequent cognitive impairments, such as memory deficits, slowed decision-making, and compromised judgment. A notable example is the 1996 Mount Everest disaster where confusion and poor decision-making, likely exacerbated by severe hypoxia, were identified as contributing factors [107]. In response to these risks, climbers now incorporate cognitive training and acclimatization strategies into their preparation for high-altitude expeditions. These preparations involve exercises aimed at enhancing cognitive resilience, focusing on maintaining mental acuity under oxygen-deprived conditions. Additionally, practical methods such as the use of supplemental oxygen, the establishment of staging camps at progressively higher altitudes for better acclimatization, and frequent cognitive assessments during climbs are applied to help preserve cognitive function. These interventions are critical for ensuring that climbers can safely navigate the mental and physical demands of extreme high-altitude environments, where sound decision-making is essential for survival.

9.9.3 High-Altitude Research and Work Environments

Researchers and workers operating at high altitudes, such as in the Andes, Himalayas, or Antarctic plateaus, often encounter cognitive difficulties as a result of hypoxia. Studies have shown that individuals working above 3500 m frequently experience impairments in cognitive functions such as attention, memory, and executive decision-making. For example, research conducted by Willie et al. [108] on scientists at the Pyramid International Laboratory in Nepal, situated at an altitude of 5050 m, demonstrated significant declines in cognitive tasks related to sustained attention and working memory. Similarly, a study by Barkaszi et al. [109] on workers in Antarctica reported diminished cognitive performance, particularly in tasks involving complex problem solving and reaction time, under hypoxic conditions. To counteract these cognitive challenges, high-altitude workers implement acclimatization protocols and engage in cognitive training exercises both prior to and during their deployments. Practical applications include gradual altitude exposure to facilitate acclimatization and continuous monitoring of cognitive performance to detect early signs of cognitive decline. Additionally, the use of supplemental oxygen and medications such as acetazolamide are employed to help sustain cognitive function during extended periods at high altitudes. These interventions are vital for ensuring that researchers and workers can maintain the necessary cognitive abilities to perform effectively in these extreme environments.

9.10 Future Directions and Research Needs

Despite the significant progress that has been made in the field of high-altitude physiology and medicine, notable research gaps persist. The available research primarily focuses on physiological responses to hypoxia and not on individual cognitive function. Data is limited in the case of long-term studies and post-exposure cognitive improvements. Wearable technologies for real-time monitoring are limited. Another important aspect is biological variations among individuals, such as genetic variations, which plays an important role in defining such adverse effects of hypoxia on physiology and are relatively lesser explored. Comprehensive research is needed to identify the most effective acclimatization strategies and personalized interventions. Addressing these gaps is essential for developing targeted approaches to prevent and manage cognitive dysfunction in high-altitude environments, thereby enhancing safety and performance for those operating in such challenging conditions.

9.11 Conclusion

This chapter has provided an in-depth review of the various aspects of high-altitude hypoxia-induced neuronal damage, potential mechanisms, and the effect on cognitive response. Specifically, how an increase in altitude affects neuronal and cognitive processes due to changes in hemodynamics, microvascular, disturbed autoregulation, neuroinflammation, mitochondrial disturbances, neurotransmitter imbalance, and impaired ion channel functioning leading to impairments in memory, attention, and executive function necessary for the survival at high altitudes. It also emphasized how different factors, whether individual or environmental, can influence the extent of hypoxia's effects on cognitive processes, as well as the positive impacts of cognitive training, pre-acclimatization, and gradual ascent. In recent times, our world has seen a tremendous increase in activities involving high altitude like military, mountaineering, and research and for that there is a strong need to develop guidelines and strategies for countering such cognitive impacts of high altitude on cognitive responses. Another important need is to understand the mechanism of hypoxia-induced cognitive dysfunction and the need for real-time cognition monitoring technologies. By advancing this knowledge, we can improve the tools used to assess cognitive function, refine preventive measures, and develop more effective intervention protocols. Ultimately, these efforts will help to ensure greater safety, enhance performance, and support the overall well-being of those exposed to high-altitude conditions.

References

1. Tremblay JC, Ainslie PN. Global and country-level estimates of human population at high altitude. Proc Natl Acad Sci. 2021;118(18):e2102463118. https://doi.org/10.1073/pnas.2102463118.

2. Prince TS, Thurman J, Huebner K. Acute mountain sickness. In: StatPearls. StatPearls Publishing; 2023. http://www.ncbi.nlm.nih.gov/books/NBK430716/.
3. Rink C, Khanna S. Significance of brain tissue oxygenation and the arachidonic acid cascade in stroke. Antioxid Redox Signal. 2011;14(10):1889–903. https://doi.org/10.1089/ars.2010.3474.
4. Laland K, Seed A. Understanding human cognitive uniqueness. Annu Rev Psychol. 2021;72(1):689–716. https://doi.org/10.1146/annurev-psych-062220-051256.
5. Su R, Jia S, Zhang N, Wang Y, Li H, Zhang D, Ma H, Su Y. The effects of long-term high-altitude exposure on cognition: a meta-analysis. Neurosci Biobehav Rev. 2024;161:105682. https://doi.org/10.1016/j.neubiorev.2024.105682.
6. Aboouf MA, Thiersch M, Soliz J, Gassmann M, Schneider Gasser EM. The brain at high altitude: from molecular signaling to cognitive performance. Int J Mol Sci. 2023;24(12):10179. https://doi.org/10.3390/ijms241210179.
7. Shi QH, Ge D, Zhao W, Ma X, Hu KY, Lu Y, Liu ZX, Ran JH, Li XL, Zhou Y, Fu JF. A computerized evaluation of sensory memory and short-term memory impairment after rapid ascent to 4280 m. Biomed Environ Sci. 2016;29(6):457–60. https://doi.org/10.3967/bes2016.060.
8. Lefferts WK, DeBlois JP, White CN, Day TA, Heffernan KS, Brutsaert TD. Changes in cognitive function and latent processes of decision-making during incremental ascent to high altitude. Physiol Behav. 2019;201:139–45. https://doi.org/10.1016/j.physbeh.2019.01.002.
9. Grocott MPW, Martin DS, Levett DZH, McMorrow R, Windsor J, Montgomery HE. Arterial blood gases and oxygen content in climbers on Mount Everest. N Engl J Med. 2009;360(2):140–9. https://doi.org/10.1056/NEJMoa0801581.
10. West JB. Human responses to extreme altitudes. Integr Comp Biol. 2006;46(1):25–34. https://doi.org/10.1093/icb/icj005.
11. Garrido E, Segura R, Capdevila A, Aldomá J, Rodríguez FA, Javierre C, Ventura JL. New evidence from magnetic resonance imaging of brain changes after climbs at extreme altitude. Eur J Appl Physiol Occup Physiol. 1995;70(6):477–81. https://doi.org/10.1007/BF00634375.
12. Kottke R, Pichler Hefti J, Rummel C, Hauf M, Hefti U, Merz TM. Morphological brain changes after climbing to extreme altitudes—A prospective cohort study. PLoS One. 2015;10(10):e0141097. https://doi.org/10.1371/journal.pone.0141097.
13. West JB. The physiologic basis of high-altitude diseases. Ann Intern Med. 2004;141(10):789. https://doi.org/10.7326/0003-4819-141-10-200411160-00010.
14. Ainslie PN, Subudhi AW. Cerebral blood flow at high altitude. High Alt Med Biol. 2014;15(2):133–40. https://doi.org/10.1089/ham.2013.1138.
15. Hochachka PW, Clark CM, Brown WD, Stanley C, Stone CK, Nickles RJ, Zhu GG, Allen PS, Holden JE. The brain at high altitude: hypometabolism as a defense against chronic hypoxia? J Cereb Blood Flow Metab. 1994;14(4):671–9. https://doi.org/10.1038/jcbfm.1994.84.
16. Vaccari A, Brotman S, Cimino J, Timiras PS. Adaptive changes induced by high altitude in the development of brain monoamine enzymes. Neurochem Res. 1978;3(3):295–311. https://doi.org/10.1007/BF00965576.
17. Petajan JH. Neuropsychological acclimatization to high altitude. J Hum Evol. 1973;2(2):105–15. https://doi.org/10.1016/0047-2484(73)90058-4.
18. Davis JN, Carlsson A. Effect of hypoxia on tyrosine and tryptophan hydroxylation in unanaesthetized rat brain. J Neurochem. 1973;20(3):913–5. https://doi.org/10.1111/j.1471-4159.1973.tb00055.x.
19. Raymond R, Millhorn D. Regulation of tyrosine hydroxylase gene expression during hypoxia: role of Ca2+ and PKC. Kidney Int. 1997;51(2):536–41. https://doi.org/10.1038/ki.1997.74.
20. Shimada M. Alteration of acetylcholine synthesis in mouse brain cortex in mild hypoxie hypoxia. J Neural Transm. 1981;50(2–4):233–45. https://doi.org/10.1007/BF01249145.
21. Zhong Z, Zhou S, Xiang B, Wu Y, Xie J, Li P. Association of peripheral plasma neurotransmitters with cognitive performance in chronic high-altitude exposure. Neuroscience. 2021;463:97–107. https://doi.org/10.1016/j.neuroscience.2021.01.031.

22. Kanekar S, Sheth CS, Ombach HJ, Olson PR, Bogdanova OV, Petersen M, Renshaw CE, Sung Y-H, D'Anci KE, Renshaw PF. Hypobaric hypoxia exposure in rats differentially alters antidepressant efficacy of the selective serotonin reuptake inhibitors fluoxetine, paroxetine, escitalopram and sertraline. Pharmacol Biochem Behav. 2018;170:25–35. https://doi.org/10.1016/j.pbb.2018.05.002.
23. Ji W, Zhang Y, Ge R, Wan Y, Liu J. NMDA receptor-mediated excitotoxicity is involved in neuronal apoptosis and cognitive impairment induced by chronic hypobaric hypoxia exposure at high altitude. High Alt Med Biol. 2021a;22(1):45–57. https://doi.org/10.1089/ham.2020.0127.
24. Shushanyan RA, Avtandilyan NV, Grigoryan AV, Karapetyan AF. The role of oxidative stress and neuroinflammatory mediators in the pathogenesis of high-altitude cerebral edema in rats. Respir Physiol Neurobiol. 2024;327:104286. https://doi.org/10.1016/j.resp.2024.104286.
25. Hatch K, Lischka F, Wang M, Xu X, Stimpson CD, Barvir T, Cramer NP, Perl DP, Yu G, Browne CA, Dickstein DL, Galdzicki Z. The role of microglia in neuronal and cognitive function during high altitude acclimatization. Sci Rep. 2024;14(1):18981. https://doi.org/10.1038/s41598-024-69694-9.
26. Sharma R, Cramer NP, Perry B, Adahman Z, Murphy EK, Xu X, Dardzinski BJ, Galdzicki Z, Perl DP, Dickstein DL, Iacono D. Chronic exposure to high altitude: synaptic, astroglial and memory changes. Sci Rep. 2019;9(1):16406. https://doi.org/10.1038/s41598-019-52563-1.
27. Engelhardt S, Patkar S, Ogunshola OO. Cell-specific blood-brain barrier regulation in health and disease: a focus on hypoxia. Br J Pharmacol. 2014;171(5):1210–30. https://doi.org/10.1111/bph.12489.
28. Bailey DM, Rimoldi SF, Rexhaj E, Pratali L, Salinas Salmòn C, Villena M, McEneny J, Young IS, Nicod P, Allemann Y, Scherrer U, Sartori C. Oxidative-nitrosative stress and systemic vascular function in highlanders with and without exaggerated hypoxemia. Chest. 2013;143(2):444–51. https://doi.org/10.1378/chest.12-0728.
29. Schoch HJ. Hypoxia-induced vascular endothelial growth factor expression causes vascular leakage in the brain. Brain. 2002;125(11):2549–57. https://doi.org/10.1093/brain/awf257.
30. Picón-Pagès P, Garcia-Buendia J, Muñoz FJ. Functions and dysfunctions of nitric oxide in brain. Biochim Biophys Acta (BBA) - Mol Basis Dis. 2019;1865(8):1949–67. https://doi.org/10.1016/j.bbadis.2018.11.007.
31. Li Y, Wang Y. Effects of long-term exposure to high altitude hypoxia on cognitive function and its mechanism: a narrative review. Brain Sci. 2022;12(6):808. https://doi.org/10.3390/brainsci12060808.
32. Ogoh S. Cerebral blood flow regulation during hypoxia. Exp Physiol. 2015;100(2):109–10. https://doi.org/10.1113/expphysiol.2014.084202.
33. Chen R, Lai UH, Zhu L, Singh A, Ahmed M, Forsyth NR. Reactive oxygen species formation in the brain at different oxygen levels: the role of hypoxia inducible factors. Front Cell Dev Biol. 2018;6:132. https://doi.org/10.3389/fcell.2018.00132.
34. Beckhauser TF, Francis-Oliveira J, De Pasquale R. Reactive oxygen species: physiological and physiopathological effects on synaptic plasticity. J Exp Neuroscie. 2016;10(Suppl 1):23–48. https://doi.org/10.4137/JEN.S39887.
35. Massaad CA, Klann E. Reactive oxygen species in the regulation of synaptic plasticity and memory. Antioxid Redox Signal. 2011;14(10):2013–54. https://doi.org/10.1089/ars.2010.3208.
36. Redza-Dutordoir M, Averill-Bates DA. Activation of apoptosis signalling pathways by reactive oxygen species. Biochim Biophys Acta Mol Cell Res. 2016;1863(12):2977–92. https://doi.org/10.1016/j.bbamcr.2016.09.012.
37. Liu Y, Xue C, Lu H, Zhou Y, Guan R, Wang J, Zhang Q, Ke T, Aschner M, Zhang W, Luo W. Hypoxia causes mitochondrial dysfunction and brain memory disorder in a manner mediated by the reduction of Cirbp. Sci Total Environ. 2022;806:151228. https://doi.org/10.1016/j.scitotenv.2021.151228.

38. Heerlein K, Schulze A, Hotz L, Bärtsch P, Mairbäurl H. Hypoxia decreases cellular ATP demand and inhibits mitochondrial respiration of A549 cells. Am J Respir Cell Mol Biol. 2005;32(1):44–51. https://doi.org/10.1165/rcmb.2004-0202OC.
39. Sharma C, Kim S, Nam Y, Jung UJ, Kim SR. Mitochondrial dysfunction as a driver of cognitive impairment in Alzheimer's disease. Int J Mol Sci. 2021;22(9):4850. https://doi.org/10.3390/ijms22094850.
40. Arias-Cavieres A, Fonteh A, Castro-Rivera CI, Garcia AJ. Intermittent Hypoxia causes targeted disruption to NMDA receptor dependent synaptic plasticity in area CA1 of the hippocampus. Exp Neurol. 2021;344:113808. https://doi.org/10.1016/j.expneurol.2021.113808.
41. Khuu MA, Pagan CM, Nallamothu T, Hevner RF, Hodge RD, Ramirez J-M, Garcia AJ. Intermittent hypoxia disrupts adult neurogenesis and synaptic plasticity in the dentate gyrus. J Neurosci. 2019;39(7):1320–31. https://doi.org/10.1523/JNEUROSCI.1359-18.2018.
42. Zhao F, Yang J, Cui R. Effect of hypoxic injury in mood disorder. Neural Plast. 2017;2017:1–10. https://doi.org/10.1155/2017/6986983.
43. Mukandala G, Tynan R, Lanigan S, O'Connor JJ. The effects of hypoxia and inflammation on synaptic signaling in the CNS. Brain Sci. 2016;6(1):6. https://doi.org/10.3390/brainsci6010006.
44. Shimoda LA, Polak J. Hypoxia. 4. Hypoxia and ion channel function. Am J Physiol Cell Physiol. 2011;300(5):C951–67. https://doi.org/10.1152/ajpcell.00512.2010.
45. Dallas M, Boycott HE, Atkinson L, Miller A, Boyle JP, Pearson HA, Peers C. Hypoxia suppresses glutamate transport in astrocytes. J Neurosci. 2007;27(15):3946–55. https://doi.org/10.1523/JNEUROSCI.5030-06.2007.
46. Richalet J-P, Letournel M, Souberbielle J-C. Effects of high-altitude hypoxia on the hormonal response to hypothalamic factors. Am J Phys Regul Integr Comp Phys. 2010;299(6):R1685–92. https://doi.org/10.1152/ajpregu.00484.2010.
47. Osborne DM, Pearson-Leary J, McNay EC. The neuroenergetics of stress hormones in the hippocampus and implications for memory. Front Neurosci. 2015;9 https://doi.org/10.3389/fnins.2015.00164.
48. Bian S-Z, Zhang L, Jin J, Zhang J-H, Li Q-N, Yu J, Chen J-F, Yu S-Y, Zhao X-H, Qin J, Huang L. The onset of sleep disturbances and their associations with anxiety after acute high-altitude exposure at 3700 m. Transl Psychiatry. 2019;9(1):175. https://doi.org/10.1038/s41398-019-0510-x.
49. Iwasaki K-I, Zhang R, Zuckerman JH, Ogawa Y, Hansen LH, Levine BD. Impaired dynamic cerebral autoregulation at extreme high altitude even after acclimatization. J Cereb Blood Flow Metab. 2011;31(1):283–92. https://doi.org/10.1038/jcbfm.2010.88.
50. Lucas SJE, Burgess KR, Thomas KN, Donnelly J, Peebles KC, Lucas RAI, Fan J, Cotter JD, Basnyat R, Ainslie PN. Alterations in cerebral blood flow and cerebrovascular reactivity during 14 days at 5050 m. J Physiol. 2011;589(3):741–53. https://doi.org/10.1113/jphysiol.2010.192534.
51. Lafuente JV, Bermudez G, Camargo-Arce L, Bulnes S. Blood-brain barrier changes in high altitude. CNS Neurol Disord Drug Targets. 2016;15(9):1188–97. https://doi.org/10.2174/1871527315666160920123911.
52. Yoon S, Zuccarello M, Rapoport RM. pCO2 and pH regulation of cerebral blood flow. Front Physiol. 2012;3 https://doi.org/10.3389/fphys.2012.00365.
53. Wang S, Tang C, Liu Y, Border JJ, Roman RJ, Fan F. Impact of impaired cerebral blood flow autoregulation on cognitive impairment. Front Aging. 2022;3:1077302. https://doi.org/10.3389/fragi.2022.1077302.
54. Yang L, Su J, Zhang X, Jiang C. Hypercapnia modulates synaptic interaction of cultured brainstem neurons. Respir Physiol Neurobiol. 2008;160(2):147–59. https://doi.org/10.1016/j.resp.2007.09.006.
55. Falla M, Micarelli A, Hüfner K, Strapazzon G. The effect of cold exposure on cognitive performance in healthy adults: a systematic review. Int J Environ Res Public Health. 2021;18(18):9725. https://doi.org/10.3390/ijerph18189725.

56. Nerín MA, Palop J, Montaño JA, Morandeira JR, Vázquez M. Acute mountain sickness: influence of fluid intake☆. Wilderness Environ Med. 2006;17(4):215–20. https://doi.org/10.1580/1080-6032(2006)17[215:AMSIOF]2.0.CO;2.
57. Pun M, Guadagni V, Bettauer KM, Drogos LL, Aitken J, Hartmann SE, Furian M, Muralt L, Lichtblau M, Bader PR, Rawling JM, Protzner AB, Ulrich S, Bloch KE, Giesbrecht B, Poulin MJ. Effects on cognitive functioning of acute, subacute and repeated exposures to high altitude. Front Physiol. 2018;9:1131. https://doi.org/10.3389/fphys.2018.01131.
58. Hencz A, Magony A, Thomas C, Kovacs K, Szilagyi G, Pal J, Sik A. Mild hypoxia-induced structural and functional changes of the hippocampal network. Front Cell Neurosci. 2023;17 https://doi.org/10.3389/fncel.2023.1277375.
59. De Wachter J, Roose M, Proost M, Habay J, Verstraelen M, De Bock S, De Pauw K, Meeusen R, Van Cutsem J, Roelands B. Prefrontal cortex oxygenation during a mentally fatiguing task in normoxia and hypoxia. Exp Brain Res. 2024;242(7):1807–19. https://doi.org/10.1007/s00221-024-06867-y.
60. Wei X, Ni X, Zhao S, Chi A. Influence of exposure at different altitudes on the executive function of plateau soldiers—evidence from ERPs and neural oscillations. Front Physiol. 2021;12 https://doi.org/10.3389/fphys.2021.632058.
61. Li XY, Wu XY, Fu C, Shen XF, Yang CB, Wu YH. Effects of acute exposure to mild or moderate hypoxia on human psychomotor performance and visual-reaction time. Space Med Med Eng. 2000;13(4):235–9.
62. Davis JE, Wagner DR, Garvin N, Moilanen D, Thorington J, Schall C. Cognitive and psychomotor responses to high-altitude exposure in sea level and high-altitude residents of Ecuador. J Physiol Anthropol. 2015;34(1):2. https://doi.org/10.1186/s40101-014-0039-x.
63. Jansen GFA, Basnyat B. Brain blood flow in Andean and Himalayan high-altitude populations: evidence of different traits for the same environmental constraint. J Cereb Blood Flow Metab. 2011;31(2):706–14. https://doi.org/10.1038/jcbfm.2010.150.
64. Zhang J-H, Shen Y, Liu C, Yang J, Yang Y-Q, Zhang C, Bian S-Z, Yu J, Gao X-B, Zhang L-P, Ke J-B, Yuan F-Z-Y, Pan W-X, Guo Z-N, Huang L. EPAS1 and VEGFA gene variants are related to the symptoms of acute mountain sickness in Chinese Han population: a cross-sectional study. Mil Med Res. 2020;7(1):35. https://doi.org/10.1186/s40779-020-00264-6.
65. Beall CM. Two routes to functional adaptation: Tibetan and Andean high-altitude natives. Proc Natl Acad Sci. 2007;104(suppl_1):8655–60. https://doi.org/10.1073/pnas.0701985104.
66. Simonson TS. Altitude adaptation: a glimpse through various lenses. High Alt Med Biol. 2015;16(2):125–37. https://doi.org/10.1089/ham.2015.0033.
67. Su R, Wang C, Liu W, Han C, Fan J, Ma H, Li H, Zhang D. Intensity-dependent acute aerobic exercise: effect on reactive control of attentional functions in acclimatized lowlanders at high altitude. Physiol Behav. 2022;250:113785. https://doi.org/10.1016/j.physbeh.2022.113785.
68. Walsh J, King T, Drouin P, D'Urzo K, Tschakovsky M, Day T. Acute exercise impairs cognitive function at high altitude. FASEB J. 2019;33 https://doi.org/10.1096/fasebj.2019.33.1_supplement.738.8.
69. Furian M, Bitos K, Hartmann SE, Muralt L, Lichtblau M, Bader PR, Rawling JM, Ulrich S, Poulin MJ, Bloch KE. Acute high altitude exposure, acclimatization and re-exposure on nocturnal breathing. Front Physiol. 2022;13 https://doi.org/10.3389/fphys.2022.965021.
70. Basnyat B, Murdoch DR. High-altitude illness. Lancet (London, England). 2003;361(9373):1967–74. https://doi.org/10.1016/S0140-6736(03)13591-X.
71. Lieberman HR, Shukitt Hale B, Banderet LE, Glenn JF, Lieberman HR, Shukitt Hale B, Banderet LE, Glenn JF. Exposure to hypobaric hypoxia a method for the comparison of neuropsychologic tests in humans. Soc Neurosci Abstracts. 1991;17(1–2):1–2.
72. Taylor A. High-altitude illnesses: physiology, risk factors, prevention, and treatment. Rambam Maimonides Med J. 2011;2(1):e0022. https://doi.org/10.5041/RMMJ.10022.
73. Institute of Medicine (US) Committee on Military Nutrition Research, Marriott BM, Carlson SJ. The effect of altitude on cognitive performance and mood states. In: Nutritional needs in cold and in high-altitude environments: applications for military personnel in field operations. National: Academies Press (US); 1996. https://www.ncbi.nlm.nih.gov/books/NBK232882/.

74. Scott SB, Graham-Engeland JE, Engeland CG, Smyth JM, Almeida DM, Katz MJ, Lipton RB, Mogle JA, Munoz E, Ram N, Sliwinski MJ. The effects of stress on cognitive aging, physiology and emotion (ESCAPE) project. BMC Psychiatry. 2015;15:146. https://doi.org/10.1186/s12888-015-0497-7.
75. Bliemsrieder K, Weiss EM, Fischer R, Brugger H, Sperner-Unterweger B, Hüfner K. Cognition and neuropsychological changes at altitude—A systematic review of literature. Brain Sci. 2022;12(12):1736. https://doi.org/10.3390/brainsci12121736.
76. San T, Polat S, Cingi C, Eskiizmir G, Oghan F, Cakir B. Effects of high altitude on sleep and respiratory system and theirs adaptations. TheScientificWorldJOURNAL. 2013;2013:241569. https://doi.org/10.1155/2013/241569.
77. Frost S, Orr JE, Oeung B, Puvvula N, Pham K, Brena R, DeYoung P, Jain S, Sun S, Malhotra A, Heinrich EC. Improvements in sleep-disordered breathing during acclimatization to 3800 m and the impact on cognitive function. Physiol Rep. 2021;9(9):e14827. 10.14814/phy2.14827.
78. Aksel G, Çorbacıoğlu ŞK, Özen C. High-altitude illness: management approach. Turk J Emerg Medi. 2019;19(4):121–6. https://doi.org/10.1016/j.tjem.2019.09.002.
79. Burtscher M, Millet GP, Burtscher J. Hypoxia conditioning for high-altitude pre-acclimatization. J Sci Sport Exerc. 2022;4(4):331–45. https://doi.org/10.1007/s42978-021-00150-0.
80. Wang J, Ke T, Zhang X, Chen Y, Liu M, Chen J, Luo W. Effects of acetazolamide on cognitive performance during high-altitude exposure. Neurotoxicol Teratol. 2013;35:28–33. https://doi.org/10.1016/j.ntt.2012.12.003.
81. Lafleur J, Giron M, Demarco M, Kennedy R, BeLue R, Shields C. Cognitive effects of dexamethasone at high altitude. Wilderness Environ Med. 2003;14(1):20–3. https://doi.org/10.1580/1080-6032(2003)014[0020:ceodah]2.0.co;2.
82. Hu S, Shi J, Xiong W, Li W, Fang L, Feng H. Oxiracetam or fastigial nucleus stimulation reduces cognitive injury at high altitude. Brain Behavior. 2017;7(10).e00762. https://doi.org/10.1002/brb3.762.
83. Bouzat P, Séchaud G, Banco P, Davranche K, Casini L, Baillieul S, Manhes P, Botrè F, Mazzarino M, De La Torre X, Robach P, Verges S. The effect of zolpidem on cognitive function and postural control at high altitude. Sleep. 2018;41(10) https://doi.org/10.1093/sleep/zsy153.
84. Muthuraju S, Maiti P, Solanki P, Sharma AK, Singh SB, Prasad D, Ilavazhagan G. Cholinesterase inhibitors ameliorate spatial learning deficits in rats following hypobaric hypoxia. Exp Brain Res. 2010;203(3):583–92. https://doi.org/10.1007/s00221-010-2266-7.
85. Uema S, Horita M, Takadera T. Protective effects of calcium ions via L-type calcium channels and NMDA receptors on prostaglandin E2-induced apoptosis in rat cortical cells. Mol Biol Rep. 2021;48(5):4517–25. https://doi.org/10.1007/s11033-021-06472-0.
86. Ji W, Zhang Y, Luo J, Wan Y, Liu J, Ge R-L. Memantine ameliorates cognitive impairment induced by exposure to chronic hypoxia environment at high altitude by inhibiting excitotoxicity. Life Sci. 2021b;270:119012. https://doi.org/10.1016/j.lfs.2020.119012.
87. Kauser H, Sahu S, Kumar S, Panjwani U. Guanfacine ameliorates hypobaric hypoxia induced spatial working memory deficits. Physiol Behav. 2014;123:187–92. https://doi.org/10.1016/j.physbeh.2013.10.028.
88. Prasad J, Baitharu I, Sharma AK, Dutta R, Prasad D, Singh SB. Quercetin reverses hypobaric hypoxia-induced hippocampal neurodegeneration and improves memory function in the rat. High Alt Med Biol. 2013;14(4):383–94. https://doi.org/10.1089/ham.2013.1014.
89. Jiang S, Fan F, Yang L, Chen K, Sun Z, Zhang Y, Cairang N, Wang X, Meng X. Salidroside attenuates high altitude hypobaric hypoxia-induced brain injury in mice via inhibiting NF-κB/NLRP3 pathway. Eur J Pharmacol. 2022;925:175015. https://doi.org/10.1016/j.ejphar.2022.175015.
90. Chen C, Li B, Chen H, Qin Y, Cheng J, He B, Wan Y, Zhu D, Gao F. Epigallocatechin-3-gallate ameliorated iron accumulation and apoptosis and promoted neuronal regeneration and memory/cognitive functions in the hippocampus induced by exposure to a chronic high-alti-

tude hypoxia environment. Neurochem Res. 2022;47(8):2254–62. https://doi.org/10.1007/s11064-022-03611-2.
91. Jing L, Wu N, Zhang J, Da Q, Ma H. Protective effect of 5,6,7,8-Tetrahydroxyflavone on high altitude cerebral edema in rats. Eur J Pharmacol. 2022;928:175121. https://doi.org/10.1016/j.ejphar.2022.175121.
92. Zhou B, Li M, Cao X, Zhang Q, Liu Y, Ma Q, Qiu Y, Luan F, Wang X. Phenylethanoid glycosides of Pedicularis muscicola Maxim ameliorate high altitude-induced memory impairment. Physiol Behav. 2016;157:39–46. https://doi.org/10.1016/j.physbeh.2016.01.037.
93. Zhang X, Zhang X, Xv J, Jia W, Pu X, Wang H, Liang H, Zhuoma-Lamao, & Lu, D. Crocin attenuates acute hypobaric hypoxia-induced cognitive deficits of rats. Eur J Pharmacol. 2018;818:300–5. https://doi.org/10.1016/j.ejphar.2017.10.042.
94. Du X, Liu T-L, Tao W, Li M-X, Li X-L, Yan L. Effect of aqueous extract of Astragalus membranaceus on behavioral cognition of rats living at high altitude. J Trad Chin Med. 2022;42(1):58–64. https://doi.org/10.19852/j.cnki.jtcm.2022.01.005.
95. Zhu A, Wu Z, Zhong X, Ni J, Li Y, Meng J, Du C, Zhao X, Nakanishi H, Wu S. Brazilian green propolis prevents cognitive decline into mild cognitive impairment in elderly people living at high altitude. J Alzheimers Dis. 2018;63(2):551–60. https://doi.org/10.3233/JAD-170630.
96. Zheng H, Su Y, Sun Y, Tang T, Zhang D, He X, Wang J. Echinacoside alleviates hypobaric hypoxia-induced memory impairment in C57 mice. Phytother Res. 2019;33(4):1150–60. https://doi.org/10.1002/ptr.6310.
97. Alonso-Alconada D, Álvarez A, Arteaga O, Martínez-Ibargüen A, Hilario E. Neuroprotective effect of melatonin: a novel therapy against perinatal hypoxia-ischemia. Int J Mol Sci. 2013;14(5):9379–95. https://doi.org/10.3390/ijms14059379.
98. Kishore K, Ray K, Kumar S, Anand JP, Thakur L, Ravi D, Suresh A, Kolpe R, Kizhekkedath J, Kumar B, Panjwani U. Tyrosine supplementation A nutraceutical approach to counter heat stress induced cognitive decline. Def Life Sci J. 2021;6(3):205–13. https://doi.org/10.14429/dlsj.6.15870.
99. Yu L, Chen Y, Wang W, Xiao Z, Hong Y. Multi-vitamin B supplementation reverses hypoxia-induced tau hyperphosphorylation and improves memory function in adult mice. J Alzheimers Dis. 2016;54(1):297–306. https://doi.org/10.3233/JAD-160329.
100. Lassen NA, Friberg L, Kastrup J, Rizzi D, Jensen JJ. Effects of acetazolamide on cerebral blood flow and brain tissue oxygenation. Postgrad Med J. 1987;63(737):185–7. https://doi.org/10.1136/pgmj.63.737.185.
101. Swenson ER. Pharmacology of acute mountain sickness: old drugs and newer thinking. J Appl Physiol. 2016;120(2):204–15. https://doi.org/10.1152/japplphysiol.00443.2015.
102. Kushwah N, Jain V, Kadam M, Kumar R, Dheer A, Prasad D, Kumar B, Khan N. Ginkgo biloba L. prevents hypobaric hypoxia–induced spatial memory deficit through small conductance calcium-activated potassium channel inhibition: the role of ERK/CaMKII/CREB signaling. Front Pharmacol. 2021;12:669701. https://doi.org/10.3389/fphar.2021.669701.
103. Dupuy O, Gauthier CJ, Fraser SA, Desjardins-Crèpeau L, Desjardins M, Mekary S, Lesage F, Hoge RD, Pouliot P, Bherer L. Higher levels of cardiovascular fitness are associated with better executive function and prefrontal oxygenation in younger and older women. Front Hum Neurosci. 2015;9 https://doi.org/10.3389/fnhum.2015.00066.
104. Bircher HP, Eichenberger U, Maggiorini M, Oelz O, Bärtsch P. Relationship of mountain sickness to physical fitness and exercise intensity during ascent. J Wilderness Med. 1994;5(3):302–11. https://doi.org/10.1580/0953-9859-5.3.302.
105. Jaeggi SM, Buschkuehl M, Jonides J, Perrig WJ. Improving fluid intelligence with training on working memory. Proc Natl Acad Sci USA. 2008;105(19):6829–33. https://doi.org/10.1073/pnas.0801268105.
106. Zeidan F, Johnson SK, Diamond BJ, David Z, Goolkasian P. Mindfulness meditation improves cognition: evidence of brief mental training. Conscious Cogn. 2010;19(2):597–605. https://doi.org/10.1016/j.concog.2010.03.014.

107. Mangione L, Nelson D. The 1996 Mount Everest tragedy: contemplation on group process and group dynamics. Int J Group Psychother. 2003;53:353–73. https://doi.org/10.1521/ijgp.53.3.353.42824.
108. Willie CK, Stembridge M, Hoiland RL, Tymko MM, Tremblay JC, Patrician A, Steinback C, Moore J, Anholm J, Subedi P, Niroula S, McNeil CJ, McManus A, MacLeod DB, Ainslie PN. UBC-Nepal expedition: an experimental overview of the 2016 University of British Columbia Scientific Expedition to Nepal Himalaya. PLoS One. 2018;13(10):e0204660. https://doi.org/10.1371/journal.pone.0204660.
109. Barkaszi I, Takács E, Czigler I, Balázs L. Extreme environment effects on cognitive functions: a longitudinal study in high altitude in Antarctica. Front Hum Neurosci. 2016;10 https://doi.org/10.3389/fnhum.2016.00331.

Part II

Desert, Marine, Jungle

Soldier Health Under Desert Conditions 10

Abhishek Bharadwaj, Pooja Chaudhary, Rajinder K. Gupta, Sachin Kumar, and Rajeev Varshney

Abstract

Stress has been described as a measure of cognitive, physical and biological response to any stressor. The stressor can be an event or stimuli which are perceived as a threat. Environmental stressors include a variety of physical stimuli, noise, temperature (heat and cold), pressure (hypobaric [altitude] and hyperbaric [underwater]), photoperiods (24 h darkness or light), and vibration alters a person's ability to function. Impact of thermal stress is predominant on soldiers under desert conditions affecting their performance. Desert and jungle warfare and operational aspects have certain physiological implications. Man cannot adapt to low quantities of water permanently. The adapted soldier needs an equal quantity of water or at times more than the unacclimatized one. Detrimental effects of heat exposure are mainly attributed to excessive elevations in core body temperature. Normally, body temperature is maintained stable relatively by balancing body heat production [due to metabolism] with heat loss. Operating in desert is very difficult task to achieve. But, when there is an available chance to train in the desert this will be the best preparation for operating in this climate. In spite of being fit, soldier still need at least 10 to 12 days to get used to the desert climate so he can be ready to carry on the operation well. Heat stress management is critical for preventing heat casualties. Most importantly, leadership must fully inform personnel of the risk associated with heat stress.

A. Bharadwaj (✉) · P. Chaudhary · R. K. Gupta · S. Kumar · R. Varshney
Defence Institute of Physiology and Allied Sciences (DIPAS), Defence Research and Development Organization (DRDO), Timarpur, Delhi, India
e-mail: abhishek.dipas@gov.in

S. Srivastava et al. (eds.), *Health and Sustenance in Extreme Environment*,
https://doi.org/10.1007/978-981-96-8642-1_10

Keywords

Desert warfare · Heat stress · Dehydration · Acclimatization

10.1 Introduction

Wars are synonymous with unrest, danger, and uncertain situations, causing different types and levels of stress that impact soldiers' performance.

Stress is described as a measure of cognitive, physical, and biological responses to any stressor. A stressor can be an event or stimuli perceived as a threat. It should be perceived strongly enough to elicit a response that affects performance both cognitive and physical performances, thus affecting the ultimate well-being [1, 2].

Like operations, during training, exposure to extremes of weather and situations (viz. temperature, humidity, barometric pressure, vibrations, noise, etc.) induce physiological strain degrading cognitive and physical performances.

Any stressor can affect the performance of an individual through physiological responses. Environmental stressors affect performance differently and the resultant effect gets exaggerated even leading to fatality when clubbed with non-environmental stressors. Environmental stressors include physical stimuli such as noise, temperature (heat and cold), pressure (hypobaric [altitude] and hyperbaric [underwater]), photoperiods [24 h darkness or light], and vibration, which affect an individual's functioning. It is interesting and important to note that a certain amount of stress can sometimes be beneficial to human performance. It has been seen in certain studies that performance can be positively correlated with stress up to a point, however, beyond that threshold performance starts to decline [3]. Moderate stress has proved to be beneficial in increasing performance by providing optimal stimulation, thus eliciting alertness, individual effort, and motivation, whereas excessive or too little can be associated with decreases in performance [1].

10.2 Background

Reports of death and injury in military populations due to exertional heat illness (EHI), and its most severe form, heat stroke, have been reported from antiquity. Heat exposure when beyond the tolerant limit can be debilitating also leading to fatality. This impacts battlefield outcomes and thus success. This aspect has been recognized by commanders for millennia [4–6]. Scientific studies pertaining to heat on military personnel began during British and European colonialism; heat became a major source of illness among troops deployed in tropical and desert climates [7]. Since then, studies have been conducted to understand and enhance soldiers' tolerance to ambient stress, with the intention of optimizing effectiveness and reducing heat casualties [8, 9].

Desert and jungle warfare and operational aspects have certain physiological implications. Land desert warfare has been characterized by fighting vehicles and mobile infantry in "hard" or "soft" vehicles who rarely march long distances. Dust and haze impair visibility during the day, so cooling vehicles in hot ambient temperatures may be more urgent than cooling the man. Nights are seldom completely dark and hence allow maneuvers.

10.3 Desert Regions Have Typical Characteristics

1. Characteristics of different desert areas are:

   (a) Large areas with hard surfaces, having a small percentage of sand. These areas are ideal for transportation.

   (b) Extended huge sand hills. Blowing wind and storms form such structures reaching heights of several feet. The possibility of transportation through these contours is very limited.
   (c) Small sand hills. Wind blows covering a large area with a few inches of sand hampering transport.
   (d) Volcanic area. A widespread area of volcanic stones, thus hindering maneuvering and transportation through such terrain.
   (e) Muddy areas. These areas exist in several parts of the desert. Transportation when wet is extremely difficult but easy when dry.
   (f) Roads and channels. Very rarely found in deserts and it usually connects villages and oases.

2. Weather.

   (a) Heat. Desert is characterized by high-temperature days with low-temperature nights. Summers are harsh while winters in the desert have temperatures reaching subzero. This huge span of temperature range and alterations affects military personnel and operations.
   (b) Wind. The direction of the wind is changeable; when it increases it carries with it sand forming large and small sand dunes/hills. Under extreme heat, these winds blow in the form of "loo."
   (c) ain. The percentage of rain in the desert is less than 10 inches yearly and storms come without prior warning where within a few minutes dry valleys fill with water.

3. Water.
   Water wells in the desert are rare and only found in an oasis. Even then water needs to be distilled before use. So, the problem of water (potable) is a major problem during desert operations.
4. Plants.
   Usually scarce in the desert, mainly in the oasis where grass and palm trees exist along with spiny small trees in some other areas.

5. Desert Dwellers. In the desert, there are no populated areas except for some groups of natives and village dwellers near an oasis.

The desert is described as a castle for the experienced and knowledgeable, and a grave for the ignorant and naive. To thrive and perform in this environment, a soldier needs to be well prepared so he can work at full capacity. Keeping high morale of personnel in the desert is of special importance because of the harsh climate, dust storms, and the way of living.

10.4 Acclimatization

A period of adaptation to the desert conditions is an essential requirement to avoid casualties specifically related to heat. An acclimatized soldier is capable to accept changes in the weather and adjusts better than an unacclimatized soldier. Where situations do not permit a long enough acclimatization period and the troops have to carry out large assignments, strenuous work is advisable to be done during comparatively cooler periods of the day.

In the desert, water wells are very few, and most are not drinkable because of high salinity. Therefore, water must be checked before drinking. The human need for water varies according to the amount of exercise, type of food intake, and ambient temperature. Though it is preferable to take a small amount of water intermittently, during comparatively cooler times a considerable amount of water should be consumed. Soldiers should adhere to a strict system of water usage. Man cannot adapt to low quantities of water permanently. The adapted soldier needs an equal quantity of water or at times more than the unacclimatized one. In case there is not enough potable water, the rate of activity must be reduced till the coolest times of the day. If water is restricted, sometimes the efficiency of personnel decreases gravely even if they are relaxed. An appropriate amount of salt is needed to keep the body balanced, salt should be distributed with food and additional quantities may be provided during adaptation and heavy work. Medical consultation should be taken about the amount of salt taken.

10.5 Health

With heat-related maladies, soldiers are susceptible to other desert-related diseases. A strict regime of physical training, hygiene, washing, and adhering to guidelines of health precautions is of utmost importance. Soldiers to avoid intestinal diseases need careful food distribution and clean cooking habits and practices. Much effort has been devoted toward understanding soldiers' tolerance to heat stress, with the goal of maximizing effectiveness and minimizing heat casualty risk [8, 9].

The detrimental effects of heat exposure are mainly attributed to excessive elevations in core body temperature. Normally, body temperature is maintained relatively stable by balancing body heat production [due to metabolism] with heat loss. Heat stress occurs when this homeostatic balance gets altered such that heat production exceeds heat loss, causing a rise in body temperature. This occurs with physical exertion and/or if heat loss is hindered due to environmental factors (high ambient temperature, radiation, relative humidity) and clothing.

Military exercises and operations often create heat stress. High temperatures with humidity are encountered during training exercises or deployments in tropical and dry climate systems and even during temperate summer months. Rapid deployment under heat stress conditions is particularly challenging without prior heat acclimatization. Similarly, divers performing tasks encounter very warm shallow-water temperatures. Vehicles, vessels, and aircraft contribute as a major source of environmental heat stress [10]. High temperatures in armored vehicles and sea-crafts produce extremely high humidity levels as unevaporated sweat accumulates inside crew compartments [11, 12] along with lack of proper ventilation. Thermal radiation from jet engines, avionics, and sun exposure through aircraft canopies greatly elevate cockpit temperatures [13]. Training and operational tasks can be physically demanding. Personal protective clothing and equipment, such as body armor, flight ensembles (G-suits, immersion protection), and chemical, biological, radiological, and nuclear (CBRN) garments, can be heavy and impede body heat loss. Physiological response to heat stress initiates by redistributing blood flow to skin and increased sweating, thus promoting evaporative heat loss. These responses thermoregulate body temperature, though increased skin blood flow and sweat rate induce strain on the body. Redistribution of blood to the skin, along with sweat-induced dehydration, confines the amount of blood returning to the heart. This threatens the supply of blood to vital organs and the brain. Due to these factors, the ability to manage other stressors [as prolonged sentry duty standing, acceleration (e.g., tactical aviation), and hemorrhage] which reduces blood flow to the brain gets seriously impaired by additional heat stress [14, 15].

In combination with exercise and environmental heat stress, the body's capacity to lose heat gets inundated, causing body temperature to rise persistently until a serious heat illness occurs. The most common being heat exhaustion. This is characterized by the lack of ability to continue physical activity resulting in extreme cardiovascular strain and dehydration. The most severe heat illness, heatstroke, is defined by extremely high body temperatures and impaired mental status. Heatstroke requires immediate and aggressive cooling to avoid organ/tissue damage and death. The risk of heat-related illness and degradation in performance may not be equal between soldiers. This is because the physiological responses to a given level of heat stress are highly variable between individuals. Numerous factors contribute to this variability; the most common are summarized in Table 10.1 below.

Table 10.1 Risk factors and their implications on soldiers' health

Risk factors	Implications
Personal factors	
Low fitness levels	Better aerobic fitness improves evaporative capacity Reduces cardiovascular strain
Unacclimatized	Acclimatization helps improves sweating and evaporative cooling Reduces cardiac workload
Hydration levels	Dehydration reduces sweating rate compromises thermoregulation
Motivation	Willingness to compete and complete pushes the tolerance limit
Physical ailments	
Infections (viral/bacterial, etc.)	Hyperthermia Immune response compromised
Pyrexia	Accentuates hyperthermia
Skin grafting	Reduces sweat production
Medication	Different medications have varied effects

Prolonged heat stress impairs complex mental tasks requiring memory, interpreting information, and reasoning whereas on the contrary, simple reaction time tasks as target detection, largely remain unaffected. Prolonged heat exposure also degrades coordination and steadiness thus affecting marksmanship. Heat-related deteriorations in steadiness and attention affect tracking ability, which is required for piloting, target acquisition, and weapons alignment.

Understanding the gravity and intricacies of heat injuries along with their risk factors is extremely important. Heat-related incidences such as heat cramps, heat exhaustion, and heat stroke are common due to the physically demanding activities and operations. These exertional heat ailments if not managed meticulously can lead to fatality. Susceptibility to these ailments depends on certain intrinsic and extrinsic factors, including CNS dysfunction and ambient heat stress.

10.6 Risk Factors for Heat Injuries

Factors governing the risk of heat illness in the military are:

- Age.
- Physical fitness.
- Environmental conditions.
- The nature of military training and operations.

Considering these risk factors measures should be planned toward implementing effective preventive schedules, preserving health, and maintaining well-being of the soldier.

10.6.1 Age and Physical Fitness

Younger and physically lesser fit individuals are at higher risk for heat-related injuries. This is due to the body's compromised ability to thermoregulate and adapt to hot environments. Better fitness and heat acclimatization can help prevent such injuries by:

- Decreasing resting core temperature.
- Augmenting sweat rate.
- Reducing physiological strain.
- Improving exercise capacity in the heat.

Higher physical fitness and heat acclimation helps military personnel improve their resilience to heat stress thus reducing their chances of heat injuries.

10.7 Heat Casualty

Hydration plays an important role in maintaining the temperature of the body, which loses efficiency with lost fluids. Ensuring ample hydration along with nourishment can prevent heat injuries by maintaining electrolyte balance and thermoregulatory processes. The recommended hydration strategy includes ensuring your body is properly nourished prior to exercise and replenishing a portion of sweat losses during exercise if feasible. Emphasizing proper hydration and nutrition can help military personnel assist their bodies in dealing with heat stress and cut down the risk of heat injuries.

Heat casualties are classified under three types:

1. Heat exhaustion. This occurs when not enough fluid is consumed. Dehydration through sweat and simultaneous not enough fluid intake to compensate, causes drowsiness, lassitude, and feeling of weakness.
2. Salt depletion. Similar to heat exhaustion for unacclimatized individuals. Salt depletion in the body happens after sweating, this causes gradual exhaustion, vomiting, headache, cramping, and stiffening of muscles in the hand, legs, and belly. This situation can be resolved by drinking water with salt but strictly in according to the physician's supervision.
3. Sunstroke. This is one of the most perilous situations encountered due to heat exposure. This occurs when the thermoregulatory system stops regulating body temperature and the body loses its ability to regulate its temperature by sweating. The core body temperature increases and if not attended immediately chances of fatality are very high.

This can be encountered by soldiers who are acclimatized and immediate treatment decreases the body temperature. This can be done by putting the patient in a shack and taking his clothes off to cool down.

Operating in a desert is very challenging and difficult to achieve, however, training under such desert conditions would best prepare for activities in such a climate. In spite of being fit, soldiers need a minimum period of 8 days to acclimatize to the desert climate and be operation ready. Heat acclimatization and adequate water/fluid intake are pivotal for optimizing the effectiveness and functionality of a desert warrior. Adherence to prescribed work/rest cycles based on environment, intensity of work, and clothing effectively mitigates the cumulative amount of heat stress and strain.

Commanders and medical officers must be educated and made aware of these risk factors to adequately manage the risk of heat illness and casualties. Heat stress management is critical for preventing heat casualties. Most importantly, leadership must be fully informed and should take the onus to inform and educate the personnel of the risk associated with heat stress and their management.

References

1. Kavanagh J. Stress and performance: a review of the literature and its applicability to the military. Santa Monica: RAND Corporation; 2005.
2. Lloyd A, Havenith G. Interactions in human performance: an individual and combined stressors approach. Temperature (Austin, Tex). 2016;3(4):514–7.
3. Tiwari GK. Stress and human performance. Indo-Indian J Soc Sci Res. 2011;7(1):40–9.
4. Goldman RF. Introduction to heat-related problems in military operations. In: Pandolf KB, Burr RE, editors. Medical aspects of harsh environments, vol. 1. Washington, DC: Borden Institute, Walter Reed Army Medical Center; 2001. p. 3–49.
5. Jarcho S. A Roman experience with heat stroke in 24 B.C. Bull N Y Acad Med. 1967;43(8):767–8.
6. Olsen EW. The Battle of Hattin, 1187 (Master's Thesis). U.S. Army Command and General Staff College; 1997.
7. Bricknell M. Heat illness—a review of military experience (part 1). J R Army Med Corps. 1995;141(3):157–66.
8. Ashworth ET, Cotter JD, Kilding AE. Methods for improving thermal tolerance in military personnel prior to deployment. Mil Med Res. 2020;7(1):58.
9. Nye NS, O'Connor FG. Exertional heat illness considerations in the military. In: Adams WM, Jardine J, editors. Exertional heat illness: a clinical and evidence-based guide. Switzerland: Springer; 2020. p. 181–209.
10. Kozlowski F. U.S. marines in battle: an-Najaf. Washington, DC: United States Marine Corps History Division; 2009.
11. Jacobs I, Michas R, Limmer R, Kerrigan-Brown D, McLellan T. Heat stress mitigation for Leopard 2C tank crew. Defence Research and Development Canada TR 2007-082; 2007. p. 1–34.
12. Schlader ZJ, Gagnon D, Adams A, Rivas E, Cullum CM, Crandall CG. Cognitive and perceptual responses during passive heat stress in younger and older adults. Am J Physiol Regul Integr Comp Physiol. 2015;308(10):R847–54. https://doi.org/10.1152/ajpregu.00010.2015.
13. Kenefick RW, Cheuvront SN, Castellani JW, O'Brien C. Thermal stress. In: Davis JR, Johnson R, Stepanek J, Fogarty JA, editors. Fundamentals of aerospace medicine. 4th ed. Philadelphia: Lippincott Williams & Wilkins; 2008. p. 206–20.
14. Crandall CG, Rickards CA, Johnson BD. Impact of environmental stressors on tolerance to hemorrhage in humans. Am J Physiol Regul Integr Comp Physiol. 2019;316(2):R88–R100.
15. Nunneley SA, Stribley RF. Heat and acute dehydration effects on acceleration response in man. J Appl Physiol Respir Environ Exerc Physiol. 1979;47(1):197–200.

Progression of Heat Illness to Heat Stroke: Race Against Time

11

Kalpana Barhwal, Anshika Baranwal, and Abhinav Dixit

Abstract

Heat stroke represents a spectrum of pathophysiological conditions progressing from heat illness to heat exhaustion, thereby encompassing central nervous system dysfunction, multi-organ failure, and metabolic disturbances leading to morbidity and mortality. Heat illness and heat stroke continue to be a formidable challenge not only for the civilian population but also for the military and paramilitary operations in India. Heat-related illnesses represent a spectrum of disorders that include minor heat cramps, heat exhaustion, and life-threatening heat stroke. While the human body tightly regulates core temperature through physiological mechanisms that include alterations in blood circulation and regulation of evaporative heat loss through sweating, extreme temperatures override these regulatory mechanisms. During high heat stress index, the core body temperature rises above 40°C resulting in the involvement of multiple organ systems and triggering pathophysiological mechanisms of heat illness.

Depending on its cause, heat stroke could be non-exertional/classic heat stroke due to climatic factors and heat waves, or exertional, resulting from exercise and dehydration. The pathophysiology of heat stroke involves a complex interplay of thermoregulatory and cardiovascular factors, leading to catastrophic systemic inflammatory response, eventually causing multi-organ failure. The clinical diagnosis of heat stroke is extremely important and is primarily based on the triad syndrome, viz., evidence of hyperthermia, CNS alterations, and history of exposure to environmental heat.

K. Barhwal
Department of Physiology, All Indian Institute of Medical Sciences (AIIMS), Bhubaneswar, India

A. Baranwal · A. Dixit (✉)
Department of Physiology, All Indian Institute of Medical Sciences (AIIMS), Jodhpur, India
e-mail: dixit@aiimsjodhpur.edu.in

S. Srivastava et al. (eds.), *Health and Sustenance in Extreme Environment*,
https://doi.org/10.1007/978-981-96-8642-1_11

Despite advancements in understanding heat stroke, effective treatment strategies for its management and treatment are still obscure. A comprehensive multistep prevention protocol could assist in preventing the symptoms associated with heat stroke. Effective preventive and mitigation strategies are also required to make the susceptible populations aware of heat-related illnesses.

Keywords

Heat stroke · Pathophysiology · Diagnosis · Prevention · Mitigation strategies

11.1 Introduction

Global climate change has resulted in an upsurge in the number and intensity of heat waves even in areas that were not predominantly associated with hot temperatures. Temperature increases due to climate changes pose an increasing risk of heat-related illnesses not only to the civilian population but also affects the military training operations at large. It poses a '*noncombatant threat*' to the military personnel, thereby encumbering various military activities. Although the guidelines for preventing heat injuries are different for civilian and military personnel, understanding of the pathophysiology and progression from heat illness to heat stroke will reinforce treatment and management strategies.

Heat-related illnesses (HRI) due to high ambient temperatures pose a major health problem, and several studies have indicated a significant increase in both mortality and morbidity [1, 2]. HRI represents a spectrum of disorders including minor heat cramps, heat exhaustion, and life-threatening heat stroke, which increase the vulnerability of individuals with cardiovascular, respiratory, and/or cerebrovascular diseases [3]. Heat stroke is clinically defined as a condition in which the core temperature of the body rises above 40°C, resulting in the involvement of multi-organ systems [4]. Heat stroke often occurs as a progression from milder heat-related illnesses such as heat cramps, heat syncope (fainting), and heat exhaustion. Heat stroke, which is life-threatening, is an outcome of prolonged exposure to high ambient temperatures, often in combination with dehydration, leading to failure of the body's thermostat system. It could be exertional, resulting from exercise and dehydration or non-exertional due to climatic factors and heat waves.

While a diagnosis of heat stroke is highly based on the heat index due to the absence of clinically definitive diagnostic markers other than an increase in body temperature above 104°F, headache, weakness, fatigue, etc., which could also happen in cases of fever, the treatment is symptomatic and largely biased toward electrolyte infusion and reducing body temperature. Other treatment modalities are based on the extent of damage to particular organs. Interestingly, pathophysiological events leading to the transition of heat illness to heat stroke, which also hold the secret for individual susceptibility to heat stroke and early markers for diagnosis still remain to be investigated. This is particularly relevant in the Indian scenario, from both a military and civilian prospective. The Indian forces guard long borders along the western frontiers, where temperatures can soar above 45°C. Besides, specific military forces engaged in operations employing tanks, naval vessels, etc., are exposed to high temperatures in the work environment, making them more susceptible to heat stress. Previous studies by researchers at the Defence Institute of Physiology and Allied Sciences (DIPAS) show a significant decrease in body mass, peak expiratory flow rate, and 24-h urinary output in crew of armored vehicles. The study demonstrated severe hypohydration of 3% body mass in the crew due to strenuous heat load, leading to physiological strain. The large number of heat stroke-related deaths reported annually from different parts of the country, particularly Andhra Pradesh and Odisha, also makes research on the progression of heat illness to heat stroke socially relevant.

Despite the relevance of the subject, research on heat stroke in general, and on blood and circulatory physiology during heat illness in particular, is scarce. While one study by DIPAS shows a significant decrease in plasma volume at 3% and 4% hypohydration using radiolabeled techniques, there is a single report on heat stroke-induced bleeding in sheep that was related to changes in blood platelets [5]. In light of the above, there is a necessity to undertake directed basic research not only to understand the differential susceptibility of individuals to heat stroke but also to identify suitable therapeutic and prophylactic targets that can modulate the progression of heat illness to heat stroke.

11.2 Epidemiology

According to the statistical approximations by the WHO, a large population is exposed to extreme heat due to global warming and climate change. It is also predicted that the frequency and intensity of extreme heat waves will continue to rise exponentially in the twenty-first century due to climate change. The highest global temperatures in over 100,000 years were recorded in 2023, contributing to climate-related disaster mortalities in many developing countries like South Asia, the Middle East, Africa, and South America [6]. The elderly, infants, and poor are the most susceptible segments of the population that is severely affected by heat stress owing to existing health conditions, lack of resources, and health care. Recent statistics reveal an increase in heat-related deaths in the elderly population over 65 years of age by 85% between the periods 2000–2004 and 2017–2021 [7].

Asia leads in heat-related deaths, with over 45% mortality due to heat-related illnesses between 2000 and 2019. Previous studies report an increase of 0.5°C in the mean temperature across India between 1971 and 2011 [8]. It is further projected to rise by 2.2–5.5°C by the end of the twenty-first century [9]. Europe stands at 36% of heat-related mortalities, with an estimated 61,672 heat-related deaths in 2022. The incidence of heat stroke in the United States varied from 17.6 to 26.5 cases per 100,000 population [10]. In Saudi Arabia, the incidence varies seasonally from 22 to 250 cases per 1000,000 population. The heat stroke-related mortality in Saudi Arabia is estimated at 50% [11] (Fig. 11.1).

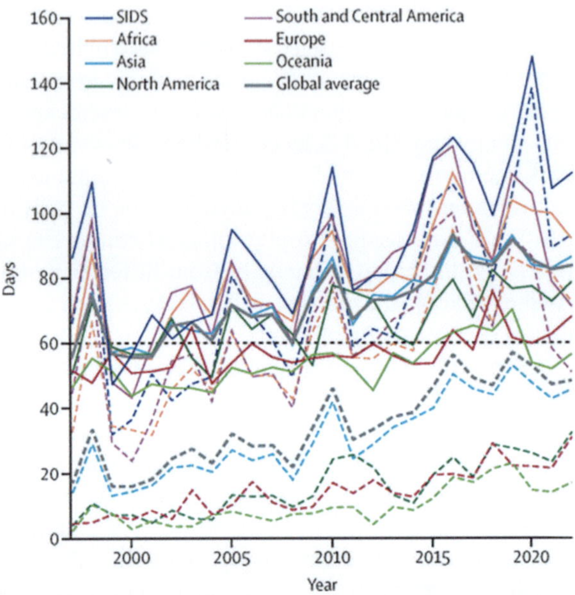

Fig. 11.1 Population-weighted days of exposure to temperatures above the 84th percentile for 1986–2005. The number of days of exposure to warm temperatures for different regions is displayed using solid lines; the heavy solid line represents the global average. The number of days of exposure to warm temperatures made at least twice as probable due to climate change is plotted with dashed lines; the heavy dotted line represents the global average. *SIDS* = Small Island Developing States. (Adapted from: Romanello et al. [7])

According to epidemiological data, it is evident that susceptibility to heat illness and heat stroke varies among individuals, with some developing heat exhaustion while others developing heat stroke. Susceptibility to heat illness and heat stroke depends on genetic predisposition to certain genes, viz., heat shock proteins (HSPs), genes encoding coagulation proteins, and certain cytokines [12]. However, additional studies are required to find putative genes responsible for the pathogenesis of heat stroke and to understand the underlying cellular and systemic responses to heat stress.

11.3 Classification, Clinical Features, and Risk Factors

Heat-related illnesses represent a spectrum of disorders ranging from minor heat cramps, heat exhaustion, and life-threatening heat stroke [3]. While the human body tightly regulates core body temperature through physiological mechanisms that include alterations in blood circulation and regulation of evaporative heat loss through sweating, extreme temperatures can override these regulatory mechanisms. During high heat stress index, core body temperature rises above 40°C, resulting in the involvement of multiple organ systems and triggering pathophysiological mechanisms of heat illness [4]. Heat stroke often occurs as a progression from milder heat-related illnesses such as heat cramps, heat syncope (fainting), and heat exhaustion. Clinically, heatstroke is characterized by central nervous system (CNS) dysfunction, multi-organ failure, and extreme hyperthermia (usually >40.5 °C) [3].

Depending on its cause, heat stroke could be classified as non-exertional/classic heat stroke due to climatic factors and heat waves or exertional, resulting from exercise and dehydration [14]. (i) Classic heat stroke frequently occurs as an epidemic arising in individuals at the extremes of age, related to either limited mobility/chronic diseases or in predisposed individuals upon exposure to excessive environmental temperature and humidity. Besides the elderly, prepubertal children are also regarded as at risk owing to underdeveloped thermoregulatory mechanisms and a high surface area to mass ratio, leading to poor heat dissipation and increased heat absorption. It presents with a rapid elevation of core temperature exceeding 40°C and is associated with a low sweating rate and CNS dysfunction. Patients with classic or non-exertional heat stroke often hyperventilate and develop respiratory alkalosis. (ii) Exertional heat stroke occurs in active individuals, viz., athletes, laborers, construction workers, military personnel, and others who overexert themselves in the heat due to their occupational needs. It is sporadic in nature, with drug abuse, alcohol or both being the major risk factors. It is often characterized by excessive heat production that overwhelms heat-loss mechanisms. Sweating and CNS dysfunctions are common along with metabolic acidosis, hypocalcemia, and hyperkalemia as depicted in Fig. 11.2. The mortality rates of exertional heat stroke are, however, low because it usually affects the health of young persons, and its recognition and treatment are usually very prompt [13].

The predisposition to heat stroke, be it classic or exertional, often depends on environmental exposure, metabolic demands, and several risk factors, as depicted in Fig. 11.3. Probable risk factors include age, acute/chronic medical conditions, body composition, sex differences, dehydration, and drug and alcohol abuse. Aging is considered a primary risk factor, as it is known to impede behavioral thermoregulatory responses, impair cardiovascular function, and decrease sudomotor and immune

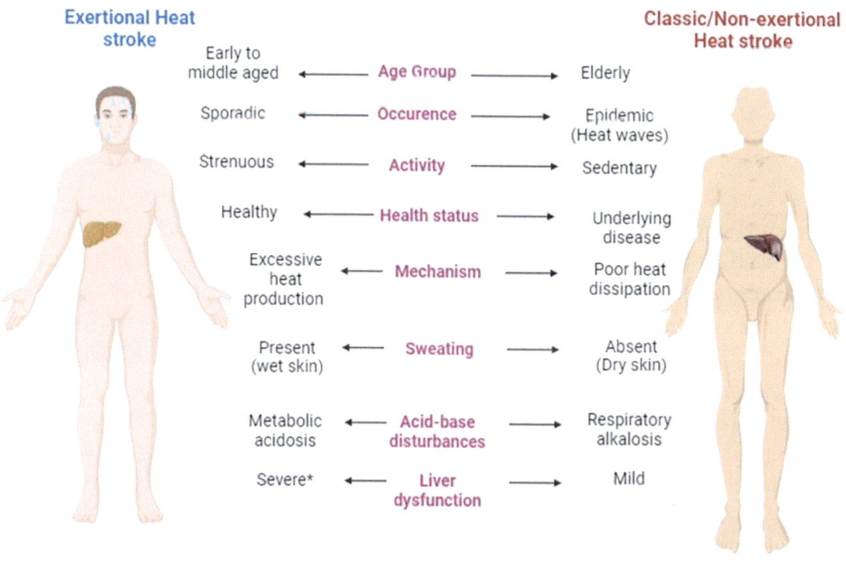

Fig. 11.2 Clinical features of exertional and classic/non-exertional heat stroke

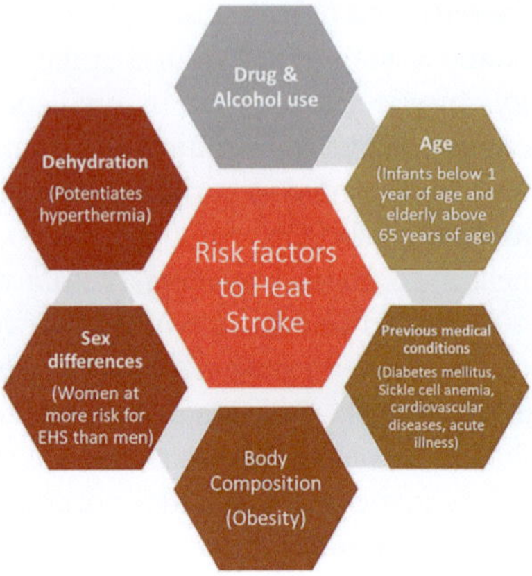

Fig. 11.3 Potential risk factors associated with heat stroke

functions. These risk factors contribute to increased mortality and morbidity in the elderly as well as in prepubertal children, particularly infants below 1 year of age [15]. Besides aging, any debilitating illness such as diabetes mellitus, cardiac disease, congenital disorders egg sickle cell trait, previous heat injury, or skin abnormalities like psoriasis, eczema, burns, etc., poses a major risk factor for susceptibility to heat illness and progression to heat stroke. Studies by Dervis et al. suggest impaired sudomotor and thermoregulatory responses in individuals with high fate mass [16]. This could be due to impaired microvascular functions of the skin, leading to diminished sweat production, thereby compromising evaporative heat loss mechanisms. In addition, high adiposity results in decreased ability to dissipate heat, posing a threat to hyperthermia. Obesity is also associated with chronic inflammation, metabolic disturbances, and decreased cardiovascular fitness, culminating as risk factors toward predisposition to exertional heat stroke.

Physical differences between men and women also influence thermoregulation. However, the physiological basis for these sex differences is uncertain and the evidence is contradictory. Some studies have shown presumptions of women having increased risk of exertional heat stroke [17, 18] while recent studies indicate men to have increased risk during self-paced activities [19]. Though women are lean and have a smaller body surface area (BSA) they tend to have a larger BSA to body mass ratio. These morphological features contribute to lower metabolic heat production and greater heat dissipation as compared to men. However, Notley et al. suggested similar metabolic heat loads between men and women during a cycling task. However, at high-intensity workload, women tend to display reduced evaporative heat loss response when compared to men. Additional studies are therefore required to understand the discrepancy and possible influence of female sex hormones on exertional heat stroke.

Inadequate hydration also poses a risk particularly for soldiers and military personnel to exertional heat stroke. During excessive exercise, plasma volume decreases and plasma osmolality and blood viscosity increase, much of which is attributed to increased metabolic demand. The increased viscosity not only leads to cardiac strain but also leads to diminished cerebral blood flow and syncope if the plasma volume continues to fall under exertional heat stress [20, 21]. It is also hypothesized that dehydration-induced hyperthermia during exertional heat stroke is associated with increased reactive oxygen species (ROS) production, leading to increased intestinal permeability, but more studies are required to support the hypothesis.

Additional risk factors that could increase the susceptibility and predisposition to heat stroke include certain drugs like amphetamines, anticholinergics, antidepressants, antihistamines, antipsychotics, beta blockers, benzodiazepines, calcium channel blockers, diuretics, illicit drugs (e.g., cocaine, heroin, lysergic acid diethylamide [LSD]), etc. Dietary supplements (primarily ergogenic stimulants) are also known as potential risk factors. Alcohol use could increase hypotension and impair thermoregulatory responses too [22].

11.4 Pathophysiology

Thermoregulation, a mechanism that maintains normal core body temperature between 36.7°C and 37°C, is tightly regulated in which it can fluctuate without evoking any autonomic thermoregulatory mechanisms. These mechanisms are pertinent to maintaining several molecular and cellular functions. It is only when core body temperature gets elevated due to compromised thermoregulatory mechanisms that heat-related illnesses and heat stroke ensue. The elevation in core temperature could be due to internal metabolic heat production by any strenuous activity or external environmental heat modulated by several factors like humidity, acclimatization, solar radiation, restricted cooling mechanisms, etc. The pathophysiology of heat stroke is, therefore, a complex interplay of thermoregulatory and cardiovascular factors leading to catastrophic systemic inflammatory responses eventually causing multi-organ failure (Fig. 11.4). As body temperature rises, it is sensed by a large number of thermosensors present on the skin and/or internal organs, including transient receptor potential (TRP) channels. The signal is relayed to the somatosensory cortex and preoptic area of the hypothalamus to elicit and trigger autonomic responses, namely, cutaneous vasodilation, activation of the sympathetic nervous system, and stimulation of sweating through cholinergic activation. The cutaneous vasodilation leads to a twofold increase in cardiac output, which approximately is redistributed to skin circulation at the expense of splanchnic and renal blood flow. If the core body temperature continues to rise under heat stress, dehydration ensues due to increased loss of water and salts via sweating, clinically manifesting as muscle cramps. With decreased central venous pressure and redistribution of blood flow, thermoregulatory mechanisms get jeopardized resulting in hyperthermia and progression towards heat stroke. Hyperthermia incites damage to proteins, membrane lipids, and DNA leading to cellular dysfunction, tissue damage, and death [23]. This triggers the cellular stress response and induction of heat shock transcription factors (HSFs) that upregulate the expression levels of heat shock proteins (HSPs). HSPs are molecular chaperones that are required for preserving cellular structures and functions. Concomitant to heat stress, the elevated temperature causes proteins to

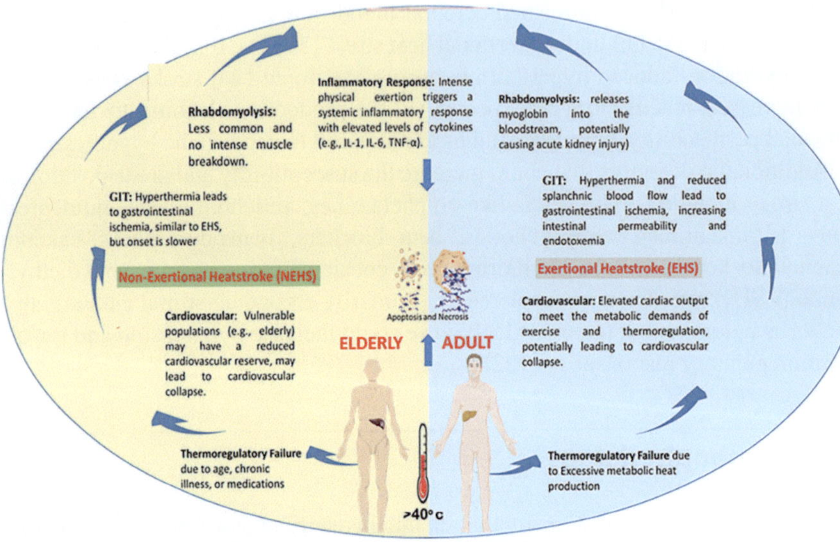

Fig. 11.4 Putative pathophysiological processes of tissue injury and multi-organ failure in exertional and non-exertional (classic) heat stroke

unfold or misfold forming aggregates that undergo proteasomal degradation. Induced HSPs recognize these misfolded proteins and help regain their native structure preventing aggregation and subsequent degradation [24]. Studies have shown high HSP expression in young soldiers during exertional heat injury indicating the central role of heat stress response in preventing heat injury [25]. However, the heat stress response fails/weakens under conditions like aging, lack of heat acclimatization, and epigenetic and genetic factors that predispose individuals to debilitating heat stroke.

Both classical and exertional heat stroke share fundamental pathophysiologic processes, viz., hyperthermia, multi-organ failure, endothelial activation triggering coagulation pathways, and inflammatory responses, but the key distinction lies in the role of skeletal muscle in exertional heat stroke that is often pronounced as rhabdomyolysis, kidney injury, and lactic acidosis [26]. During exertional heat stroke, elevated muscle tension increases muscle breakdown which is often associated with a high incidence of rhabdomyolysis [27]. This could also be due to decreased muscle blood flow due to competition for accessible cardiac output between skin and muscles resulting in skeletal muscle ischemia. Ischemia decreases the ATP turnover in skeletal muscle thereby increasing the intramuscular calcium (Ca^{2+}), which in turn, activates the proteolytic enzymes particularly caspases, contributing to muscle breakdown as seen in exertional rhabdomyolysis.

Thus, the underlying pathophysiological mechanisms leading to heat stroke encompass a transition from a reversible, compensable thermoregulatory state to a nonreversible state [28]. The non-compensable state involves a cascade of events associated with inflammatory and noninflammatory responses. The inflammatory responses include activation of innate immune response characterized by leukocytosis, elevation of pro-inflammatory cytokines like tumor necrosis factor-α (TNF-α), interleukin-1β, interferon γ, and anti-inflammatory cytokines like interleukin-6, interleukin-10, and TNF receptors p55 and p75 [29, 30]. The increased levels of cytokines from skeletal muscle particularly interleukin-1 and interleukin-6 lead to widespread inflammation manifesting as vascular congestion and migration from vascular beds to nearby interstitial tissues [32]. In human and animal models of heat stroke, activation of innate immunity has been associated with the release of intracellular molecules, viz., pathogen-associated molecular patterns (PAMPs) and damage-associated molecular patterns (DAMPs) [30]. Elevated cytokines, PAMPs, and DAMPs breach intestinal permeability leading to leakage of bacteria and endotoxins from the gut lumen (*Leaky gut hypothesis*). Studies by Pires et al. and others have documented increased intestinal permeability during exercise with and without heat stress [31] due to loss of tight junction integrity. This facilitates the leakage of endotoxins in the systemic circulation leading to oxidative and nitrosative stress. During exertional heat stroke, the liver becomes severely affected, evidenced by high levels of lipopolysaccharides, thereby compromising the reticuloendothelial system. Under these catastrophic conditions, liver necrosis and endotoxemia occur leading to sepsis.

The inflammatory responses subsequently contribute to endothelial cell injury by activating coagulation pathways, as evidenced by the appearance of thrombin–antithrombin III complexes, soluble fibrin monomers, and low levels of protein C, protein S, and antithrombin III. Thus, hyperthermia during heat stroke evokes a prothrombotic state, leading to microvascular thrombosis and disseminated intravascular coagulation (DIC). Circulating levels of adhesion molecules like soluble endothelial cell leukocyte adhesion molecule 1 (ELAM1), intercellular adhesion molecule 1 (ICAM1), E-selectin, and metabolites of nitric oxide are also elevated in patients with heat stroke [33]. Studies by [37] in the baboon model have shown profound fibrinolytic activity demonstrated by increased levels of tissue plasminogen activator, which gets rapidly inhibited by an early rise in plasminogen activator inhibitor-1 activity, thereby evoking a biphasic response of fibrinolysis activation.

11.5 Diagnosis of Heat Stroke

The clinical features as depicted in Fig. 11.1 differ in classic and exertional heatstroke and form the basis for diagnosing heat stroke. The clinical diagnosis of heat stroke is primarily based on the triad syndrome, viz., evidence of hyperthermia, CNS alterations, and history of exposure to environmental heat. While, hyperthermia is common in both ranging between 40°C and 43.6°C, relying on increased core temperature as a diagnostic yardstick could be misleading. This is because core body temperature could fall during transit to the hospital or primary health care centers and should not lead to the exclusion of the diagnosis of heat stroke. Rectal temperature measurement is the most accurate method in a clinical setting and should be considered for the diagnosis of heat stroke. While dry skin, reflecting a decrease in sweat gland response, is a characteristic feature of non-exertional heat stroke seen in elderly patients; profuse sweating and wet skin are typical of classic/exertional heat stroke. Since the brain is extremely sensitive to hyperthermia, CNS alterations are inevitable in heatstroke and may include delirium, slurred speech, unconsciousness, and post-cooling seizures. The symptoms closely resemble stroke and should be ruled out during the diagnosis of heat stroke [34]. Other cardiovascular alterations include hypotension, arrhythmia, and tachycardia, which can lead to circulatory shock in severe cases.

In principle, the diagnosis of heat stroke can be confirmed with laboratory investigations that include complete blood count (CBC), kidney function tests, liver function tests, and serum electrolytes. Kidney function tests portray renal failure as a common feature in patients with exertional heat stroke, accounting for 25–30%, whereas it is less common in classic heat stroke (<5%). The major contributors to kidney dysfunction in heat stroke include hypovolemia, rhabdomyolysis, and disseminated intravascular coagulopathy. Rhabdomyolysis is often detectable by reddish-brown urine, along with clinical features like muscle aches, muscle weakness, and muscle stiffness. Respiratory disorders are common in both classic and exertional heat stroke, with symptoms including hypoventilation, cyanosis, and acute respiratory distress syndrome (ARDS). Exertional heat stroke is characterized by metabolic acidosis, whereas blood gas analysis in classic heat stroke shows respiratory alkalosis. Other diagnostic criteria in exertional heat stroke include marked liver dysfunction, elevated creatine kinase, hypocalcemia, and hyperkalemia. Calcium and phosphate levels remain normal; however, there is a mild increase in creatine kinase levels in patients with classic heat stroke. Experimental data suggests that these traditional clinical biomarkers may not be sensitive diagnostic indicators of tissue injury in all heat stroke patients, and the search for precise and sensitive biomarkers is still underway. The alarming global rise in temperatures necessitates further research so as to identify precision-based diagnostic biomarkers for early recognition of heat stroke.

11.6 Management and Treatment

Despite the advancements in understanding heat stroke, effective treatment strategies are still obscure. Nevertheless, the primary objective of treating heat stroke is to alleviate hyperthermia through effective cooling strategies. Evaporative cooling is preferred for non-exertional heat stroke, while conductive cooling is preferred for exertional heat stroke. Evaporative cooling is a noninvasive, efficient, and rapid, and involves positioning the undressed patient laterally or elevating them on hands and knees to expose the maximum skin surface, applying warm water (20 °C) to the entire undressed body, and using a large circulating fan to blow room air for cooling. Supplementing with cool air or oxygen can assist but should not be the sole method. Conductive cooling methods include injection of cold fluids, lavage of the stomach or bladder, application of ice packs, and immersion in ice or cold water. For on-the-job heat stroke, ice or cold-water immersion is recommended, and ice packs should be applied to the whole body rather than just the armpits and groin to enhance their effectiveness. Advanced cooling techniques include intravascular heat exchange catheter devices and hemodialysis with cold dialysate (30–35 °C). Continuous monitoring of the patient is essential, utilizing a pulse oximeter and heart monitor, and regularly assessing core body temperature, fluid intake, and fluid output.

Fluid intake guidelines are another critical aspect of managing heat stress. It is recommended to increase fluid intake before, during, and after physical activity in hot conditions to maintain hydration and prevent dehydration. Maintaining a balance of electrolytes and water is essential, as excessive water intake without electrolyte replacement can lead to hyponatremia, a condition characterized by low sodium levels in the blood. Salt tablets should be avoided due to the risk of hypertonic hypernatremia, where the body's sodium concentration becomes excessively high. Monitoring and intervention are necessary to ensure individuals maintain proper fluid and electrolyte intake. During physical exertion in hot environments, it is crucial to monitor for signs of heat stress and intervene promptly if symptoms appear. This can involve moving the individual to a cooler environment, providing fluids, and using cooling techniques to lower body temperature.

To prevent the progression of heat stroke to multi-organ failure, it is pertinent to prioritize and stabilize the patient's airway, breathing, and circulation (ABCs). For the treatment of heat cramps, heat exhaustion, and heat syncope, oral isotonic or hypertonic rehydration solutions have been found effective. Antipyretics such as aspirin and acetaminophen are ineffective for hyperthermia and should not be used. Treatment should be continued until the core body temperature reaches 39 °C [35–38]. Patients should rest for at least 2 days and continue taking dietary supplements before returning to work or engaging in strenuous physical activity in hot weather. Figure 11.5 summarizes the various first aid measures to combat symptoms of heat cramps, heat exhaustion, and eventually heat stroke. However, large studies involving both humans and animal models are required to evaluate the existing treatment strategies and to develop personalized precision-based biomarkers, drugs, and effective treatment and therapies.

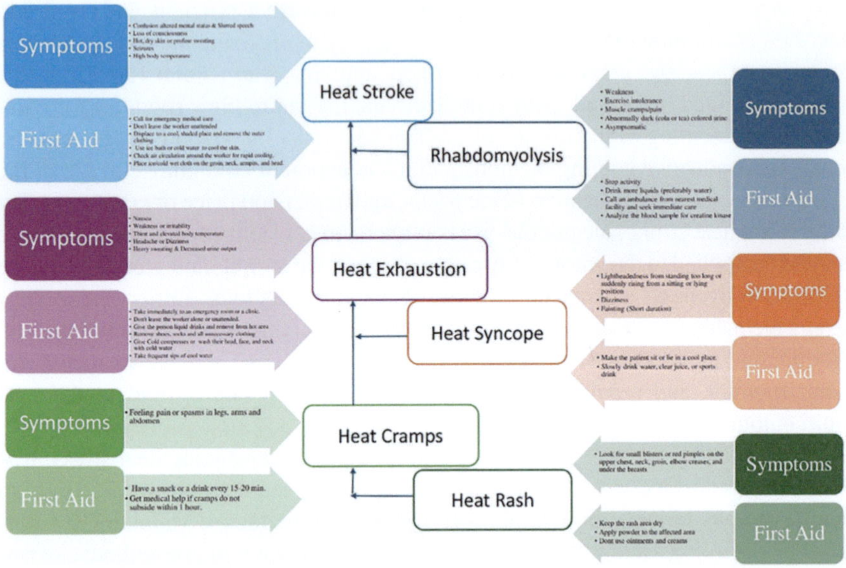

Fig. 11.5 First aid measures to combat symptoms of heat stroke

11.7 Prevention Strategies

Heat stroke is more preventable than treatable because therapeutic options for multi-organ failure are limited. Thorough awareness of heat-related illnesses can reduce mortality and morbidity among populations at risk, viz., elderly, military personnel, athletes, football players, those with preexisting health conditions, or those who are not acclimatized to heat. They should be well informed about the importance of prevention, the dangers of heat exposure, the symptoms to watch for, and the appropriate treatment methods. Mitigation and adaptation strategies are also crucial, as people can gradually reduce their risk factors and acclimate to heat over time, thereby enhancing their resilience to heat stress. Monitoring climatic conditions, especially during periods of extreme heat, can also help in planning and adjusting activities to reduce heat exposure.

A comprehensive multistep prevention protocol could assist in preventing the symptoms associated with heat stroke. Primarily, it is essential to assess individual risk factors, such as age, fitness level, medical history, and current medications, which influence a person's susceptibility to heat-related illnesses. Considering workload and acclimatization status of individuals is also critical, as those who are not accustomed to heat or are engaged in strenuous activities are at a higher risk. Recognizing early symptoms of heat stress, such as excessive sweating, dizziness, nausea, and headache, allows for timely intervention and treatment.

Thus, by integrating education, assessment, fluid management, and vigilant monitoring, the risk of heat-related illnesses can be significantly reduced, and individuals can safely engage in physical activities even in hot weather conditions [39].

11.8 Mitigation and Adaptation Strategies

Heat stress can be effectively mitigated by modifying factors such as metabolic heat production, convective heat exchange, radiative heat exchange, and evaporative heat exchange. Alleviating variables such as ventilation, air conditioning, screening, insulation, and process or operation modifications can significantly alter environmental heat load. Additionally, protective clothing and equipment, along with labor-saving devices, can reduce metabolic heat production. Convective heat control involves managing environmental variables like air temperature and air movement speed. While engineering solutions can enhance convective heat loss, radiative heat exchange can be accomplished by shielding workers from heat sources or using reflective shielding, which can significantly reduce radiant heat load. Reflective shielding is often the easiest and least expensive solution, requiring careful placement to avoid work interference. Increasing air movement with fans or blowers and decreasing ambient water vapor pressure through air conditioning can enhance evaporative cooling. Eliminating water vapor sources, such as steam leaks and wet floors, also improves evaporative heat loss.

Limiting or adjusting exposure duration could help in managing daily exposure time and temperature under heat stress conditions. This can be achieved by scheduling hot jobs during cooler parts of the day, redesigning workspaces to increase distance from radiant heat sources, adjusting work/rest schedules, and allowing work interruption if a worker feels heat discomfort. Other strategies include mechanizing physical job components, reducing work time, utilizing planned exposure times, and increasing the workforce to distribute the workload.

Heat acclimatization is an effective mitigation strategy to enhance a worker's capacity in enduring hot conditions. The effectiveness of a heat acclimatization protocol depends on variables such as heat modality, ambient conditions, session frequency, and duration, all of which impact physiological responses to heat and exercise. Although these factors are tailored for athletic performance, they are often impractical for military settings due to challenges like limited access to heat chambers and the necessity for additional exercise sessions, which can lead to overtraining. Short-term heat acclimatization protocols are used to reduce barriers and prevent interference with other training priorities. Despite adding extra training sessions that may cause fatigue and overtraining, short-term heat acclimatization often requires access to heat chambers and results in limited adaptation. Passive heat exposure methods, such as saunas and hot-water immersion, allow groups to acclimate simultaneously and are straightforward to install. These techniques are typically employed postexercise to assist in maintaining core temperature, potentially minimizing disruption to essential training. Implementing a 7–14-day heat-acclimatization program can reduce the risk of heat-related illnesses.

Screening workers for low heat intolerance is also crucial for preventing heat-related illnesses. Heat intolerance varies widely among individuals due to congenital, functional, or acquired factors such as physical fitness, acclimatization, and health conditions. Medical screening for heat intolerance should include a history of previous heat-related illnesses and a heat tolerance test to assess heat susceptibility.

A Heat Alert Program is extremely vital during heat waves to prevent heat-related illnesses. Operational tasks include preparing facilities for heat emergencies, ensuring ventilation efficiency, and maintaining cooling equipment. Criteria for declaring a Heat Alert are set based on weather forecasts. During a Heat Alert, non-essential tasks are postponed, teams are augmented to reduce individual heat exposure, and rest periods in air-conditioned areas are increased. Core temperature monitoring, hydration monitoring, and medical evaluations for affected workers are also implemented, along with exercising caution during shift changes.

During extreme heat waves and increased severity of heat stress, it is typical to modify either the work environment or the protective clothing of the workers. Impermeable clothing inhibits heat exchange with the environment, thereby increasing the metabolic burden and subsequent heat production. An alternative strategy could be supplementary body cooling technologies, viz., water-cooled garments, air-cooled garments, cooling vests, and wetted overgarments. These systems aim to reduce heat stress and conserve drinking water, enhancing performance in challenging environmental conditions. Water-cooled garments use conductive heat exchange through tubing to cool the body, requiring external devices like batteries and pumps. Their effectiveness depends on weight, mobility constraints, and environmental conditions, with condensation potentially affecting performance. Air-cooled garments provide cooling by distributing air next to the skin, with cooling effectiveness influenced by airflow rate and air inlet temperature. Cooling vests, which use ice or phase-change material packs, provide adequate cooling for 2–4 h, depending on environmental conditions, without noise or mobility limitations. However, practical limitations exist for cooling systems in work settings. Phase-change and ice vests lack temperature control and do not stay cool long enough. Water cooling systems tether the user, restricting movement. Although auxiliary cooling can reduce heat injury risk and recovery time, wearing protective clothing still impairs performance due to mechanical and communication barriers, even without heat stress [40, 41].

11.9 Conclusion

Globally, heat waves are increasing in intensity, frequency, and duration due to climate change, posing a fatal threat through heat stroke. Risk factors include age, sex differences, dehydration, body composition, chronic medical illnesses, and drug and alcohol abuse. However, the susceptibility of some individuals to heat stroke warrants further research. The pathophysiological processes underlying heat stroke involve a catastrophic interplay of inflammatory and noninflammatory responses, compromising thermoregulatory and cardiovascular functions. The failure of the heat stress response (mediated by HSFs and HSPs), concomitant with inflammatory events triggers endotoxemia, subsequent liver dysfunction, and coagulopathy, resulting in tissue injury, multi-organ dysfunction, and cellular death. Effective preventive and mitigation strategies are required to make susceptible populations aware of heat-related illnesses. Increased mortality and morbidity due to heat stroke also necessitate the development of novel biomarkers, and underlying cellular and molecular responses to ameliorate the symptoms associated with heat stroke.

References

1. Abidine El Morjani ZE, Ebener S, Boos J, Abdel Ghaffar E, Musani A. Modelling the spatial distribution of five natural hazards in the context of the WHO/EMRO Atlas of Disaster Risk as a step towards the reduction of health impact related to disasters. Int J Health Geogr. 2007;6:8.
2. Basu R, Samet JM. Relation between elevated ambient temperature and mortality: a review of the epidemiologic evidence. Epidemiol Rev. 2002;24:190–202.
3. Bouchama A, Knochel JP. Heat stroke. N Engl J Med. 2002;346(25):1978–88.
4. Dematte JE, O'Mara K, Buescher J, Whitney CG, Forsythe S, McNamee T, et al. Near-fatal heat stroke during the 1995 heat wave in Chicago. Ann Intern Med. 1998;129:173–81.
5. Mohanty D, Gomez J, Mustafa KY, Khogali M, Das KC. Pathophysiology of bleeding in heat stress: an experimental study in sheep. Exp Hematol. 1997;25(7):615–9.
6. Diffenbaugh NS, Scherer M. Observational and model evidence of global emergence of permanent, unprecedented heat in the 20th and 21st centuries. Clim Chang. 2011;107:615–24.
7. Romanello M, Napoli CD, Green C, Kennard H, Lampard P, et al. The 2023 report of the Lancet Countdown on health and climate change: the imperative for a health-centered response in a world facing irreversible harms. Lancet. 2023;402(10419):2346–94.
8. Kothawale DR, Munot AA, Kumar KK. Surface air temperature variability over India during 1901–2007, and its association with ENSO. Clim Res. 2010;42:89–104.
9. Dholakia HH, Mishra V, Garg A. Predicted increases in heat related mortality under climate change in urban India. IIMA Working Papers WP2015-05-02. Indian Institute of Management Ahmedabad, Research and Publication Department; 2015.
10. Jones TS, Liang AP, Kilbourne EM, et al. Morbidity and mortality associated with the July 1980 heat wave in St. Louis and Kansas City, Mo. JAMA. 1982;247:3327–31.
11. Ghaznawi HI, Ibrahim MA. Heat stroke and heat exhaustion in pilgrims performing the Haj (annual pilgrimage) in Saudi Arabia. Ann Saudi Med. 1987;7:323–6.
12. Moseley PL. Heat shock proteins and heat adaptation of the whole organism. J Appl Physiol. 1997;83:1413–7.
13. Shapiro Y, Seidman DS. Field and clinical observations of exertional heat stroke patients. Med Sci Sports Exerc. 1990;22:6–14.
14. Glazer JL. Management of heatstroke and heat exhaustion. Am Fam Physician. 2005;71(11):2133–40.
15. Millyard A, Layden JD, Pyne DB, et al. Impairments to thermoregulation in the elderly during heat exposure events. Gerontol Geriatr Med. 2020;6:233372142093243.
16. Dervis S, Coombs GB, Chaseling GK, et al. A comparison of thermoregulatory responses to exercise between mass-matched groups with large differences in body fat. J Appl Physiol. 2016;120:615–23.
17. Carter R, Cheuvront SN, Williams JO, Kolka MA, Stephenson LA, Sawka MN, et al. Epidemiology of hospitalizations and deaths from heat illness in soldiers. Med Sci Sports Exerc. 2005;37:1338–44.
18. Marsh SA, Jenkins DG. Physiological responses to the menstrual cycle. Sports Med. 2002;32(10):601–14.

19. Belval LN, Giersch GEW, Adams WM, Hosokawa Y, Jardine JF, Katch RK, et al. Age- and sex-based differences in exertional heat stroke incidence in a 7-mile road race. J Athl Train. 2020;55(12):1224–9.
20. Notley SR, D'Souza AW, Meade RD, Richards BJ, Kenny GP. Whole-body heat exchange in women during constant-and variable-intensity work in the heat. Eur J Appl Physiol. 2020;120(12):2665–75
21. Leyk D, Hoitz J, Becker C, et al. Health risks and interventions in exertional heat stress. Dtsch Arztebl Int. 2019;116:537–44.
22. Vescovi PP, Coiro V. Hyperthermia and endorphins. Blood. Pharmacol Ther. 1993;47:301–4.
23. Kültz D. Molecular and evolutionary basis of the cellular stress response. Annu Rev Physiol. 2005;67:225–57.
24. Gomez-Pastor R, Burchfiel ET, Thiele DJ. Regulation of heat shock transcription factors and their roles in physiology and disease. Nat Rev Mol Cell Biol. 2018;19:4–19.
25. Sonna LA, et al. Exertional heat injury and gene expression changes: a DNA microarray analysis study. J Appl Physiol. 2004;96:1943–53.
26. Laitano O, Oki K, Leon LR. The role of skeletal muscles in exertional heat stroke pathophysiology. Int J Sports Med. 2021;42(8):673–81.
27. Knochel JP. Exertional rhabdomyolysis. N Engl J Med. 1972;287:927–9.
28. Epstein Y, Yanovich R. Heatstroke. N Engl J Med. 2019;380(25):2449–59.
29. Chang DM. The role of cytokines in heatstroke. Immunol Investig. 1993;22:553–61.
30. Bouchama A, Al-Sedairy S, Siddiqui S, Shail E, Rezeig M. Elevated pyrogenic cytokines in heatstroke. Chest. 1993;104:1498–502.
31. Pires W, Veneroso CE, Wanner SP, et al. Association between exercise-induced hyperthermia and intestinal permeability: a systematic review. Sports Med. 2017;47:1389–403.
32. Roberts GT, Ghebeh H, Chishti MA, Al-Mohanna F, El-Sayed R, Al-Mohanna F, Bouchama A. Microvascular injury, thrombosis, inflammation, and apoptosis in the pathogenesis of heatstroke: a study in baboon model. Arterioscler Thromb Vasc Biol. 2008 Jun;28(6):1130–6.
33. Shieh SD, Shiang JC, Lin YF, Shiao WY, Wang JY. Circulating angiotensin-converting enzyme, von Willebrand factor antigen and thrombomodulin in exertional heat stroke. Clin Sci (Lond). 1995;89:261–5.
34. Leon LR, Bouchama A. Heat stroke. Compr Physiol. 2015;5:611–47.
35. Wang H, Alves J. Heat related illnesses: diagnosis and treatment update. Sriwij J Forensic Medicolegal. 2023;1(2):25–9.
36. Alele FO, Malau-Aduli BS, Malau-Aduli AEO, Crowe M. Epidemiology of exertional heat illness in the military: a systematic review of observational studies. Int J Environ Res Public Health. 2020;17(19):7037.
37. Bouchama A, Al-Mohanna F, Assad L, Baturcam E, Eldali A, Owaidah T, Dehbi M. Tissue factor/factor VIIa pathway mediates coagulation activation in induced-heat stroke in the baboon. Crit Care Med. 2012;40(4):1229–36.
38. Parsons IT, Stacey MJ, Woods DR. Heat adaptation in military personnel: mitigating risk, maximizing performance. Front Physiol. 2019;10:1485.
39. Hasan F, Marsia S, Patel K, Agrawal P, Razzak JA. Effective community-based interventions for the prevention and management of heat-related illnesses: a scoping review. Int J Environ Res Public Health. 2021;18(16):8362.
40. Ashworth ET, Cotter JD, Kilding AE. Methods for improving thermal tolerance in military personnel prior to deployment. Mil Med Res. 2020;7(1):58.
41. Carton C, Oopik V, Rintamaki H, Bourdon L, Glitz KJ, Werner A, et al. Management of heat and cold stress—guidance to NATO medical personnel. NATO HFM. 2012;187

Effect of Marine Motion on Crew Performance

12

P. Gokulakrishnan, Rohit Verma, and Dilip Raghavan

Abstract

Marine motion is peculiar enough to have a disease named after its effect—seasickness. The occupation of seafaring exposes crew to the vagaries of the vessel dynamics as they ply in the watery medium. The waters of the world follow a universal dictum—irrespective of where the vessel plies, in freshwater or on the high seas, the movements affect human physiology in ways that need special attention. With the world comprising two-thirds of water and international trade largely dependent on shipping for the world economy to thrive, motion disorders in mariners warrent unique medical consideration. This chapter describes the effect of motion on the vestibular system of mariners, its implications ranging from benign discomfort to serious safety concerns, and its amelioration, both by therapeutic and by design-based preventive measures.

Keywords

Motion sickness · Seasickness · Visual display · Artificial horizon · Human performance · Operator performance · Ship designing

The only sure cure for seasickness is to sit under a tree—Spike Milligan KBE, Irish Comedian

P. Gokulakrishnan (✉)
Institute of Naval Medicine, INHS Asvini Campus, Mumbai, India
e-mail: gokulakrishnan.parth@navy.gov.in

R. Verma
Classified Specialist (Marine Medicine), Institute of Naval Medicine, INHS, Mumbai, India

D. Raghavan
Headquarters Southern Naval Command, Kochi, India

S. Srivastava et al. (eds.), *Health and Sustenance in Extreme Environment*, https://doi.org/10.1007/978-981-96-8642-1_12

Historically, men have taken to the sea in ships to forge beyond territorial boundaries, to trade, and to spread civilisations to faraway lands. Marine crafts function at optimum capability only when their crew is free to function without any limitation. There are no passengers among the crew, for any extra being carried would have a penalty on supplies, space and thus endurance of the mission, which is the critical bottom line. Each member of the crew delivers a share of the workload. Any condition that degrades the functions of even a small number of the crew would threaten the overall operations and the safety of others on board the ship.

It is well recognised that the movements of a ship in motion cause a form of sickness. This predates the causality of vehicular motions, which later on led to the recognition of the condition we now call motion sickness or kinetosis. More than 2000 years ago, Greek physician Hippocrates observed that '*sailing on the seas proves that motion disorders the body*'. Famous sailors like Admiral Nelson, Charles Darwin and Christopher Columbus suffered from seasickness. Elevated to head the Spanish Armada in 1588, the Duke of Medina Sidonia, known as a competent general, soon developed debilitating seasickness [1].

Motion sickness is a generic syndrome of an illness with associated emesis caused by motion: on ships, aircrafts, moving cars, or on a roller coaster ride. Actually, the term 'motion sickness' is incorrect from two perspectives. Firstly, it can be induced even during a virtual reality simulation without actual motion. Secondly, while sickness implies that it is a disease, this disease is actually a completelyexpected response of a human exposed to motion without any underlying disorder [2]. Although the responses in health and clinical presentation are similar across all motions, seasickness, being platform induced in the marine medium, is predominantly described in this chapter.

> The greatest and most important problems of life are all fundamentally insoluble. They can never be solved but only outgrown—Carl Jung

The symptom syndrome in excess of a dozen manifestations includes faintness, dizziness, sweating, sleepiness, headache and nausea. The word 'nausea' stems from the Greek word '*naus*', meaning ship. Close to it is also the Sanskrit word '*Nauka*'. Naval personnel traditionally have downplayed seasickness, regarding it as an occupational hazard that is to be endured and 'gotten over with'. However, statistics emerging from research conducted by world navies show that up to one-third of a crew may be incapacitated even in moderate sea conditions, and in rough seas, this number could go up to an alarming 80% of crew [3, 4].

While getting over seasickness is often assumed to be just a matter of time, and indeed many would get conditioned to it [5], sufficient data exist to show that this assumption is both erroneous and trivialised to shun sickness as a sign of weakness or unsuitability to be a mariner. Bruner's survey of 699 men sailing on US Navy warships concluded that 39% were never seasick, 38% were only occasionally sick, 10% were frequently sick and 13% were almost always sick [6]. When extrapolated to thousands of personnel, consistently deployed on seagoing vessels, most major Navies would be grappling with the enormity of the burden of this common condition.

An Indian Naval study reports that moderately severe symptoms were observed among 19.31% of personnel needing medication and rest. Only 1.91% had symptoms severe enough to be given protracted excuses from duties along with medication and rest. About 54.5% of personnel reported seasickness during the first voyage, 26.5% of personnel persisted to be symptomaticwhile being put to sea for up to 10 exposures and 13.6% continued to have symptoms even after 10 exposures to sea [7], forming the so-called chronic seasickness-prone mariner.

> Why do we get seasick?: Travel unsettles the appropriate. You're bound to be inappropriate—Lynne Tillman

The movement of a ship in water produces oscillation in three axes. The ship's motion is generated directly either by the water and wind influences, put together called wave motion, and otherforces that act against the hull, which are mechanical, platform based; or indirectly by the inertia of the ship while turning, accelerating or coming to a stop. Surge is the motion along the fore-and-aft*X*-axis. Sway is the motion along midships on the transverse *Y*-axis. Heave is the linear motion along the vertical *Z*-axis. Rolling is a rotation around the *X*-axis, pitching is a rotation around the *Y*-axis and yawing is a spinning top-like rotation around the *Z*-axis (Fig. 12.1).

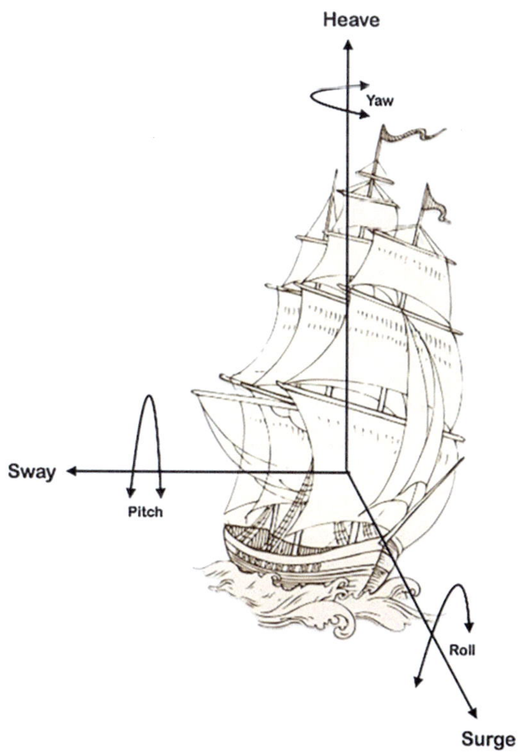

Fig. 12.1 The axes of motion on a ship

Seasickness is a physiological response to motion in anyone with an intact and functional vestibular system [8, 9]. Change in the motions in any of the axes, or rotational effect or a combination of these can cause sickness;thus the basis of motion sickness is the repetitive, persistent or distressing acceleration of the vestibular system [10].

Pathophysiology of Seasickness—'If Qi is insufficient above, the brain is not sufficiently filled by it, the ears suffer a ringing noise, the head is bent low by it, the eyes experience dizziness'—Inscription in The Huangdi Neijing

The motion-sensitive inputs to the central nervous system originate from different sensory systems viz. vestibular, visual and somatosensory. But they are not self-cancelling signals. Vision, for example, cannot distinguish the effect of perceived motion from the actual motion (akin to the illusory movement when looking at a moving train across the window) and becomes less reliable in darkness. Darkness is akin to a lack of a point of reference at a distance, say the horizon. Semicircular canals, working as gyroscopes, record changes in angular velocity in the vestibular system, but they are incapable of registering constant-velocity rotation. The otoliths behave as accelerometers, measuring the direction of accelerations, but they cannot differentiate between gravity and inertial forces. In the brainstem and cerebellum, all sensory signals converge, to be weighed according to their reliability and 'loudness' in a balance optimised by evolution to obtain the best estimate of our natural self-propelled motion. However, passive artificial motion (experienced in a transport system) induces unnatural motion inputs that lead to combinations of sensory signals that confuse our brains. Since each sensory signal is interpreted by CNS as resulting from a different natural motion, our brain tries but fails to solve a confusing problem of conflicting information, termed the sensory conflict (Fig. 12.2).

The theory of CNS-sensor mismatch as the aetiology of seasickness is further influenced by other factors: females suffer from seasickness 1.7 times more than males [11]. Pregnancy makes women more vulnerable to seasickness. Very young age bestows high tolerance to motion sickness, with cases rare below the age of two. The susceptibility appears to be highest in the decade starting second year of age. Tolerance gradually increases with every year added to age [12, 13].

Sleep deprivation exacerbates seasickness. In the maritime environment, adequate rest is a rarity; sleep is an important component of a healthy physiological state and a major determinant of crew performance at sea [14]. Conditions aboard a vessel are often not conducive to satisfying rest and sufficient sleep. Sleep deficiencies are commonly reported because of the ship's disturbing motion, noise, vibration and lack of privacy [15]. Sleep inadequacy is one of the significant risk factors for crew fatigue [16] and in the development of seasickness because it interferes with the vestibular system conditioning.

A person's personality and past memories of distressing sufferings at sea seem to have a bearing on future outlook towards overcoming their symptoms and willingness to face provoking motion stimuli [17]. Some studies suggest that motion sickness tends to be worse in introverts [18]. Genetic factors play a role in the formation of susceptibility to seasickness too [19].

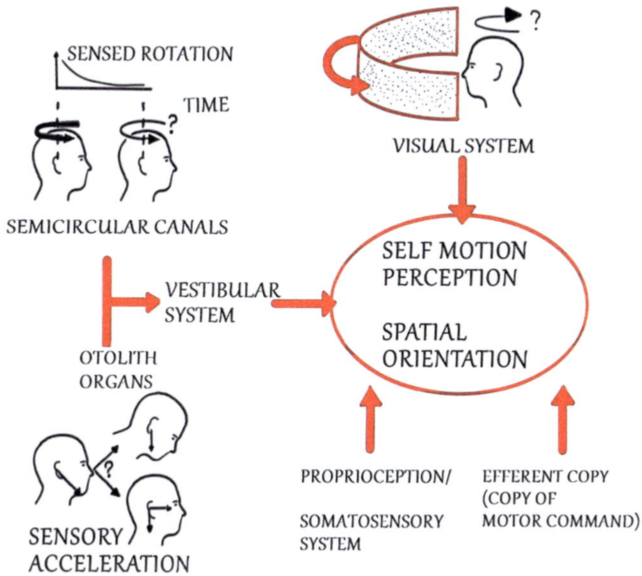

Fig. 12.2 Conflict of sensory inputs: the cocktail of motion sickness induction

Recent progress has been in identifying brain mechanisms and loci associated with motion sickness and nausea per se. Modern tools include conventional neurophysiological battery, pathway mapping, functional MRI and impulse stimuli imaging; and studies have implicated multiple regions including the cerebral cortex, brainstem and cerebellum.

12.1 Is Susceptibility to Seasickness Predictable?

How soon, how long and how seriously a person gets sick will depend on a combination of the frequency and acceleration characteristics of the vector in the three axes. Lawther and Griffin studying a wider range of provocative frequencies, proved that humans are most affected by vertical oscillations in the frequency range of 0.1–0.5 Hz (6–30 cycles per minute) with a peak around 0.16 Hz [20]. Within a given frequency, motion sickness increases with increasing acceleration [21, 22]. Golding and colleagues correlated fore and aft and vertical acceleration of similar magnitudes and frequencies and concluded that the first causes stronger motion sickness than the vertical counterpart [23]. Other studies have shown that equivalent nausea responses are evoked by oscillations in both the *X* and *Y* axes [24–30].

Sensitivity to develop motion sickness depends on the rate of change of the dynamics of motion. Reason and Brand [31] described three features that affect this rate change: receptivity, adaptability and retentiveness. Receptivity refers to a person's internal amplification of the motion trigger or the range of motion stimuli that produce a response. In some, a given motion stimuli may be significantly damped, while in others the signal may be amplified. Adaptability refers to the capability of an exposed individual to accept the motion, endure it better and have fewer symptoms. This is often referred to as the process of getting one's 'sea legs'. Receptivity and adaptability need not be related; a highly receptive person may soon adapt—quickly enough to reduce sensory dysbalance before the nausea threshold is reached. Retentiveness refers to a person's ability to retain and register the internal model of motion and adapt to the environment with repeated exposures, i.e. the case of returning to sea after several days in port and experiencing no seasickness.

12.2 How to Address the Problem of Motion Sickness in Ships

Researchers have [32] described Motion Sickness Incidence (MSI) and Motion Induced Interruptions (MII) criteria which quantify the trigger-response. The North Atlantic Treaty Organisation Standard Agreement 4154 includes both the older 'Angular and Acceleration Limits' as well as the more modern MSI and MII criteria. The MSI model is based on the theory of McCauley and O'Hanlon, who attributed seasickness primarily to yaw and heave components of ship motion [33]. The MSI model, however, does not take into consideration the habituation associated with experience at sea. A rate of 20% of a ship's crew, subjected to exposure in a 4-h watch, is considered a reasonably acceptable MSI. This figure has been arrived at by studying subjects with little or no experience at sea. For the calculated MSI, a two-fold increase in the magnitude of the motion, or a fourfold increase in exposure duration, has the effect of doubling the predicted incidence of vomiting [34]. These parameters are essential human habitability and engineering considerations in shipbuilding.

How do we address a problem as complex as this? Multifactorial causes would require multiple solutions, combined effectively. In this case, it would involve the environment (the ship itself, its operations, design changes and crew comfort measures), the susceptible patient (prevention and care of sickness using medical science) and organisational measures.

12.2.1 Design-Based Solutions

There are many design solutions to motion-induced difficulties aboard ship (handholds, anti-skid decks, raised edges on table surfaces, enveloping restraints like rings around a container, and even railings-restraints on bunk beds, etc.). All of these engineering solutions have been used aboard ships for centuries. However, these represent attempts only to cope with motion while performing gross motor tasks and may have little or no effect on fine-motor and cerebral tasks. In other words, these may reduce fall risk or physical injury hazard, but the suffering, ill feeling of the stomach churning or dizziness would continue to trouble the crew.

Motion stabilisers in ships include bilge keels, anti-roll fins and ballast tanks. These devices are used to reduce rolling. In addition, such naval design imperatives as prevention of movement of cargo, improvement in the accuracy of targeting with weapons and reduction of the stress on the vessel hull are also important considerations behind including 'stabilisers' in the ship design. One would reasonably expect that stabilisers would also alleviate motion sickness, in addition to reducing the number of MII experienced. Alas, as Griffin [22] indicates, there appears to be little evidence that damping the rollwould directly translate to mitigating motion sickness. Some stabilisers may produce jerky and unpredictable contra-motions, which may contribute to sickness instead of reducing it. Most importantly, however, roll stabilisers have little effect on motions in the other axes—heave and pitch. There is a need to more fully understand the effects of roll and pitch motions on motion sickness so as to deploy appropriate roll stabilising devices.

A ship would navigate keeping economical speed and optimal navigational safety in consideration. Avoidance of sailing, altering the course to a more placid condition or simply taking a break from watches are not envisaged! Crew comfort, unless it has a commercial gain, would always be subsumed by these important interests, especially in the operational tasking of Navies. But ship designers can't be blamed if scientific literature is unsure as to how much motion in which axis is actually related to a seasickness. Alas, as with many illnesses, lack of accuracy of predictable stimuli will preclude the development of mitigating design.

Thus, there is a need to include human factor analyses of ship arrangements. Bittner and Guignard [35] evaluated two mission-critical workstations for USCG from a human factors perspective. Their study identified five potential engineering approaches to enhance seakeeping through the prevention and mitigation of adverse motion effects on personnel. They are:

1. Locate critical stations near the ship's effective centre of rotation to reduce the metacentric height of the axes of movement.
2. Minimise head movements while at the watch station by designing the panels/workstations, and design of visual display terminals.
3. Align the operator with a principal axis of the ship's hull. The further away from the centre, the larger the amplitude of the swing.
4. Reduce provocative factors by optimising the layout of sleeping quarters to improve sleep during rough seas, improve ventilation and temperature control and lastly,
5. Provide an external visual frame of reference to the sea outside, particularly the horizon, by using even an artificial 3D Earth-fixed visual reference if natural viewing of the sea is not possible.

Research is pursued increasingly to quantify motion sickness dose value (MSDV), by complex observational parameters, so as to objectively predict the motion sickness burden on the crew in various sea states [36].

12.2.2 Pharmacological

The two most common groups of anti-seasickness drugs are anticholinergics and antihistamines.

12.2.3 Anticholinergics

Anticholinergics have antagonistic action on the parasympathetic nervous system and block muscarinic receptors, inhibiting cholinergic transmission from the vestibular nuclei to the vomiting centre. They are shorter acting making them more appropriate for shorter journeys.

Transdermal therapeutic system and intranasal spray containing scopolamine are also often used. Freedom from oral administration is indeed welcome when the crew tends to vomit! The patches are effective up to 3 days and are used both preventively and for long-term therapy. There are no prospective randomised studies on the patch's effect on motion sickness, nor any substantive comparative study with other drugs and delivery routes. However, the authors of a Cochrane review concluded that scopolamine, while not being superior to antihistamines or combination drugs in combating seasickness, does have fewer adverse effects, and can be considered suitable on that count!! [41]. The adverse effects derive from its anticholinergic properties and include, especially, mucosal dryness, mydriasis, palpitations, urinary retention and, very rarely, hallucinogenic effects, which are enough to put navy on alert while considering prescribing scopolamine to an active crew member [42]. All of these medicines have in common that they do not totally prevent or suppress seasickness and are perhaps completely acceptable to treat lay passengers but not the ship's crew.

12.2.4 Antihistamines

A typical H1- antagonist is dimenhydrinate. Other than that the first-generation drugs cyclizine, meclizine and promethazine are also used.Dimenhydrinate is often prescribed in combination with cinnarizine for short-term acute therapy. This H1-antagonist also acts as a dopamine, serotonin and bradykinin receptor antagonist, as well as as a calcium inhibitor. This is believed to result in synergistic effects in the vestibular apparatus [37].

There is considerable individual variation in response to antihistamines, some seem to be completely rendered sleepy to be of any use on a watch. Typical adverse effects of the antihistaminics, seen in about 5% of all who take these medicines, are fatigue, slowed reactions, drowsiness, impaired coordination and concentration [38]. As authors have pointed out [39, 40], the unpredictable nature of individual susceptibility to the adverse effects of these medications may not be acceptable when the user is operating sophisticated or potentially hazardous equipment, or making command and control type decisions.

12.2.5 Other Pharmacological Agents

Most experts agree that sympathomimetics and dopamine antagonists do not demonstrate any advantage over the above options, especially for prolonged usage. 5-HT 3 antagonists like ondansetron may help in faster gastric emptying reducing the quantum of vomitus and associated social issues in the workspace.

Apart from these drug therapies, natural remedies are also used to suppress seasickness. A placebo-controlled study using ginger as an antiemetic showed few adverse effects [43]. The active ingredients contained are believed to act as antagonists at the 5-HT 3 receptor, which has an important role in the vomiting centre. It is questionable if culturally, ginger, with its sharp taste and pungency, is acceptable to everyone, but if you like it spicy you might probably have one more cure for seasickness to try!!

High-dose vitamin C, another remedy that has been credited with some antihistaminic effect, was shown in a prospective, double-blind, placebo-controlled study to reduce the symptoms of seasickness without identifiable adverse effects [44]. Studies have proposed safe and delicious methods of using spearmint, chamomile, lemon and other naturally occurring plant-based substances to treat nausea [45]. If not completely curative, they would certainly provide an interesting concoction to sip while suffering!!

Though not pharmacological, and in addition to being contrary to one's intuition, eating is actually believed to prevent motion sickness. This is explained by postulations that say eating reduces the type of retrograde gastric motions that are typically associated with motion sickness [46]. Bles et al. [47] in their study conducted on the training vessel *MV Zeefakkel* proved that motion sickness is seen reduced in the afternoons after the subjects had consumed lunch, and it was hypothesised that vectin-induced gastric myoelectric dampening was at least partially responsible. Griffin [22] also advises that going without food for long periods can result in lessened resistance to motion and weaken the effort to adopt preventative measures.

Finally, consumption of alcohol is considered a temporary predisposing factor that should be limited prior to encountering a moving environment [17]. Drinks are best kept for parties in harbours!

12.2.6 Desensitisation and Biofeedback

In desensitisation therapy, the subjects are subjected to increasing motion stimulus over time with concurrent administration of psycho-therapeutic treatments to help allay their fears and anxiety. In the Royal Air Force, it is reported that this procedure effectively prevents the recurrence of air sickness upon a return to flying duties in 85% of those who have undergone desensitisation [11]. An optimum desensitisation program would entail identifying the appropriate combinations of stimuli actually felt at sea, and then exposing the susceptible crew to these combinations.

In Biofeedback (also known as autogenic feedback), the subject learns to acquire voluntary control over certain autonomic responses that characterise motion sickness, allowing 'at will' attenuation, thus increasing their tolerance to the provocative motion. The technique involves both visual and auditory feedback of several autonomic responses (e.g., pulse, respiratory rate, pulse pressure, sweating and galvanic skin response) that occur during motion sickness. Subjects are oriented toward controlling some of the responses or even reducing them [48]. Jones et al. [49] reported a 79% reduction in symptomatology among 53 aircrews of USAF treated for airsickness. However, other studies indicate that this method of treatment is not completely effective due to the inability to apply the biofeedback skills of ANS self-regulation while suffering in the distracting and difficult motion environment [50].

12.2.7 Behavioural

Keeping one's head steady is indeed a virtue!Reducing head movements to a minimum results in a reduction of excitatory stimulation of the vestibular system and may prevent its provoking [51]. Stott [47] postulates that an oscillatory motion stimulus is most well tolerated in a posture that demands the least corrections while maintaining a stable balance, i.e. recumbent position. Many sailors know the comforts of conducting watches from the forbidden bunks, but that would come with incurring the unacceptable wrath of the Officer on Deck! These measures, while theoretically effective, have limited applicability to the crew on board a ship.

A crew member who is motivated to further his career in the Navy, also positively conditions himself to service at sea, and this helps not only to keep returning to sea cheerfully but also assists his adaptation and retention.

12.3 Summary

Exposure to unfamiliar motion on a ship produces seasickness.Susceptibility to seasickness shows a large degree of variation within and between individuals and usually follows a reliable pattern of progression of symptomatology before vomiting, which offers transient relief from discomfort.

Seasickness is dependent on the vestibular apparatus of the inner ear. Having said that, symptoms result from a conflict of sensory information regarding body orientation and motion. Normal locomotor sensation produces an accustomed pattern of inputs from visual apparatus, the inner ear as well as gravireceptors of the musculoskeletal system. When there is a sustained discordance between these inputs, sensory mismatch sets in with spillage of an electrical impulse to the vomiting centre. Two main categories of neural mismatch have been identified, representing a conflict between visual and vestibular receptors or a mismatch between semicircular canals and the otoliths.

The incidence of motion sickness in a particular set of conditions can be traced to a combination of factors including the physical characteristics of the motion, innate susceptibility, and training for acclimatisation. Prevention may involve the selection and training of individuals, the use of technology to reduce provocative stimuli, ship engineering to mitigate seasickness and neuropsychological measures. Pharmaceutical agents have been found to offer some relief but may produce side effects unacceptable while performing duties at sea.

References

1. Johnson B. The Spanish Armada, 1588. https://www.historic-uk.com/HistoryUK/HistoryofEngland/Spanish-Armada/
2. Cha YH, Golding JF, Keshavarz B. Motion sickness diagnostic criteria: Consensus Document of the Classification Committee of the Bárány Society. J Vestib Res. 2021;31(5):327–44.
3. Pethybridge RJ. Sea sickness incidence in Royal Navy ships. Gosport: INM Report 37/ 82, Institute of Naval Medicine; 1982.
4. Dobie TG. The importance of the human element in ship design. Arlington VA, June 13: Presented at the Ship Structure Symposium; 2000. http://www.shipstructure.org/
5. Pethybridge RJ, Davies JW, Walters JD. A pilot study on the incidence of sea sickness in RN personnel on 2 ships. Alverstoke, Hampshire: INM Report 55/78, Institute of Naval Medicine; 1978.
6. Bruner JMR. Seasickness in a destroyer escort squadron. US Armed Forces Med J. 1955;6(4):469–90.
7. Gupta AK, Vijaya Kumar B, Rajguru R, Parate KD. Indian J Occup. Environ Med. 2021;25(2):119–24.
8. Mallinson AI, Longridge NS. Motion sickness and vestibular hypersensitivity. J Otolaryngol. 2002;31(6):381–5.
9. Kennedy RS, Graybiel A, McDonough RC, Beckwith FD. Symptomatology under storm conditions in the North Atlantic in control subjects and in persons with bilateral labyrinthine defects. Acta Otolaryngol. 1968;66(6):533–40. https://doi.org/10.3109/00016486809126317.
10. Bertolini G, Straumann D. Moving in a moving world: a review on vestibular motion sickness. Front Neurol. 2016;7:14. Published Online 2016 Feb 15. https://doi.org/10.3389/fneur.2016.00014.
11. Benson AJ. Motion sickness. In: Ernsting J, Nicholson AN, Rainford DS, editors. Aviation medicine. Oxford: Butterworth-Heinemann; 1999. p. 318–38.
12. Lawther A, Griffin MJ. A survey of the occurrence of motion sickness amongst passengers at sea. Aviat Space Environ Med. 1988;59:399–406.
13. Wertheim AH. Working in a moving environment. Ergonomics. 1998;41(12):1845–58.
14. Miller NL, Matsangas P, Kenney A. The role of sleep in the military: implications for training and operational effectiveness. In: Laurence H, Matthews MD, editors. The Oxford handbook of military psychology. Oxford: Oxford University Press; 2012. p. 262–81. https://doi.org/10.1093/oxfordhb/9780195399325.013.0084.
15. Colwell J. Modeling ship motion effects on human performance for real time simulation. Nav Eng J. 2005;117:77–90.
16. Smith A. (2007). Adequate crewing and seafarers' fatigue: The international perspective.
17. Guedry FE. Factors influencing susceptibility: individual differences and human factors. AGARD (Advisory Group for Aerospace Research and Development) Lecture Series 175: Motion Sickness: Significance in Aerospace Operations and Prophylaxis; 1991.
18. Kottenhoff H, Lindahl L. Laboratory studies on the psychology of motion-sickness. Acta Psychol. 1960;17:67–73.
19. Reavley CM, Golding JF, Cherkas LF, Spector TD, MacGregor AJ. Genetic influences on motion sickness susceptibility in adult women: a classical twin study. Aviat Space Environ Med. 2006;77(11):1148–52.

20. O'Hanlon JF, McCauley ME. Motion sickness incidence as a function of the frequency and acceleration of vertical sinusoidal motion. Aerosp Med. 1974;45(4):366–9.
21. McCauley ME, Royal JW, Wylie CD, O'Hanlon JF, Mackie RR. Motion sickness incidence: exploratory studies of habituation, pitch and roll and the refinement of a mathematical model. Goeleta, CA: Human Factor Research Inc. Technical Report; 1976. p. 1733–2.
22. Griffin MJ, editor. Handbook of human vibration. London: Academic Press; 1990.
23. Golding JF, Markey HM, Stott JR. The effects of motion direction, body axis, and posture on motion sickness induced by low frequency linear oscillation. Aviat Space Environ Med. 1995;66(11):1046–51.
24. Joseph JA, Griffin MJ. Motion sickness: effect of changes in magnitude of combined lateral and roll oscillation. Aviat Space Environ Med. 2008;79(11):1019–27. https://doi.org/10.3357/ASEM.2196.2008.
25. Beard GF, Griffin MJ. Discomfort caused by low-frequency lateral oscillation, roll oscillation and roll-compensated lateral oscillation. Ergonomics. 2013;56(1):103–14. https://doi.org/10.1080/00140139.2012.729613.
26. Golding JF, Mueller AG, Gresty MA. A motion sickness maximum around the 0.2 Hz frequency range of horizontal translational oscillation. Aviat Space Environ Med. 2001;72(3):188–92.
27. Wyllie IH, Griffin MJ. Discomfort from sinusoidal oscillation in the roll and lateral axes at frequencies between 0.2 and 1.6 Hz. J Acoust Soc Am. 2007;121(5 Pt1):2644. https://doi.org/10.1121/1.2715654.
28. Donohew BE, Griffin MJ. Motion sickness with fully roll-compensated lateral oscillation: effect of oscillation frequency. Aviat Space Environ Med. 2009;80(2):94–101. https://doi.org/10.3357/ASEM.2345.2009.
29. Griffin MJ, Mills KL. Effect of frequency and direction of horizontal oscillation on motion sickness. Aviat Space Environ Med. 2002;73(6):537–43.
30. Golding JF, Markey HM. Effect of frequency of horizontal linear oscillation on motion sickness and somatogravic illusion. Aviat Space Environ Med. 1996;67(2):121–6.
31. Reason JT, Brand JJ. Motion sickness. London: Academic Press; 1975.
32. Baitis AE, Applebee TR, McNamara TM. Human factors considerations applied to operations on the FFG-8 and LAMPS MK-III. Nav Eng J. 1984;96(3):191–9.
33. O'Hanlon JF, McCauley ME. Motion sickness incidence as a function of the frequency and acceleration of vertical sinusoidal motion. Aerosp Med. 1974;45(4):366–9.
34. Stevens SC, Parsons MG. Effects of motion at sea on crew performance: a survey. Mar Technol. 2002;39(1):29–47.
35. Bittner AC, Guignard JC. Human factors engineering principles for minimizing adverse ship motion effects: theory and practice. Nav Eng J. 1985;97(4):205–13. https://doi.org/10.1111/j.1559-3584.1985.tb01355.x.
36. Taylor NC, Bekker A, Kruger K. The operational development of diagnostic seasickness criteria through a human-cyber-physical system. Appl Ergon. 2024;119:104316. https://doi.org/10.1016/j.apergo.2024.104316. Epub 2024 May 28. PMID: 38810325
37. Koch A, Cascorbi I, Westhofen M, Dafotakis M, Klapa S, Kuhtz-Buschbeck JP. The neurophysiology and treatment of motion sickness. Dtsch Arztebl Int. 2018;115(41):687–96. https://doi.org/10.3238/arztebl.2018.0687. PMID: 30406755; PMCID: PMC6241144
38. Scholtz AW, Schwarz M, Baumann W, Kleinfeldt D, Scholtz HJ. Treatment of vertigo due to acute unilateral vestibular loss with a fixed combination of cinnarizine and dimenhydrinate: a double-blind, randomized, parallel-group clinical study. Clin Ther. 2004;26:866–77.
39. Golding JF, Gresty MA. Pathophysiology and treatment of motion sickness. Curr Opin Neurol. 2015;28:83–8.
40. Shupak A, Gordon CR. Motion sickness: advances in pathogenesis, prediction, prevention, and treatment. Aviat Space Environ Med. 2006;77:1213–23.
41. Spinks A, Wasiak J. Scopolamine (hyoscine) for preventing and treating motion sickness. Cochrane Database Syst Rev. 2011;6:CD002851.

42. Rubner O, Kummerhoff PW, Haase H. An unusual case of psychosis caused by long-term administration of a scopolamine membrane patch paranoid hallucinogenic and delusional symptoms. Nervenarzt. 1997;68:77–9.
43. Grøntved A, Brask T, Kambskard J, Hentzer E. Ginger root against seasickness. A controlled trial on the open sea. Acta Otolaryngol. 1988;105:45–9.
44. Jarisch R, Weyer D, Ehlert E. Impact of oral vitamin C on histamine levels and seasickness. J Vestib Res Equilib Orientat. 2014;24:281–8.
45. Khorasani F, Aryan H, Sobhi A, Aryan R, Abavi-Sani A, Ghazanfarpour M, et al. A systematic review of the efficacy of alternative medicine in the treatment of nausea and vomiting of pregnancy. J Obstet Gynaecol. 2020;40(1):10–9. https://doi.org/10.1080/01443615.2019.1587392. Epub 2019 Jun 19. PMID: 31215276
46. Uijtdehaage SH, Stern RM, Koch KL. Effects of eating on vection-induced motion sickness, cardiac vagal tone, and gastric myoelectric activity. Psychophysiology. 1992;29:193–201.
47. Bles W, de Graaf B, Keuning JA, Ooms J, de Vries MH, Wientjes CJE. Experiments on motion sickness aboard the M.V. "Zeefakkel". Report IZF 1991 A-34, Soesterberg, NL- TNO Institute for Perception; 1991.
48. Stott JRR. In: Crampton GH, editor. Adaptation to nauseogenic motion stimuli and its application in the treatment of airsickness, in motion and space sickness. Boca Raton: CRC Press; 1990. p. 373–90.
49. Jones DR, Levy RA, Gardner L, Marsh RW, Patterson JC. Self-control of psychophysiologic response to motion stress: using biofeedback to treat airsickness. Aviat Space Environ Med. 1985;56(12):1152–7. PMID: 3910020
50. Jozsvai EE, Pigeau RA. The effect of autogenic training and biofeedback on motion sickness tolerance. Aviat Space Environ Med. 1996;67(10):963–8. PMID: 9025819
51. Bittner AC Jr, Guignard JC. Human factors engineering principles for minimization of adverse ship motion effects: theory and practice. Nav Eng J. 1985;97(4):205–13.

Challenges and Opportunities for Soldiers Operating in Jungles

13

Sibnarayan Datta and Dev Vrat Kamboj

Abstract

Soldiers need to carry out both defensive and offensive operations in a wide variety of terrains, ranging from the depth of seas to high-altitude mountains. Of the different terrains, jungles are considered one of the most challenging environments to operate due to the diversity of prevailing conditions. On the one hand, jungles present hostile conditions in the forms of dense vegetation, hot and humid climate, biting and attacking animals, disease-carrying vectors, etc. On the other hand, many of the features of the jungles also provide ample opportunities for camouflage, protective shelters, protection from enemies and animals, and food and drink during emergency situations, which soldiers can use during jungle operations. In this chapter, we discuss these challenges and opportunities. In addition, we also discuss the R&D opportunities that can enhance the combat efficiency of the soldiers during jungle operations.

Keywords

Jungle operations · Tropics · Rainforests · Camouflage · Vector-borne diseases

13.1 Introduction

The word '*jungle*' originates from the Sanskrit word '*Jaṅgala*', which broadly referred to regions with harsh, arid or inhospitable environmental conditions. Currently, jungle denotes large areas of land dominated by huge trees, dense

S. Datta · D. V. Kamboj (✉)
Defence Research Laboratory (DRL), Defence Research and Development Organisation (DRDO), Tezpur, Assam, India
e-mail: dvkamboj.drl@gov.in

S. Srivastava et al. (eds.), *Health and Sustenance in Extreme Environment*,
https://doi.org/10.1007/978-981-96-8642-1_13

impenetrable underbrush, harshly hot and humid environment, weird noises, and a home to several hostile animals, all of which elicit a deterring, unpleasant sense of fear. Apart from adventure-seeking communities and forest security forces, soldiers are required to conduct defensive and offensive operations in the jungles.

In general terms, jungle operations denote deep penetrations of densely vegetated environments by small combat units having specialized training and tactics to conduct defensive and offensive missions. As per the classical field manual on jungle warfare [1], soldiers fight two enemies in the jungle—man and nature. Tropical jungles feature dense vegetation, gusty rivers, rough terrain, very low visibility, hot and humid environment, sudden and drastic change in climatic conditions, numerous attacking, biting animals, sharp background noises (insect buzz, animal calls, gushing wind, etc.), and odour of the wet and decaying vegetation, which together pose serious obstacles in carrying out military operations [2, 3].

Therefore, under jungle operations, strong willpower and patience are considered the most important weapons of a soldier against the enemy during a tactical mission [4, 5]. These specialist soldiers are trained to face and overcome the obstacles met in the jungles and to use the jungle as an ally. Even though soldiers are well trained to acclimatize to the jungle environment, imparted with skills for survival, jungles may present a unique set of challenges to the soldiers and may also offer certain opportunities as well, which are not available in other terrains [5]. In this chapter, some of these challenges and opportunities are discussed.

13.2 The Jungle Environment

The word 'jungle' is commonly used to refer to forests situated in the tropics. The region of Earth flanking both sides of the Equator and between the *Tropic of Cancer* and the *Tropic of Capricorn* is referred to as the *Tropics*, while some parts of the temperate zones in both hemispheres near the *Tropic of Cancer* and the *Tropic of Capricorn* is commonly termed as subtropics [6]. Tropics/subtropics hosts wide diversity of geographical feature zones ranging from snow-capped mountains and forests to swamps and barren deserts. The tropics receive more sunlight and heavy rainfall (more than 1000 cm annually) than other parts of the globe and therefore feature hot (25–35 °C), humid (85–90%) climatic conditions that support abundant growth of vegetation throughout the year [1–6].

Like the other biomes, the tropics cover approximately 6–8% of the Earth's surface and have unique environmental conditions in Asia, Africa, Oceania, and Central and South America. Due to heavy rainfall, most of the jungle regions feature deep rivers that become raging and torrent during the rainy seasons frequently causing flash floods, large water bodies, swamps, numerous ridges, and riverine networks, resulting in a complicated terrain [7]. Additionally, forests are characterized by fine-grained clay-like upland soil, with silty soil in lowlands, an abundance of rapidly growing flora. Often larger trees are connected by dense and thick networks of vines, a characteristic feature of rainforests. Another factor that makes the jungle environment extremely uncomfortable is the absence of air circulation. Even though the environmental conditions greatly differ between jungles in different climatic regions, the presence of one or more of these features poses considerable difficulty in movement and operations.

13.3 Terrain Features of Different Jungle Types

Depending upon the prevailing climatic conditions in different geographical regions, various jungle types are observed. In the following section, a broad classification of diverse types of jungles important for military operations is presented for a general understanding.

13.4 Evergreen Rainforest

Primary evergreen rainforests in the lowlands are characterized by a dense assemblage of evergreen flora with a connected 'canopy' of tree branches and dense woody vines with broad leaves that obstruct the sun from reaching the forest floor, resulting in a relatively limited abundance of shade-tolerant, shrubs. Due to heavy to very heavy rainfall and low penetration of sunlight, the forest always remains humid and the forest floor is damp and slippery throughout the year. In comparison, highland forests (also termed '*moss forests*' or '*cloud forests*') demonstrate extreme variation in terms of climate and vegetation, especially in the hilly terrains situated at an altitude of about 1000–3000 m, near the equator. These forests are characterized by the domination of large trees along with the thick growth of mosses and ferns. Due to the presence of rocky, steep, and sloppy terrains, often covered with mosses and heavy slime, movement is extremely hazardous. Despite being utterly uncomfortable terrain to deal with, such a forest provides an advantageous ground for guerrilla operations [4, 5].

On the other hand, secondary evergreen rainforest are characterized by layered vegetation including relatively lower and open top canopy of tree branches, complex mass of vines, shrubs and bushes forming the intermediate canopy, and thick mat of grass on the forest floor. Due to the presence of a tangled mass of this intermediate canopy, these forests are most difficult to penetrate. In contrast to the primary evergreen rainforests, secondary evergreen rainforests are relatively more illuminated by sunlight, but the horizontal visibility is significantly obstructed due to dense undergrowth. Despite having somewhat lower humidity, higher temperatures make these forests uncomfortable.

13.4.1 Swamps

Swamps are often connected with evergreen rainforests in the coastal areas that are inundated with tidal flooding. Most of the swamp forests are characterized by standing saline water of different depths, extensive entangled network of closely spaced mangrove vegetation with sharp protruding '*pneumataphores*' or specialized breathing roots [1]. In the relatively less-saline inland margins, swamps may harbour diverse vegetation including dense vines, which may hinder horizontal visibility. Even though these terrain features greatly restrict movement across the terrain, the same features also provide suitable grounds for covert and underwater special operations. Apart from the mangrove swamps, other types of swamps such as '*Palm Swamp*', characterized by dominant open canopy palm-type vegetation along with mixed jungle vegetation are seen at low to high flooded grounds.

13.4.2 Marsh

Marshy forests are met in regions where water is less brackish to support mangrove vegetation, and too watery for common forest trees. In general, marshes harbour thick growth of tough and tall masses of reed plants and therefore provide a suitable place for ambushes.

13.4.3 Dry Season Deciduous Forest

Dry deciduous forests are present in tropical regions experiencing long-lasting dry seasons receiving very little or no precipitation. As opposite to the evergreen rainforests, the predominant vegetation in these forests is deciduous in nature and sheds leaves at a particular time of the year. The undergrowth in dry deciduous forests depends upon the opening of the upper canopy and may feature extremely dense formation of shrubs and vines with sharp thorns and knife-like edges, saw-toothed grasses. Secondary vegetation, temperature, and humidity fluctuate depending on the prevailing climatic conditions and may resemble evergreen rainforests during the rainy season.

13.4.4 Bamboo Forests

Bamboo belongs to the hollow-stemmed grass family that is frequently found across the wet tropics. Bamboo species found in different regions vary distinctly in terms of height, often reaching several meters and growing densely over a large area of land. Making trails across Bamboo Forests for movement through is extremely difficult and noisy.

13.5 Characteristics of Jungle Operations

Any form of military operation is broadly termed a jungle operation. Although the term jungle operations is often interchangeably referred to as jungle warfare, jungle operations refer to both combat and non-combat operations, while jungle warfare deals with the tactics associated with combat operations. Colonel John M. Collins (USA), in his famous book *'Military Geography for Professionals and the Public'* [6] has mentioned the following 10 distinct characteristics of operations in jungle situations:

(i) small or medium units including special forces usually conduct jungle operations
(ii) infantry foot soldiers are better suited for jungle operations, as compared to mechanized units
(iii) special forces can use the natural features for cover and ambushes
(iv) support from heavy weapons or air support is generally limited
(v) airstrips, even if available, are small and difficult to operate
(vi) movement is slow due to thick vegetation, rivers, and other water bodies
(vii) soldiers need to be fully self-contained with communications, navigation, and logistics
(viii) soldiers operate under low-visibility conditions, even during daytime
(ix) navigation through jungle terrains require non-conventional techniques
(x) adaptability of the troops to the prevailing conditions is more important than the technology or the numbers

Jungle operations are quite distinct from conventional military operations, as jungle operations do not involve formal battle, but comprise raids, ambushes, and violent contacts. It is therefore important for the soldiers to well understand the challenges and the opportunities that jungles offer, to survive and fight in such scenarios.

13.6 Challenges in Jungle Operations

13.6.1 Climate

As mentioned earlier, tropical climate, especially in jungles is quite unpleasant. Tropics experience the most fluctuating climatic conditions, as compared to any other region of the Earth. In summer, the presence of high humidity makes the heat feel worse, as compared to the same heat in dry climatic regions. On the other hand, during winter, the considerable period also offers chilly conditions, especially in high-altitude regions. Similarly, the tropics experience the heaviest rainfall, often continuously, when the sun does not shine for days. Such a condition often led to flash floods and waterlogging.

13.6.2 Movement

Jungles due to their vegetation, climatic conditions and terrain features pose obstacle to the movement of the troops or vehicles. Due to the absence of heavy equipment, soldiers must improvise with available resources, while faced with distinct jungle features obstacles including rocks, rivers, trenches, etc. Dense shrubs and bushes, especially those having thorns or woody edges may penetrate protective clothing and inflict sharp injuries to incautious troupers, while vines and creepers may entrap and trip an experienced person. Such sharp injuries, if not attended properly, may easily get infected with bacterial or fungal organisms prevalent in humid jungle conditions. Conditions in jungles are especially difficult during or just after rainfall. In jungles with dense vegetations, troops are often recommended to follow the banks of streams, movement in these areas is also challenging due to the soft, slippery soil, and rocky riverbeds, especially during night-time when most of the special operations are conducted [1, 3, 4]. Use of trails, even if present, may not be permissible, especially during covert operations, as such routes are expected to be monitored by the enemies and may often lead to fierce ambushes. Under such offensive circumstances, soldiers must stay extremely alert for traps, snipers, and hidden enemy positions [1, 8, 9]. Alternatively, valleys and ridgelines may provide a more secure way but movement through such terrains is slower and tougher for soldiers carrying heavy equipment and ammunitions. Owing to the above-mentioned reasons, transportation of heavy military vehicles and equipment is also restricted. Transportation by pack animals and or on-foot soldiers is the choice of mode of transportation in most of the jungle conditions [3, 4, 10]. Often rivers are used for downstream movement.

13.6.3 Navigation

Another problem associated with jungle movement, especially during important missions is associated with navigation and estimation of distances. Jungles rarely do have reference points to serve as checkpoints or navigational landmarks. Even though GPS-based devices are currently being used, in certain terrains soldiers must resort to classical methods of footstep counting and magnetic compasses for navigation and distance travelled estimation [4, 6, 7]. Nevertheless, soldiers must spend a significant amount of energy to travel relatively short distances over a considerable time span.

The distance travelled is significantly different in diverse terrains, and is dependent upon various other factors such as the prevailing weather conditions, physiological status of the troops, equipment load, etc. [4, 5]. For planning missions, distance is calculated on the scale of time, where the average movement rate is about 1 km/h in lowland primary evergreen forests, while the half distance can be covered in the same time limit in a secondary evergreen forest. This rate is further slowed (100–500 m/h) in mountainous terrains and swamps, while extremely slow to almost standstill in densely growing bamboo woods.

13.6.4 Visibility

As mentioned in the preceding sections, horizontal visibility in humid tropical jungles depends upon the vegetation. In primary evergreen rainforests, due to the closed canopy of tree branches, a condition of semidarkness prevails throughout the day.

Although undergrowth is quite less in these forests, the lack of sunlight and the relatively larger diameter of the primary vegetation greatly reduce horizontal visibility. In addition, such conditions provide ideal grounds for entrapment and ambush by hostile forces having ability in guerrilla warfare. Additionally, this low-visibility situation may also be used by soldiers for their protection in events of contact with the enemy. It is estimated that in evergreen forests, getting a visibility range of 30 m is exceedingly difficult even under the best climatic conditions. Conducting military operations under low-visibility conditions is quite challenging, especially in friend & foe identification, interpersonal contact, and flow of information while maintaining silence [1, 4, 5].

The conditions are more challenging in secondary evergreen rainforests and deciduous dry forests, where dense growth of shrubs, bushes, and vines obstruct the effective fields of view beyond even 10 m. Similarly, in marshes tall reed plants also sharply obstruct the effective field of vision. In comparison, swamps offer relatively better visibility, sometimes up to 50 m. Except in the case of marshy areas and short grass savannah regions, surveillance and rescue operations using airborne platforms are also not very useful as jungle canopy masks most of the inside activities [1, 4].

13.6.5 Diseases

Tropical forests are infamous for hosting a large variety of biting and hematophagous animals, many of which are venomous (snakes, scorpions, bees, etc.) or capable of transmitting various diseases (mosquitoes, ticks, mites, fleas, etc.) [11]. Soldiers operating in jungles are often exposed to several of the biting and disease-transmitting animals that may incapacitate or restrict their combat efficiency. Vector-borne diseases such as malaria, dengue fever, and typhus are known to cause total incapacitation. Additionally, in hot and highly humid conditions, almost all the soldiers are exposed to fungal diseases, especially in the foot, nails, and ears. These fungal infections spread rapidly causing severe pain, itching, and utter discomfort [11, 12]. Therefore, soldiers must obey preventive health and hygiene measures to keep themselves well. It is well documented that during the Pacific War, casualties caused by jungle-related diseases, especially malaria outnumbered the casualties from enemy fire.

Due to high heat and humidity in the tropics, soldiers experience a tremendous amount of sweating, leading to loss of water and salt from the body [11, 12]. Serious dehydration leads to ionic imbalances, which in turn manifest symptoms as heat cramps and abdominal cramps. Additionally, due to prolonged wearing of sweat-wet/rain-wet uniforms and sleeping on wet forest floors, troops may develop complicated infections of the respiratory system.

Another crucial factor for disease among soldiers during jungle operations is water, which in most cases are polluted with detritus and animal wastes. Consumption or accidental ingestion of such contaminated waters may, therefore, lead to serious gastrointestinal problems, often resulting in severe outcomes. Additionally, extreme physiological and psychological stress associated with high heat, humidity, and/or violent combat action may also seriously compromise the combat efficiency of the soldiers in jungles [12]. Therefore, acclimatization of soldiers to such psychological and physiological stressors through extensive training is essential to continue the operational capabilities of the troops.

13.6.6 Personal Hygiene and Sanitation

Maintenance of personal hygiene is one of the biggest challenges during jungle operations, as the soldiers do not have the opportunity or access to proper sanitary conditions. During relatively longer deployment in jungles, soldiers have to resort to primal sanitary practices and are compelled to use available water resources for daily routine, which often are contaminated. Since tropical areas are known for the diversity and abundance of disease-carrying insects and environmental factors that encourage rapid growth of disease-causing pathogens, improperly disposed human wastes can contaminate nearby water sources and become a source of spread of serious illnesses including typhoid fever, bacillary dysentery, cholera, etc., among the troops [11, 12].

13.6.7 Attacking Animals

Different jungles host different faunal compositions ranging from herbivores to carnivores. Except few animals belonging to the big cat family and wild boars, most other animals attack only if provoked or frightened [1, 4, 5]. Some reptiles such as crocodiles and caymans tend to attack primarily moving targets. In general, animals are not considered a substantial hazard to soldiers operating in jungles.

13.6.8 Poisonous Plants

In many parts of the tropics, poisonous trees and plants are encountered. Plants such as nettles, have a severe sting, which on contact with the skin can cause extreme irritation [5, 7]. These plants can be avoided by carefully preventing skin (e.g. covering of bare skin, gloves, etc.) from exposure to unknown vegetations during operations.

13.7 Opportunities in Jungle

13.7.1 Camouflage and Cover

One of the best advantages that a jungle offers to soldiers is perfect camouflage or concealment. This advantage is primarily provided by the vegetation present in each jungle. Using proper camouflaging measures, soldiers can conceal themselves as well as their equipment under the dense vegetation [1, 5]. However, to effectively use vegetation for camouflaging, a proper understanding of the seasonal variation in the colour of the predominant vegetation must be considered while planning operations in a particular jungle feature. Even though camouflaging is practised since historical conflicts, its significance in modern-day jungle operations became prominent during the Vietnam War [3, 4]. Modern-day armies use uniforms with different camouflaging patterns also known as the 'combat pattern' or 'operational camouflage pattern', to improve concealment in their respective operating terrains. Such patterns include the most widely used olive drab, woodland, tiger stripe patterns, etc. Of late, disruptive pixellated patterns such as digital camouflage pattern and flecktarn are gradually replacing the older patterns due to their better performance.

In this context, it is important to understand that natural concealment or camouflaging can provide positional disguise but cannot provide any cover against ammunitions. In the primary evergreen rainforests, trees with wider trunks are abundant, which can provide personal cover against ammunition. In the rest of the forest types such kinds of trees are either not available or very widely spaced, hence cannot provide sufficient cover. In such a scenario, geographical features such as irregular surfaces, mounds, large rocks, channels, etc. can provide cover at the time of contact with the enemy.

13.7.2 Obstacles and Surprises

As mentioned in the previous sections, all the features of a jungle pose serious obstacles to movement and survival. Obstacles faced by the soldiers are also applicable to the enemies, albeit the scale may differ depending upon the training of the forces [4, 5]. Nevertheless, movement across the jungle itself is exhaustive for any individual, and extreme exhaustion results in reduced levels of alertness and psychological strength necessary for fighting. In certain conditions, enemy forces can be lured to take a jungle path with more obstacles or natural features can be augmented in a specific manner to physically exhaust the enemy before an active contact is made. Similarly, when the enemy locations are known, attacks are preferably planned during the resting period, so that the fatigued condition and the '*element of surprise*' together can inflict maximum enemy casualty. Jungle forces often plan and conduct raids during heavy rainfall in tropical forests as it provides an excellent mask for the sound of troop movement [4, 5].

13.7.3 Tracking Enemy Locations

It is often said that jungles keep a record of every movement, which simply means that it is exceedingly difficult to move in a jungle without leaving any records. However, such records are both advantageous and disadvantageous for soldiers, as they can track enemy movements, while the enemy can also track the soldiers. Soldiers experienced in jungle combat can study the foot (or boot) prints and can assess the strength of the unit, general physical characteristics of the person (an idea of friend or foe), and direction of the movement. Other cues include mark on grasses, unusual formations of vines, twigs, sap, or fluid from cuts and bruises on the trees, distribution of leaves on the forest floor, disturbance in foliage and animal habitats, the unusual appearance of wet areas in the forest floor, etc. [4]. Nevertheless, trained enemies can deliberately create some of these cues to deceive and distract soldiers to an ambush. In such conditions, the utilization of scout dogs proves to be particularly useful in search and track operations.

13.7.4 Water

Water is one of the primary requirements of the soldiers operating in the hot and humid tropical jungles. Although, jungles may have abundant water sources, but consumption of water from any untested source is strongly discouraged. In emergency conditions, free-flowing water from streams and rivers can be consumed after boiling (at least 1–2 min), filtration, or chemical treatment such as the use of purification tablets [4, 5]. In addition to these resources, vines present in the jungle can provide a source of water. It is generally believed that if the vine sap is transparent, tasteless, and odourless, it is safe for consumption. Similarly, sufficient water can be squeezed out of banana tree stalk or can be collected from the lower segments of bamboo saplings. Rainwater collected directly in containers is also considered safe to consume.

13.7.5 Shelters

During jungle deployment or extended operations, it is important for soldiers to protect themselves from different climatic factors, insects, and other animals. In certain scenarios, soldiers may get struck in the jungle and need to make temporary shelters. Jungles typically offer several materials for the construction of shelters, which soldiers can use as per the situation and protection level required. For building shelter structure good quality living, hardwood is abundantly found in the jungles, while wider leaves such as palm and banana can be used for roofing thatch [1, 4, 5]. Additionally, jungle vegetation has a variety of vines, some of which can be used as excellent lashing material. Vines are also frequently used for setting up traps.

13.7.6 Food

Different jungles have a diversity of plants that can be used as food during emergency periods of jungle operations when soldiers are forced to stay beyond their planned durations. In fact, tropical areas have been gifted with numerous varieties of plants, that bear nutritious fruits, and nuts, depending upon various climatic factors including soil, humidity, elevation, sunlight, rainfall, etc. However, before considering wild vegetation for food, it is essential to ensure that the same is edible and non-poisonous [5, 7]. With adequate training in the Universal Edibility Test, soldiers can distinguish edible and non-edible fruits. Since universal edibility tests are time consuming and include several steps, in conditions of extreme emergency, the toxicity of a fruit or vegetable could be evaluated through some rapid tests, which soldiers have evolved based upon their field experiences [1, 5, 7].

Plants or fruits that excrete milky/black sap or have an unpleasant odour (onion/garlic/almond like) are avoided in the first place. Subsequently, the taste is determined by placing a small piece on the tongue (avoiding ingestion) and material with acceptable taste is cooked by boiling with the change of water multiple times. A small amount of the food is then ingested and waited for 3–4 h before further consumption. Fruits that are eaten by birds, monkeys, or other animals are usually safer for human consumption too. Additionally, citrus, bananas, and cucurbits grow abundantly throughout the year in most of the tropical and subtropical jungles and are easily identifiable due to their similarities with fruits/vegetables of the same family commonly available in the market.

Apart from fruits and vegetables, animal life forms are also abundant in tropical regions, which include birds, fowls, fish, and fur-bearing animals, which can be boiled properly and taken as food during extreme situations and as per the preferences of the soldiers. It has been reported that meat cooked inside hollow bamboo sections that are sealed with leaves and mud, does not spoil for 3–4 days if left so.

13.7.7 Natives

Jungles are often home to several native communities, which can be of great assistance in providing information about the terrain, local intelligence, food, transport, primary treatment, and other help. However, during planning a jungle operation, it is especially important to get the soldiers know about the native populations residing there and to get the soldiers acquainted with their customs and traditions to avoid any disrespect of their culture [1].

13.8 R&D Opportunities for Supporting Soldiers in Jungle Operations

As discussed in the preceding text, jungle terrains pose unique challenges to the soldiers operating there. Therefore, the operational and support requirement of the soldiers operating in the jungles is utterly different from the ones operating in other terrains. The standard issue military gears used presently by different armies are designed based on the terrains/vegetation they operate in and do fulfil the basic requirements. However, there is ample scope for further R&D towards improvement, refinement, and addition of new items to provide additional protection and support to the soldiers. The following list indicated some of the areas for R&D

towards developing products and technologies for supporting soldiers during jungle operations:

(i) Insect-repellent uniforms—made from fabrics incorporated with insect repellents to protect the soldiers from insect bites during operations in jungles.
(ii) Quick-drying uniforms and boots—made from lightweight, breathable, quick-drying fabrics for enhanced comfort during operations in highly humid jungles to protect from ailments caused by microbial infections due to prolonged wetness and sweat.
(iii) Camouflage face paint with insect repellent—dual-use face paints for camouflage as well as protection of the exposed body surface from insect bites.
(iv) Hammock—a lightweight, durable hammock with insect and rain protection features to provide the soldiers an elevated resting place, safe from different biting and hunting animals.
(v) Multipurpose waterproof ponchos—to be used as a poncho or protective tent during heavy rainfalls.
(vi) Emergency bathing solution—a skin-safe formulation for maintaining hygiene, to be used as a body cleansing solution during emergency situations, where bathing is not possible.
(vii) Emergency water purifiers—pocket-sized, straw-type water purifiers for use during extreme emergency conditions, where drinking water from natural resources is indispensable.
(viii) Medicated chewing gum or lozenge—formulations for maintaining oral hygiene during jungle deployments.
(ix) Combat haemostats—special formulations to accelerate blood clotting to prevent blood loss from combat and other wounds.
(x) Hydration balls—small eatable balls filled with nutrient and electrolyte-rich fluid to be used for hydration during extreme conditions.

These requirements are just suggestive but not exhaustive. Researchers willing to work towards supporting soldiers must interact with the jungle operatives to better understand their specific requirements.

13.9 Conclusion

Jungles continue to be the place of choice for terrorists groups, guerrilla fighters, drug traffickers, and criminals. It is therefore obvious that jungles will remain crucial grounds for future military operations. Over the decades, rapid developments in the field of military technology have completely revolutionized conventional warfare, especially in terms of real-time intelligence collection, navigation, communication, etc. Despite these technological advances, the basics of jungle operations remain the same. Conducting jungle operations and survival still require the highest levels of alertness, skills, and creative thinking, which can only be cultivated and acquired through self-discipline and rigorous training under similar conditions. As discussed in the foregoing text, jungles can offer diverse features, several of which apparently pose challenges to the soldiers, but could be exploited in favour too. Furthermore, every jungle operation is unique in numerous aspects, and thus each of these operations generates a lot of new experiences and knowledge, which should be incorporated in subsequent trainings, manuals, and SOPs.

References

1. Anonymous. Jungle warfare. In: Field manual. Washington DC: Department of the Army; 1942. p. 72–20.
2. Anonymous. Tropical operations. In: Army field manual, operations in specific environments. London: Royal Army, Ministry of Defense; 2003.
3. Perrett B, Walker W. A history of jungle warfare: from the earliest days to the battlefields of Vietnam. South Yorkshire: Pen & Sword Military; 2016.
4. Anonymous. Jungle training and operations. In: Field manual. Washington DC: Department of the Army; 1965. p. 31–0.
5. Anonymous. Jungle operations. In: Field manual. Washington DC: Department of the Army; 1982. p. 90–5.
6. Collins JM. Military geography for professionals and the public. Dulles, VA: Potomac Books; 1998.
7. Cross JP. Jungle Warfare. Annapolis, MD: Naval Institute Press; 2008.
8. Threlfall A. Jungle warriors: from Tobruk to Kokoda and beyond, how the Australian Army became the world's Most deadly jungle fighting force. Melbourne: Allen & Unwin; 2014.
9. Anonymous (2019) Tactics of jungle war. https://anzacportal.dva.gov.au/wars-and-missions/kokoda-track-1942-1943/events/jungle-warfare/tactics-jungle-war. Accessed on 24 Apr 2024.
10. Rodriguez-Hernandez SM. Jungle operations. In: Piehler KG, editor. Encyclopaedia of military science. London: SAGE Publications; 2013.
11. Carter HP. Medical problems in jungle warfare. The Military Surgeon. 1942;91:640–7.
12. Anonymous. Medical problems in jungle warfare and the Pacific area. In: US Army command and general staff school. Fort Leavenworth, Kansas; 1945. https://cdm16040.contentdm.oclc.org/digital/collection/p4013coll8/id/505 Accessed on 30 Apr 2024.

Part III

Common Issues Across Environments

Common Neurological Ailments Across All Ranks and Maintaining Mental Health in Today's Stressful Environment

14

Pawan Dhull, Rajiv Mishra, Anirban Gupta, Gaurav Ji Rana, and Shrimukta Sahoo

Abstract

The multifaceted challenges of military service significantly impact the neurological and mental health of armed forces personnel. This chapter examines common neurological ailments and their implications within the military context, highlighting the interplay of environmental, occupational, and lifestyle factors. Prevalent disorders include headache syndromes, cerebrovascular accidents (CVAs), and high-altitude-related illnesses, and combat exposure, extreme climates, and unique operational conditions play a role in exacerbating these conditions. The chapter covers various lifestyle-induced disorders such as obesity, diabetes, and alcohol use, which further complicate health outcomes. Neurological infections, autoimmune disorders, and degenerative diseases pose long-term risks, often linked to stress and traumatic brain injuries (TBI). Mental health remains a cornerstone of operational readiness, which is affected by disorders like anxiety and post-traumatic stress disorder (PTSD). The chapter also highlights systemic approaches for prevention including vaccinations, nutritional interventions, and protective protocols, which play a crucial role in risk mitigation. Additionally, it emphasizes the importance of rehabilitation programs for neurodegenerative and traumatic conditions, aimed at enhancing recovery and

P. Dhull (✉)
Armed Forces Clinic, New Delhi, India
e-mail: dhullpawan.680w@gov.in

R. Mishra
Defence Institute of Physiology and Allied Science (DIPAS), New Delhi, India

A. Gupta · S. Sahoo
Army Hospital (R&R), New Delhi, India

G. J. Rana
Command Hospital Air Force, Bangalore, India

S. Srivastava et al. (eds.), *Health and Sustenance in Extreme Environment*,
https://doi.org/10.1007/978-981-96-8642-1_14

readiness. By integrating mental and physical health strategies, armed forces personnel can achieve optimal performance in demanding environments, ensuring both resilience and operational effectiveness.

Keywords

Neurological disorders · Military personnel · Mental health · Post-traumatic stress disorder · Traumatic brain injury · Lifestyle diseases · High-altitude illness · Cerebrovascular accidents

14.1 Introduction

In the high intensity and consequential world of armed forces, neurological ailments present distinct challenges that significantly impact the lives of military personnel. The nature of military service, characterized by rigorous physical demands, constant vigilance, and exposure to traumatic events, creates a distinct environment where neurological issues can flourish. This chapter explores the prevalent neurological disorders among armed forces personnel in India, exploring their origins, manifestations, and their critical importance in maintaining mental health in a high-pressure setting. In today's high-pressure environment, safeguarding health is vital to ensuring both the operational readiness and overall well-being of armed forces personnel. Maintaining mental health in today's stressful environment is critical for the operational readiness and overall well-being of armed personnel. The chapter will cover common neurological ailments like headache disorders, central nervous system (CNS) infections, and disease related to lifestyle, environment, and autoimmune conditions. Psychiatric issues like post-traumatic stress disorder (PTSD), traumatic brain injury (TBI), and anxiety disorders, which are particularly pervasive among military personnel, often result from combat exposure, accidents, or high-stress missions. These ailments not only affect the individuals but also have far-reaching consequences for their families and their combat units. By shedding light on the specific conditions that affect them and outlining effective strategies for maintaining mental health, we aim to equip readers with the knowledge and tools necessary to support the brave individuals who serve in the armed forces. Through increased awareness and proactive measures, we can address the unique needs of military members in a systematic way, ensuring health and readiness in the face of today's demanding environments.

14.2 Common Neurological Disorders in Armed Forces Personnel

The common neurological disorders in the armed forces include headache disorders, environment-related disorders, epilepsy and seizure disorders, vascular disorders, neurological disorders related to lifestyle and occupation, and autoimmune

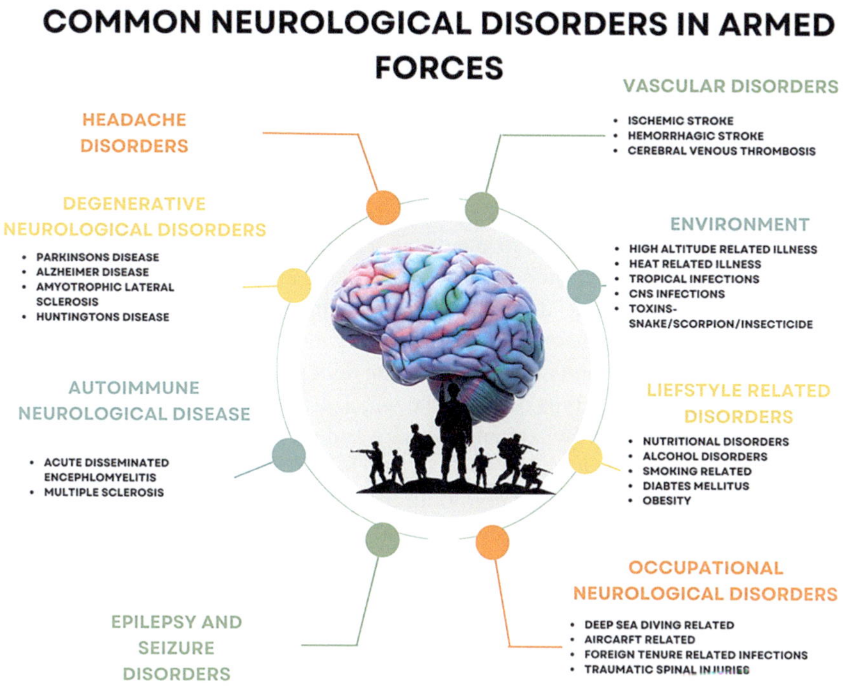

Fig. 14.1 Common neurological disorders in armed forces personnel

and degenerative neurological disorders. Figure 14.1 summarizes the major neurological disorders faced by armed forces personnel.

14.2.1 Headache Disorders

Headache disorders pose a significant concern within the armed forces, with an elevated prevalence among military personnel, in particular those with combat exposure. Combatants have significantly higher odds of developing any new-onset headache disorders than non-deployed personnel [1]. The increase in new-onset headaches and migraines among deployed soldiers is multifactorial. The key contributory factors include combat exposure, which significantly elevates the risk. Traumatic brain injury (TBI), particularly concussions, is universal among combat-deployed personnel and has a strong association with an overall increased incidence of headache. PTSD is another critical factor; the nervous tension and anxiety inherent to PTSD frequently precipitate migraine and various other headache symptoms [2]. Additionally, anxiety and depression, present alongside with headache disorders, particularly migraines, exacerbate these conditions and significantly reduce the quality of life. Environmental factors of military deployments, including excessive noise, extreme heat, and high humidity, further contribute to the development of headache disorders. Headache disorders are classified into several types, each with distinct pathophysiological mechanisms and clinical manifestations. Migraine is a complex neurological disorder, which is characterized by recurrent, severe, and prolonged headaches, often accompanied by nausea, dizziness, photophobia, and phonophobia. Migraine is categorized into migraine with or without aura where visual disturbances precede the headache in the former and there is a lack of these prodromal symptoms in the latter. Vestibular migraine, primarily affecting the vestibular system, manifests as vertigo and disequilibrium and may be accompanied by headache. Basilar-type migraine involves brain stem dysfunction, presenting with symptoms such as diplopia, dysarthria, and paresthesia. Hemiplegic migraine is characterized by transient hemiparesis or hemiplegia, often accompanied by a typical migrainous headache and other neurological deficits.

Cluster headaches are a type of headache which are typified by severe, unilateral periorbital pain, described by patients as sharp and stabbing, occurring in cyclical patterns with periods of remission. Tension-type headache is one of the most prevalent forms which presents as bilateral, dull, aching pain with a band-like distribution, often triggered due to stress, muscular tension, and poor posture. Sinus headaches are characterized by facial pain, accompanied by nasal congestion and discharge, resulting from sinus inflammation or infection. Rebound headaches, or medication-overuse headaches, arise from the chronic use of analgesics, paradoxically exacerbating headache frequency and severity. Chronic daily headaches are defined as headaches occurring on more than 2 weeks per month, and are often secondary to medication overuse, stress, or other underlying medical conditions.

Some headaches can be severe and dangerous and necessitate immediate medical attention due to their intensity, sudden onset, or association with other alarming symptoms. These headaches indicate a life-threatening condition and require prompt intervention to prevent serious complications or death. The key symptoms that signal the need for urgent medical care include a sudden, severe headache described by the individual as the worst in their life, often referred to as a thunderclap headache, which reaches peak intensity within 1 min and lasts for at least 5 min. Additional warning signs include headaches accompanied by fever, chills, night sweats, or unexpected weight loss suggestive of an underlying inflammation or infection. Neurological symptoms such as confusion, loss of consciousness, seizures, or personality changes, as well as weakness, numbness, or tingling in the face, arms, or legs, are also critical indicators. Vision changes, including blurry vision, double vision, or blind spots, and headaches accompanied by seizures further signify the need for immediate medical evaluation. Dangerous headaches are a critical condition, and hence, experiencing any of these symptoms necessitates prompt medical attention to prevent severe outcomes or death.

14.2.2 Vascular Disorders

Cerebrovascular accidents (CVAs), commonly known as strokes, represent a significant concern to military personnel with higher prevalence than anticipated due to unique risk factors such as combat exposure, environmental factors like high-altitude areas, hypoxic conditions, and extreme temperatures [3]. Elevated blood concentration, particularly higher hemoglobin levels, and certain medications further elevate the risk. Stress and anxiety inherent to PTSD can precipitate stroke symptoms. Modifiable risk factors like obesity, hypertension, smoking, insufficient physical activity, illicit drug use, unrestricted alcohol consumption, and altered sleep pattern also predispose an individual to strokes. Other risk factors encompass chronic medical ailments like kidney disease, heart disease, carotid artery disease, diabetes mellitus (DM), migraine headaches, sleep apnea, inflammatory conditions, high stress, anxiety, or depression and non-modifiable factors such as age, race, and air pollution exposure.

CVAs can be classified into ischemic and hemorrhagic strokes. Ischemic strokes constitute approximately 87% of all strokes, which are a consequence of an obstruction in a blood vessel supplying the brain, leading to insufficient oxygenation of the brain parenchyma. This obstruction can result from blood clots, plaque buildup, or other substances in the bloodstream. Hemorrhagic strokes, accounting for about 13% of strokes, occur when the brain artery ruptures or leaks, causing raised intracranial pressure and subsequent brain damage. These strokes are often associated with high blood pressure or aneurysms. Symptoms of CVA manifest suddenly, these include symptoms like numbness or weakness in the arm, leg, face or unilateral side of the body, difficulty walking, inability to maintain balance posture, or dizziness, vision problems in one or both eyes, confusion, difficulty in speaking or understanding speech, and severe headache. Immediate medical intervention is critical, particularly for ischemic strokes, where thrombolytic medications like tissue plasminogen activator (tPA) can dissolve clots, and for hemorrhagic strokes, where medical or surgical methods are used to control bleeding. Preventive strategies are essential, given that approximately one in four stroke survivors will experience another stroke within 5 years. Cerebral venous thrombosis (CVT) is an uncommon condition in the general population but more frequent among military personnel in high-altitude areas (HAA). CVT involves thrombotic occlusion of the cerebral venous system, involving deep and superficial venous systems and dural venous sinuses. CVT is a silent contributor that is a commonly disregarded and unacknowledged cause of intracranial hypertension and stroke in the young with an underlying procoagulant state [4]. High-altitude exposure significantly increases CVT risk due to hypoxia-induced hemostatic changes. Other risk factors include specific injury patterns like lower extremity fractures, underlying hypercoagulable states, and lifestyle factors such as smoking, alcohol use, and oral contraceptive use. Demographic factors like male gender and older age also elevate CVT risk. Clinical manifestations of CVT include fever, headache, focal neurological deficits, seizures, nausea, vomiting, and altered mental status [5].

14.2.3 Environment-Related Neurological Disorders

Environment-related neurological disorders are further divided into HAA-related conditions, heat-related conditions, neurological infections, conditions related to foreign tenures, and conditions related to bites and toxins.

14.2.4 High-Altitude-Related Neurological Disorders

The Indian Army deploys troops to high-altitude regions, such as the Siachen Glacier and other areas in western and eastern Himalayas and the Pir Panjal ranges. These deployments occur at elevations ranging from 9000 to 22,000 ft. (3500–6700 m), presenting substantial environmental, logistical, and medical challenges. Environmental challenges include extreme cold, high winds, and hypobaric hypoxia, which can result in hypoxia and frostbite. Logistically, the remoteness and inaccessibility of these regions complicate the transport of supplies, equipment, and personnel. Medically, high altitudes increase the risk of conditions such as acute mountain sickness (AMS), high-altitude cerebral edema (HACE), and high-altitude pulmonary edema (HAPE).

Neurological manifestations at high altitudes include alterations in cerebral blood flow, resulting in altered mental status, headaches, and fatigue. Sleep disturbances such as insomnia and daytime fatigue are prevalent and high altitudes can exacerbate preexisting neurological disorders. Acute mountain sickness (AMS) is notably common among soldiers in these regions, who rapidly ascend to HAA (3700 m) without acclimatization. Risk factors for AMS encompass genetic predispositions such as polymorphisms in the EDN1 gene and environmental factors like rapid induction to high altitude. The clinical presentation of AMS includes symptoms such as headache, dizziness, gastrointestinal upset, and fatigue. The Lake Louise Scoring System is frequently employed for recognizing AMS based on the symptoms. Treatment typically involves the administration of acetazolamide. High-altitude cerebral edema (HACE) is a rare but serious condition affecting fewer than 1% of individuals who ascend to altitudes exceeding 4000 m. Pathophysiologically, HACE is characterized by disturbances in cerebral blood flow, resulting in increased cerebral blood volume and pressure. This condition involves disruption of the blood–brain barrier (BBB), allowing fluid to permeate the brain and induce edema. Mitochondrial dysfunction leads to the heightened production of reactive oxygen species (ROS), leading to the development of HACE. Imbalance in reduction–oxidation (REDOX) homeostasis plays a critical role, activating microglia abnormally and compromising vascular endothelial cell tight junctions. Treatment strategies for HACE encompass immediate administration of supplemental oxygen to enhance brain oxygenation. Dexamethasone is employed to mitigate cerebral edema and inflammation. Urgent descent to lower altitudes is imperative to mitigate risks of severe outcomes like coma and mortality [6].

14.2.5 Heat-Related Neurological Illnesses

Heat-related neurological illnesses are predominantly seen in environments characterized by intense heat and extreme humidity. The incidence of these illnesses ranges from 0.2 to 10.5 per 1000 person-years, with prevalence rates spanning from 0.3% to 9.3% [7]. Exertional heat illness (EHI) occurs when the body fails to adequately cool itself during physical exertion. It encompasses conditions like heat cramps, marked by muscle cramps, and weakness due to dyselectrolytemia and heat exhaustion, which manifests with features such as headache, fatigue, nausea, and confusion, along with physiological signs like a rapid pulse and low blood pressure. In a study involving EHI-affected paratroopers at Agra, the clinical presentations included a confusional state in 58.97% of cases, seizures in 33.33%, coma in 29.48%, violent behavior in 20.51%, and vomiting in 26.92%. Laboratory findings showed disturbances such as hyponatremia, hypophosphatemia, and hypokalemia, along with elevated levels of transaminases, creatinine kinase, and lactate dehydrogenase [8]. The most severe form, heat stroke, occurs when the core body temperature reaches 104 °F (40 °C) or higher, leading to confusion, slurred speech, and potential loss of consciousness. Hyponatremia, where excessive water intake dilutes blood sodium levels, can also complicate cases. Risk determinants include advanced age, poor physical fitness, and ecological conditions like extreme heat and humidity and arbitrary factors such as hydration status and prior history of heat illness. Elevated biomarkers such as creatine phosphokinase are indicative of muscle damage. Prevention and deterrent strategies emphasize hydration, electrolyte balance, and avoiding strenuous activity during peak heat periods, while treatment involves rapid cooling and fluid/electrolyte replenishment to mitigate heat-related complications.

14.2.6 Neurological Infections

Neurological tropical infections present significant health risks for Indian army personnel deployed in high-risk environments. Infectious agents have been developed as weapons of mass destruction by governments and terrorists. While respiratory diseases are often prioritized, CNS infections can be equally critical. Aerosolized encephalitis viruses and CNS involvement pose significant bioterrorism threats [9]. CNS manifestations of bacterial pathogens like Brucella, Plague, Q fever, Leptospira, and Scrub typhus complicate treatment.

Though battles have been fought and won against malaria, the tide of the war now threatens to turn, with the disease looming ever closer to reclaiming its hold. It is caused primarily by mosquito species, mainly *Plasmodium falciparum* and *Plasmodium vivax*, which are endemic in tropical regions and pose a substantial threat due to prospective severe neurological complications. Malaria statistics within military forces fluctuate annually across adjacent units and even within the same unit. These variations may be linked to the influx of new soldiers or the relocation/establishment of new units from different regions. Despite troops typically being administered chemoprophylaxis, malaria cases persist due to the high parasite burden in environments where a high malaria load is seen [10]. Cerebral malaria (CM), primarily associated with *P. falciparum* infection, is the most dreaded manifestation, characterized by the parasite's cytoadherence to endothelial cells in brain capillaries, which leads to microvascular obstruction, cerebral edema, and ultimately, neurological symptoms such as altered mental status, seizures, and coma. Inflammatory responses encompass the cytokine release syndrome, which further exacerbates the blood–brain barrier disruption and neurovascular pathology, contributing to the severity of neurological outcomes. Post-Malaria Neurological Syndrome (PMNS), acute disseminated encephalomyelitis (ADEM), Guillain–Barré Syndrome (GBS), and cerebellar ataxia are additional neurological complications documented in malaria cases, affecting both the acute and postinfection phases.

Scrub typhus, caused by *Orientia tsutsugamushi* and transmitted through the bites of infected chiggers (Leptotrombidium mites), exhibits neurological manifestations in approximately 20% of cases. The neurological complications of scrub typhus encompass a spectrum from acute encephalitis to meningitis, meningoencephalitis, peripheral neuropathy and rarer presentations like optic neuritis or opsoclonus-myoclonus syndrome [11]. The pathophysiology involves the bacterium's invasion of the CNS, triggering an inflammatory response and direct neuronal injury, leading to various neurological deficits.

Japanese Encephalitis (JE), caused by the Japanese encephalitis virus, is a leading form of viral encephalitis in Asia. Spread mainly by Culex mosquitoes, it primarily affects rural agricultural regions in northern India. The infection typically presents with acute encephalitis syndrome, marked by fever, headache, and altered mental status, which can advance to severe neurological damage. The virus invades the CNS, triggering neuro-inflammation, neuronal death, and the potential for long-term neurological impairments. Up to 50% of JE survivors can develop significant neurological sequelae including cognitive impairment, movement disorders, and seizures impacting their cognitive and motor functions [12].

Neurocysticercosis (NCC), caused by the tapeworm T. Solium, is a concern in regions with poor sanitation practices where pig-raising is common. In India, NCC forms 1.2–2.5% of all intracranial space-occupying lesions seen in specialized centers. In slums where pigs are raised, there is a higher prevalence of NCC (12–15%) in the community. At the National Institute of Mental Health and Neurosciences Bangalore, NCC accounted for 0.12% of neurological disorders among 96,413 patients screened in 20 years [13]. NCC occurs when humans inadvertently ingest T. Solium eggs, leading to larval migration and cyst formation within the brain. Neurological manifestations include seizures, focal deficits, and increased intracranial pressure due to cystic growths. NCC can impair cognitive functions critical for military operations, such as spatial working memory, attention, and executive function.

Preventive strategies against these infections include rigorous antimalarial chemoprophylaxis, vaccination against JE virus, and improving hygiene and sanitation practices to prevent NCC transmission. Early diagnosis through comprehensive clinical evaluation and imaging studies is essential for timely intervention and effective management of these infections' neurological complications.

14.2.7 Foreign Tenure-Related Neurological Disorders

When Indian Army personnel are deployed on foreign assignments in regions like Sudan, Congo, Bhutan, Lebanon, and other similar contexts, they confront a spectrum of neurological challenges due to prevalent regional diseases and environmental factors [14]. In countries like Sudan and Congo, malaria presents a significant threat. Severe illnesses can be seen in the form of cerebral malaria which is characterized by neurological symptoms such as seizures, coma, and cognitive impairment. Outbreaks of bacterial meningitis are also common, potentially causing meningitis-associated encephalitis and lasting neurological deficits. Lassa fever, native to West Africa, including parts of Sudan, can lead to complications like encephalitis or meningo-encephalitis. In Bhutan and surrounding regions, JE, borne by mosquitoes, may trigger inflammation of the brain, with symptoms spanning from mild fever and headaches to more severe outcomes like seizures and paralysis, casting a shadow over those it afflicts.

NCC leads to cyst formation in the brain and spinal cord causing seizures and neurological deficits. Lyme disease, although more common in temperate regions, can occur in Bhutan due to tick bites, potentially causing neurological manifestations like meningitis, facial nerve palsy, and encephalomyelitis. In Lebanon and similar Middle Eastern settings, Leishmaniasis, a parasitic infection transmitted by sandflies, can cause Visceral Leishmaniasis affecting the CNS, leading to neurological complications if left untreated. Brucellosis, endemic in the Middle East, can cause in neurobrucellosis affecting the nervous system with symptoms such as meningitis, encephalitis, or neurovascular complications. Invasive fungal infections such as Cryptococcal meningitis or Aspergillosis pose additional risks, especially in immunocompromised individuals in military settings. Yellow fever, transmitted by mosquitoes in tropical regions of Africa and South America, can also pose risks during deployments. Severe cases may lead to yellow fever-associated encephalitis or meningo encephalitis, causing neurological symptoms such as confusion, seizures, and coma. Moreover, Dengue fever, prevalent in tropical and subtropical regions, can progress to severe forms like Dengue hemorrhagic fever or Dengue shock syndrome, which may involve neurological complications such as encephalopathy or Guillain–Barré syndrome. These diverse regional diseases underscore the complex neurological risks faced by army personnel during foreign deployments. Implementing comprehensive preventive measures, including vaccinations, prophylactic medications, and robust surveillance for early detection and treatment, is crucial to safeguarding the neurological health and well-being of deployed personnel in these challenging environments.

14.2.8 Bites and Toxin-Related Neurological Disorders

Neurotoxic snake bites and scorpion stings present significant health risks to Indian Army personnel deployed in regions endemic to venomous species. In areas such as the Western and Eastern Ghats, northeastern states, and rural sectors across India, soldiers face heightened exposure to venomous snakes like cobras, kraits, and sea snakes [15]. These snakes inject potent neurotoxins that interfere with nerve function, leading to symptoms ranging from localized pain and swelling to neurological manifestations causing paralysis and respiratory failure. Given the remote and rugged terrains where military operations often occur, rapid medical evacuation and access to specialized antivenom treatments are crucial for mitigating the potentially life-threatening consequences of envenomation.

Similarly, scorpion stings pose a significant threat to military personnel, especially in arid and rural environments such as Rajasthan and forested regions, where scorpions like the Indian red scorpion and black scorpion are prevalent. These stings inject venom that can produce immediate excruciating pain accompanied by localized swelling. In severe cases, systemic reactions may occur, leading to neurological symptoms, cardiovascular issues, and respiratory distress. The unpredictability of encounters with scorpions during operational duties, such as night patrols and field exercises, underscores the importance of preventive measures. Effective strategies include the use of protective clothing and thorough inspections of living quarters and equipment to minimize the risk of stings, thereby safeguarding military readiness and operational capabilities.

In contrast to venomous bites and stings, chemical poisoning from insecticides and rodenticides, particularly organophosphorus (OP) and carbamate compounds, presents occupational hazards for military personnel engaged in pest control and environmental sanitation tasks. Deployment in pesticide-intensive areas heightens exposure risks with OP and carbamate poisoning manifesting through symptoms such as nausea, blurred vision, and respiratory distress. Given the military's role in diverse operational environments, rigorous adherence to safety protocols, including routine use of personal protective equipment (PPE) and comprehensive training in chemical handling practices, is imperative. Effective preventive measures and swift medical responses are essential to mitigate the impact of chemical exposures on military health and operational readiness.

14.2.9 Lifestyle-Related Neurological Disorders

Armed forces personnel are often considered the healthiest citizens due to their youth and disciplined lifestyle, which includes regular physical exercise, proficiency training, balanced diet, and routine medical examinations. However, the demands of military service also involve long working hours, strict discipline, stressful situations, separation from family, harsh climatic conditions, and significant mental stress. Additionally, the constant readiness for potential threats at the borders contributes to various biological and lifestyle-related diseases across different ages and roles within the military [16]. Lifestyle-related disorders include nutrition-related, alcohol-related, and smoking-related disorders and conditions related to obesity and DM.

14.2.10 Nutritional Neurological Disorders

Nutritional health among military personnel is profoundly shaped by the unique demands of military life and the challenges in maintaining adequate dietary intake. Military training and operations subject soldiers to rigorous physical exertion, exposure to harsh environmental conditions, and prolonged stress, significantly elevating their nutritional requirements [17]. These conditions increase metabolic rates and energy expenditure, necessitating a diet rich in essential nutrients to sustain optimal performance and overall health.

Logistical constraints and limited food options during training and deployment often lead to dietary inadequacies. Irregular meal timings and disrupted eating patterns further prevent meeting heightened nutrient demands, potentially resulting in deficiencies of crucial vitamins and minerals essential for various physiological functions. Prevalent deficiencies observed among military populations include vitamin D, Iron, and B-complex vitamins (thiamine, riboflavin, niacin, pyridoxine). Vitamin D deficiency arises from reduced sunlight exposure during deployments and inadequate dietary intake, impacting bone health and immune function. Iron deficiency is notable, particularly among female soldiers and those undergoing intense physical training, affecting oxygen transport and energy metabolism and contributing to fatigue and reduced physical performance. B-complex vitamins play a vital role in energy metabolism, brain health, and cardiovascular wellness. Deficiencies in these essential nutrients can result in fatigue, cognitive decline, and various other health problems. Factors such as gender, age, and genetic variations significantly contribute to nutritional inequalities among military personnel. For example, female soldiers may need increased iron intake to compensate for menstrual losses, while genetic predispositions can alter individual nutrient requirements. Additionally, environmental factors, including extreme temperatures or high altitudes at deployment sites, can further impact nutritional health by affecting nutrient absorption, utilization, and metabolic needs. These environmental stressors heighten the risk of deficiencies, compromising overall health and operational readiness. Nutritional deficiencies can manifest clinically through a spectrum of neurological symptoms and complications. Vitamin B12 deficiency, for example, is associated with neurological manifestations such as subacute combined degeneration of cord, peripheral neuropathy, and optic neuropathy. Copper deficiency, although historically linked primarily to hematologic abnormalities, can also lead to myelo-neuropathy characterized by both upper and lower motor neuron signs. Thiamine deficiency is notorious for causing conditions like Wernicke encephalopathy and Korsakoff syndrome which affect cognitive function and memory. Moreover, deficiencies in vitamin D, folate, and vitamin E can contribute to a range of neurological disorders including cognitive impairment and various neuropathies.

14.2.11 Disorders Related to Alcohol and Smoking

The unique nature of deployment of troops in armed forces leads to prolonged periods of separation from family and society. This, coupled with ease of access to alcohol, can cause serious alcohol-related problems in armed forces personnel. The neurological ramifications of alcohol consumption among military personnel are profound and carry substantial long-term implications for both health and operational readiness. Scientific research has demonstrated that alcohol abuse can induce structural and functional changes in the brain, significantly affecting cognitive and motor abilities. Heavy alcohol consumption is strongly linked to neurological impairments, including cognitive deficits, memory loss, and reduced motor skills. These effects are intensified in individuals with PTSD, as alcohol can further disrupt brain structure and function, worsening cognitive difficulties. Research has shown that the combination of heavy drinking and PTSD can create a cycle of cognitive decline, where alcohol exacerbates the already present neurological issues. Military deployment presents unique psychological stressors that amplify the risk of alcohol dependence among troops. Combat exposure, trauma, and high levels of operational stress, separation from family are robust predictors of alcohol dependency. Soldiers may turn to alcohol as a coping mechanism to alleviate symptoms of PTSD, depression, and anxiety, thereby self-medicating their psychological distress. Moreover, the social environment within the military can perpetuate alcohol use through norms and peer influence reinforcing patterns of excessive drinking. The Army Order 3&11/2001 was introduced to curb alcoholism and to weed out habitual alcoholics [18]. Neurologically, chronic alcohol consumption disrupts neurotransmitter balance in the brain, leading to profound changes in mood regulation, motivation, and cognitive decline. These disruptions also pose long-term risks to brain health and resilience. On the other hand, the combined use of tobacco and alcohol exacerbates these neurological impacts synergistically. The inhalation of tobacco smoke triggers chemical changes in the brain inducing oxidative stress and inflammation. These processes contribute to observable abnormalities in brain morphology, neurochemistry, cerebral blood flow, and overall neurocognitive function over time. Chronic smoking is notably associated with brain atrophy, particularly affecting regions responsible for controlling memory and executive functions such as the hippocampus and prefrontal cortex. Moreover, the inflammatory response triggered by smoking, known as neuro-inflammation, further disrupts brain structure and function. Neurologically, smoking disrupts the delicate balance of neurotransmitters in the brain, altering mood, motivation, and cognitive processes. This imbalance manifests as reduced attention span, memory deficits, and slower processing speeds. Additionally, smoking interacts with various drugs, compounding its neurological effects and potentially intensifying cognitive impairments.

14.2.12 Diseases Related to Obesity and Diabetes Mellitus

Obesity significantly impacts neurological health through various interconnected mechanisms. The pathophysiology of these neurological complications involves chronic inflammation, insulin resistance, vascular damage, and hormonal changes induced by excessive adiposity, leading to a range of debilitating complications. Excess body weight leads to chronic inflammation, metabolic disturbances, and changes in vascular function, all of which significantly elevate the risk of neurological disorders. One of the most pressing concerns is the increased likelihood of both ischemic and hemorrhagic strokes, primarily due to obesity-related conditions such as hypertension, diabetes, dyslipidemia, and atherosclerosis. Studies have demonstrated that obesity not only contributes to these risk factors but also directly affects brain health, leading to poorer outcomes in individuals affected by these conditions. Furthermore, obesity is strongly associated with cognitive impairment and an increased risk of dementia, including Alzheimer's disease. Chronic inflammation and insulin resistance, prevalent in obese individuals, contribute to brain changes that impair memory, concentration, and overall cognitive function. Obstructive sleep apnea (OSA), another common consequence of obesity, disrupts normal breathing during sleep due to fat deposits around the upper airway, leading to symptoms like loud snoring, daytime sleepiness, and cognitive difficulties. Idiopathic intracranial hypertension (IIH), characterized by increased pressure in the brain, is also linked to obesity and is manifested by symptoms such as headaches, visual disturbances and neck pain. Additionally, obesity has been implicated in increasing the risk of multiple sclerosis (MS), possibly through the inflammatory effects of excess adipose tissue, leading to symptoms such as muscle weakness, balance problems, and cognitive impairment.

DM, specially related to obesity or metabolic syndrome, presents a spectrum of neurological complications, collectively known as diabetic neuropathies, arising from prolonged elevated blood sugar levels that inflict damage on nerves in the entire body. Peripheral neuropathy is the most prevalent form, affecting peripheral nerves in the legs and feet and often causing symptoms such as numbness, tingling, burning sensations, pain, and reduced sensation in extremities. Autonomic neuropathy targets the autonomic nervous system, leading to dysfunction in involuntary bodily functions like heart rate, blood pressure, digestion, bladder control, sexual function, and sweating. Diabetic amyotrophy affects nerves in the thighs, hips, or buttocks, resulting in severe pain, muscle weakness, and mobility challenges, typically on one side of the body. Focal neuropathy manifests as sudden, localized nerve damage, potentially causing symptoms like acute weakness or pain in specific nerves, leading to conditions such as Bell's palsy or double vision. Cranial neuropathy, which affects cranial nerves, presents symptoms like eye pain, vision changes, and eyelid drooping. The pathophysiology of diabetic neuropathies involves several mechanisms triggered by chronic hyperglycemia. These include metabolic and vascular changes that damage nerves and their blood supply, heightened oxidative stress damaging nerve tissues, chronic inflammation exacerbating nerve damage, and reduced nerve blood flow due to vascular complications. Management of diabetic neuropathies primarily focuses on controlling blood sugar levels to slow disease progression and alleviate symptoms. Lifestyle modifications such as regular exercise, a balanced diet and smoking cessation and proper foot care are recommended to support overall health and mitigate neuropathic symptoms.

14.2.13 Epilepsy and Seizure Disorders

Epilepsy and seizure disorders among military personnel stem from a complex interplay of factors uniquely associated with their service environment. Traumatic brain injury (TBI) stands out as a primary cause, resulting from incidents like blast injuries and penetrating head wounds encountered in combat zones. These injuries disrupt normal brain function and can lead to seizures due to altered electrical activity in the brain. Psychological stress and chronic sleep deprivation, prevalent in military settings, also contribute significantly. High stress levels and irregular sleep patterns can lower the seizure threshold, making seizures more likely in individuals predisposed to epilepsy, exacerbated further by operational demands and night missions. Infections such as meningitis and encephalitis pose another risk, particularly in regions with endemic diseases and limited medical resources during deployments. These infections can induce seizures by causing inflammation and functional disruption in the brain. Substance abuse and its withdrawal add to the complexity, as both acute intoxication and withdrawal states from substances like alcohol and benzodiazepines can trigger seizures. The pressures inherent in military life may lead some personnel to engage in illicit substance abuse as a coping mechanism, thereby increasing the risk of seizure episodes. Moreover, genetic factors play a significant role. Individuals with a family history of epilepsy or genetic disorders may have an elevated susceptibility to seizures. During medical assessments conducted at recruitment, screening for these genetic predispositions is routinely performed to identify potential vulnerabilities. Environmental exposures to neurotoxic chemicals and hazardous substances encountered during military operations, such as pesticides and nerve agents, pose additional seizure risks by directly damaging brain tissue or disrupting metabolic processes. Moreover, preexisting conditions including undiagnosed epilepsy or other neurological disorders may be exacerbated by the physical and psychological demands of military service. Metabolic imbalances such as electrolyte disturbances and dehydration, exacerbated by intense physical exertion and heat exposure in field conditions, can also precipitate seizures. Lastly, PTSD, prevalent among military personnel, contributes to seizure risk through chronic stress and anxiety, which can further lower the seizure threshold and exacerbate epilepsy symptoms.

14.2.14 Occupational Neurological Disorders

Deep-sea diving and submarine operations are pivotal for the Indian military but entail profound physiological stresses that significantly impact neurological health.

The primary challenge stems from barotrauma, where rapid changes of ambient pressure during descent and ascent exert mechanical stress on tissues, including the CNS. As divers descend, elevated pressures cause gases, mainly nitrogen, to dissolve into body tissues. Upon ascent, rapid decompression can lead to the formation of nitrogen bubbles, which can occlude blood vessels and directly affect CNS microvasculature. This phenomenon, known as decompression sickness (DCS), manifests with neurological symptoms such as headache, vertigo, confusion, and in severe cases, focal deficits resembling stroke due to embolic events or direct neuronal injury from gas bubbles [19]. The management of DCS include hyperbaric oxygen therapy (HBOT) to reduce bubble size and recompression to facilitate the elimination of nitrogen. In addition to DCS, exposure to elevated partial pressures of oxygen at depth can lead to oxygen toxicity, which causes the generation of reactive oxygen species and subsequent neuronal damage. This condition presents with symptoms such as visual disturbances, seizures, and other neurobehavioral abnormalities. Furthermore, high-pressure nervous syndrome (HPNS) may occur during deep dives, characterized by CNS hyperexcitability, leading to symptoms like tremors, myoclonic jerks, and altered consciousness. These neurological effects are compounded by psychological stressors inherent to submarine environments, including confinement, isolation, irregular sleep patterns, and prolonged operational durations, which can further compromise cognitive function and decision-making abilities. To mitigate these risks, rigorous predive assessments, comprehensive medical surveillance, and real-time monitoring of neurological parameters are imperative. These measures enable early detection of neurological impairments and facilitate prompt intervention including appropriate medical evacuation, if necessary. Technological advancements in diving equipment, gas formulations, and simulation training programs play a crucial role in improving safety protocols and preparing personnel to effectively manage the complex neurological challenges associated with deep-sea diving and submarine operations in the Indian military.

Fighter pilots and workforce aboard aircraft carriers in armed forces face significant neurological challenges due to their unique operational environments, which can lead to various complications including spinal injuries. High gravitational forces (G-force) experienced during aerial maneuvers impose considerable stress on both the cardiovascular and neurological systems, potentially resulting in G-force induced loss of consciousness (G-LOC), visual disturbances, and cerebral hypoperfusion. To mitigate these risks, preventive measures such as anti-G suits, rigorous physical conditioning, and specialized training are employed to enhance the pilot's tolerance to G-forces. Hypoxia represents a significant hazard during high-altitude flights or emergency situations, adversely impacting cognitive function, decision-making abilities, and reaction times. Implementing countermeasures such as the use of supplemental oxygen systems and adherence to altitude protocols is vital for ensuring adequate cerebral oxygenation and minimizing the risk of neurological impairments associated with hypoxia. These strategies are essential for maintaining operational effectiveness in challenging environments where oxygen levels may be compromised.

On aircraft carriers, the operational environment introduces additional neurological considerations. Prolonged exposure to engine noise and vibration can contribute to neurological fatigue and sensory impairment among personnel. Psychological stressors stemming from confined spaces, constant motion, and operational demands further exacerbate neurological responses. Moreover, the potential for spinal injuries during catapult launches, arrested landings (traps), or accidents on deck poses a significant risk. These injuries, which include compression fractures, herniated discs, and spinal cord injuries, can lead to neurological deficits which vary in severity and can profoundly impact sensory, motor, and autonomic functions depending on the extent of spinal cord involvement necessitating urgent stabilization and evacuation to medical facilities equipped to provide specialized neurosurgical care [20]. Beyond the acute phase, individuals may experience chronic sequelae from spinal injuries, including persistent pain syndromes, neurogenic bladder and bowel dysfunction due to disrupted autonomic pathways, and secondary complications such as pressure ulcers, osteoporosis, and increased susceptibility to respiratory infections, owing to reduced mobility. To mitigate these risks, comprehensive safety protocols encompass rigorous training in ejection procedures, deck operations, and emergency responses aimed at minimizing accidents leading to spinal injuries. Furthermore, ongoing advancements in technical infrastructural design, including enhanced seat configurations, restraint systems, and protective gear, are pivotal in reducing the incidence and severity of spinal trauma during high-risk maneuvers. Medical preparedness involves rapid deployment of specialized medical teams trained in spinal injury management to ensure early stabilization and critical care interventions. Rehabilitation strategies are equally crucial, focusing on restoring mobility through comprehensive rehabilitation programs that incorporate adaptive techniques and provide psychosocial support to aid personnel in coping with the long-term impacts of spinal cord injuries.

14.2.15 Autoimmune Neurological Disorders

Acute disseminated encephalomyelitis (ADEM) and MS present distinct neurological challenges for military personnel due to their potential impact on health and operational readiness. ADEM, characterized by widespread inflammation and demyelination of the central nervous system, often follows infections or, rarely, vaccinations. In military settings, where exposure to infectious agents and vaccinations is common, the risk of triggering ADEM is heightened. Symptoms can include sudden neurological deficits such as headache, fever, confusion, weakness, and seizures, necessitating prompt diagnosis through MRI and clinical evaluation. Treatment typically involves high-dose corticosteroids and supportive care, with implications for fitness for duty and potential long-term neurological consequences. Conversely, MS, a chronic autoimmune disease affecting the myelin sheath of nerve fibers, poses challenges due to its unpredictable course and varying symptomatology. Military personnel may face heightened risk factors such as genetic predisposition, environmental exposures, and the physical and psychological stressors inherent in military service. Over a span of 10 years, an analysis involving the US military identified 1827 new cases of MS, leading to an overall incidence rate of 12.9 cases per 100,000 person-years [21]. The disease manifests in different forms, from relapsing-remitting to progressive forms, each influencing the management and prognosis. Diagnosis relies on clinical evaluation, magnetic resonance imaging (MRI), and other tests to assess disease activity and guide treatment with disease-modifying therapies aimed at reducing relapses and managing symptoms. MS can impact fitness for duty, necessitating careful evaluation of personnel capabilities, and potential restrictions from high-demand roles or deployments. Long-term management involves regular monitoring, lifestyle adjustments, and access to comprehensive healthcare to support military personnel in maintaining optimal function and quality of life amid the challenges posed by MS.

14.2.16 Degenerative Neurological Disorders

Degenerative neurological disorders pose significant challenges for military personnel in India, affecting both their operational capabilities and long-term health. Parkinson's disease (PD), characterized by the progressive degeneration of dopaminergic neurons in the substantia nigra of the brain, presents with motor symptoms such as tremors, bradykinesia, and rigidity. These symptoms can severely impact the fine motor skills necessary for precise movements and tasks crucial in military operations, such as handling weapons or equipment with accuracy and efficiency.

According to Scott et al. [22], TBI and PTSD pose an increased risk for development of PD later in life. These conditions, particularly when combined with chronic pain and migraines, act synergistically, significantly raising the risk. The study provides compelling evidence that TBI and PTSD serve as long-term risk factors for PD, often preceding its onset by decades [22]. Alzheimer's disease (AD), another prevalent neurodegenerative disorder, is marked by the accumulation of amyloid plaques and neurofibrillary tangles in the brain, leading to a gradual decline in cognitive functions. Military personnel face elevated risk factors for AD due to tobacco use, TBI, depression, and PTSD. Additionally, they are exposed to other nonspecific risk factors for AD, such as vascular risk factors like obesity, and related conditions such as metabolic syndrome, as well as factors like education level and physical fitness. Combat exposure also plays a significant role in contributing to these risks. Military personnel affected by AD experience deficits in memory retention, decision-making abilities, and situational awareness, all of which are critical in operational settings where split-second judgments and recall of complex information are essential for mission success and safety. The delayed neurological effects of environmental exposures during military service have often been overlooked. With the advancing understanding of AD pathogenesis, it is feasible to quantify some of the potential long-term health risks associated with military service [23]. Amyotrophic lateral sclerosis (ALS), also known as Lou Gehrig's disease, results from the progressive degeneration of motor neurons in the brain and spinal cord. This degeneration leads to muscular weakness, spasticity, and eventually paralysis, impacting mobility and physical capabilities crucial for military duties such as marching, physical training, and responding swiftly to emergencies. Huntington's disease (HD), though less common, presents unique challenges due to its hereditary nature and progressive course. It is characterized by involuntary movements (chorea), cognitive decline, and psychiatric symptoms, including mood disturbances and behavioral changes. Military personnel affected by HD may struggle with maintaining motor coordination and cognitive flexibility required in dynamic and high-stress operational environments. Overall, these degenerative neurological disorders not only affect the physical and cognitive abilities of military personnel but also pose ongoing challenges that require specialized support, management, and adaptation strategies to maintain their well-being and operational effectiveness over time.

14.2.17 Common Mental Health Disorders in Armed Forces Personnel

Mental health disorders are a significant concern in the armed forces. Stressful environments significantly elevate the risk of psychiatric disorders among defense personnel. Psychiatric illnesses carrying high morbidity lead to the underutilization of available manpower and reduction in number of fighting fit individuals. Various factors contributing toward a higher stress level in military jobs are Operational deployment and combat, separation from family, harsh environmental conditions and terrain, civil administrative hassles at home, inability to get leave at required times, early retirement ages, and communication gap in a hierarchical structure. Combat psychiatry demands specialized skills and approaches due to unique stressors on clients, organizational requirements, and frequent transfers of both psychiatrists.

and patients. The Armed Forces have developed an efficient and well-established psychiatric service delivery system, featuring comprehensive referral, management, disposal, and documentation procedures. This system goes beyond curative medicine, engaging in community outreach to offer preventive and promotive mental health services [24]. Figure 14.2 summarizes the common mental health disorders in armed forces personnel.

14.2.18 Anxiety Disorders

Anxiety disorders are abnormal states in which the most striking features are mental and physical symptoms of anxiety. These symptoms include autonomic symptoms (palpitations, sweating, trembling, dry mouth), chest and abdomen symptoms (difficulty breathing, chest or abdominal discomfort, nausea), brain and mind symptoms (dizziness, feeling of surrounding being unreal, fear of losing control over self), and general symptoms like hot flushes or cold chills and numbness or tingling sensations.

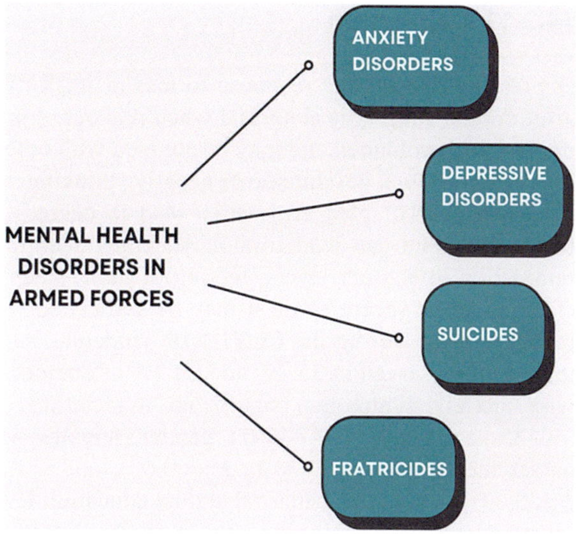

Fig. 14.2 Common mental health disorders in armed forces personnel

Phobic anxiety disorders are characterized by anxiety symptoms leading to severe distress, but restricted to a specific feared situation. Panic disorder is characterized by recurrent, intense episodes of anxiety that occur unpredictably, accompanied by overwhelming fears of dying or losing control, despite the absence of real danger. These episodes are often linked to anticipatory anxiety and can lead to significant dysfunction in daily life. Generalized anxiety disorder (GAD) is a profoundly debilitating condition, especially in its severe manifestations. Individuals with GAD experience persistent and excessive anxiety that is often free-floating, meaning it is not confined to specific situations or triggers. This constant state of worry can encompass a range of everyday life events, including health, work, and social interactions, leading to a pervasive sense of fear and unease. The anxiety associated with GAD is characterized by its relentlessness. Individuals may find it challenging to control their worry, leading to significant distress and impairment in daily functioning. This disorder can manifest physically, resulting in symptoms such as fatigue, restlessness, muscle tension, and difficulty concentrating. The cumulative effects can severely impact an individual's quality of life, making it challenging to engage in routine activities. In obsessive compulsive disorder, obsessive thoughts (words, ideas, and beliefs that are recognized by patients as their own) intrude forcibly into the mind. They are usually unpleasant, and attempts are made to exclude them, and compulsive actions are the hallmark. It is difficult to treat the illness and causes severe stress and dysfunction in the patient's life. In an acute stress reaction, an exceptional mental or physical stressor leads to sudden onset anxiety symptoms. These symptoms reduce in intensity within 48 hours of the event. PTSD is caused by an exceptionally threatening or catastrophic situation. In this disorder, the patient persistently remembers or "relives" the stressor by intrusive flashbacks, vivid memories, and recurring dreams. It is also characterized by avoidance of circumstances associated with the stressor [25].

14.2.19 Depression

Depression is a normal response to loss or misfortune, when it may be called grief or mourning. It is abnormal when it is out of proportion to the misfortune, or is unduly prolonged and may be coupled with other changes, notably a lowering of self-esteem, pessimistic or negative thinking, and a reduction in or loss of the experience of pleasure, which lead to decreased social and occupational functioning and can lead to suicide. Other features include lack of interest, retardation in activity levels, feeling of guilt, sleep disturbances, and appetite changes. In its severe form, it may be associated with delusions and impaired reality testing. During the COVID-19 pandemic, anxiety and depressive symptoms were observed in 35.2% and 28.2% of doctors, respectively. Among those with anxiety symptoms, significant associations were identified with age (20–35 years: 39.4%, $P = 0.01$), gender (females: 44.6%, $P < 0.001$), duration of service (0–10 years: 38%, $P = 0.03$), and specialty (nonclinical: 41.3%, $P < 0.001$). In contrast, marital status, education level, and current department of work were not significantly associated with anxiety symptoms [26]. Depression may necessitate inpatient care in severe cases. Treatment includes the use of medicines, psychotherapy, and social interventions. Severe patients may require electroconvulsive therapy (ECT).

14.2.20 Suicide

Suicide is a severe consequence of depression. The availability of lethal firearm in the armed forces makes the environment high risk for suicide. The presence of other psychiatric illnesses like depression, psychosis, bipolar affective disorder, or PTSD further increases the risk of suicide. Genetic loading, impulsivity, early life traumatic events, neurobiological disorders, physical disorders, psychosocial crisis, availability of means of psychoactive substance abuse have been identified as risk factors for suicide and is preventable.

14.2.21 Fratricide

Usually, the armed forces are at a higher risk of fratricide due to the availability of lethal weapons, difficult work conditions, and lack of family support in high-risk and hardship locations. Data available in the public domain for the Indian Armed Forces reveal five cases between 2016 and 2018 [27]. Certain personality traits like high impulsivity, antisocial traits, and the presence of psychiatric disorders like psychosis increase the risk of fratricide.

14.2.22 Maintaining Mental Health in Today's Stressful Environment: An Armed Forces Perspective

Maintaining optimal mental health is paramount for armed forces personnel and veterans who confront distinct stressors that can profoundly impact their psychological well-being. Evidence-based strategies are essential in navigating these challenges effectively. For armed forces personnel, recognizing symptoms such as persistent sadness, anxiety, suicidal thoughts, and changes in sleep or appetite is crucial for early intervention, which has been shown to mitigate the severity and duration of mental health conditions. Seeking timely professional help through confidential counseling services provided by military organizations is pivotal, as significant improvements in mental health outcomes with professional intervention have transpired in some studies [28]. Cultivating a strong support network among family, friends, and colleagues also plays a crucial role, providing both emotional comfort and practical assistance that can mitigate stress-related mental health challenges. Engaging in self-care practices like physical exercise, meditation, or hobbies is beneficial, with exercise particularly noted for its effectiveness in reducing symptoms of anxiety and depression. A recent study involving 722 BSF personnel was conducted to investigate the outcome of a 9-day residential yoga training program on self-rated sleep, state anxiety, and performance in a vigilance test among BSF personnel. The study involved all male participants, with an average age of 30.9 years (±7.4 years). Before and after the yoga training, the participants were assessed for self-rated sleep, state anxiety, and vigilance. The results of the study indicated significant benefits of yoga for BSF personnel. Specifically, the participants showed a visible progress in scores in the vigilance test, indicating enhanced alertness and attention. Additionally, they also experienced a decrease in state anxiety, suggesting a reduction in overall anxiety levels. Furthermore, the participants reported improved self-rated sleep, indicating enhanced sleep quality. The study highlights the positive effects of yoga on the mental and cognitive well-being of BSF personnel. Yoga can be used as a tool to enhance vigilance, reduce anxiety, and improve sleep quality among military personnel involved in high-stress roles and can be a valuable component of training programs for BSF personnel, helping to improve their mental and cognitive abilities and enhance their overall performance [29].

For veterans, fostering connectivity with fellow veterans who share similar experiences offers invaluable peer support, positively influencing mental health outcomes. Thoughtfully planning transitions to civilian life with access to career counseling and transition support programs can reduce the risk of mental health issues during this period of adjustment. Continuous monitoring of health by regular evaluations helps in the early detection and intervention for potential mental health concerns. Keeping oneself occupied in meaningful activities post-service, whether through employment, volunteerism, or hobbies, fosters a sense of purpose that enhances mental well-being. Supporting others involves staying informed about prevalent mental health conditions among armed forces personnel and veterans, which aids in reducing stigma and promoting help-seeking behaviors. Encouraging individuals to seek professional assistance, assuring confidentiality and support without career repercussions, aligns with effective organizational measures and enhances mental health outcomes. Administering empathetic emotional support as well as practical assistance to those suffering from mental health challenges significantly improves their well-being, reinforcing the importance of a supportive environment.

In order to reduce the prevalence of mental health disorders, significant measures have been put in place in Armed forces across the world. Building resilience in individuals is a primary preventive measure employed at various levels, e.g., at the Individual level (positive coping, positive thinking, realism, and altruism), family level (emotional ties, communications, and adaptability), unit level (positive command climate, teamwork, and cohesion), and community level (belongingness and cohesion). Situations like deployment in active combat or counterinsurgency (CI) ops require much intensive measures. Pre-deployment measures like screening for existing mental health issues, pre-deployment briefing, pre-induction training help in reducing the risk of mental health issues. Employment of field psychiatric teams, limiting frontline exposure, and peer support are measures which can be used pre-deployment. Psychological decompression and post-deployment screening should be used as post-deployment measures.

The Indian Armed forces have already deployed multiple measures to help personnel cope with stress. Interviews by the medical officer and the Commanding Officer before proceeding and after return from leave, along with periodic interviews with superiors, has been the practice in the Indian Armed Forces, which has helped in the early identification and timely referral of at-risk individuals to the service psychiatric establishment. Employment of a psychological counsellor during the primary deployment stage by the Indian Air Force proved beneficial and resulted in easy access to mental healthcare; this has also reduced the stigma related to psychiatric illnesses in society. Training of nonmedical personnel in identifying the red flag signs and early referral to psychiatric services has been proven to be beneficial in providing early intervention.

Various other programs have been put in place by the armed forces as preventive measures. The US Armed Forces has developed and employed Deployment Anxiety Reduction Training (DART), Post-deployment Adaptation Programs, and "Battlemind" (a 7-day reintegration program presently also includes pre-deployment sessions) [30]. Mindfulness training in the Indian Armed Forces aims to enhance mental resilience, focus, and overall well-being among personnel. Given the high-stress environments and unique challenges faced by military members, mindfulness practices offer significant benefits. Structured mindfulness training programs are integrated into the routine training schedules of military personnel, including meditation, breathing exercises, and techniques to enhance awareness and emotional regulation. Regular workshops and seminars led by trained instructors or psychologists further introduce and reinforce these practices. Additionally, mindfulness techniques are incorporated into physical training sessions to strengthen the mind-body connection and improve performance under stress. To ensure accessibility, mobile apps and online platforms provide mindfulness exercises and guided meditation sessions, allowing personnel to practice mindfulness at their convenience. Senior officers and leadership within the Armed Forces endorse and participate in mindfulness training, promoting a culture of mental well-being and resilience. The impact of mindfulness training is substantial. It enhances attention and focus, crucial for operational effectiveness and decision-making in high-pressure situations. Regular practice of mindfulness techniques reduces stress and anxiety, contributing to better mental health among personnel. It also fosters emotional regulation, helping individuals manage their reactions to stressful or traumatic events more effectively. Furthermore, mindfulness practice is associated with various physical health benefits, including lower blood pressure, improved sleep, and reduced symptoms of chronic pain. By cultivating a mindful approach to challenges and adversities, personnel develop greater resilience, enabling them to better cope with the demands of military life. Mindfulness also promotes empathy and better communication, enhancing teamwork and camaraderie among military personnel. The introduction of mindfulness training in the Indian Armed Forces represents a proactive approach to enhancing the well-being and operational fitness of military personnel. By integrating mindfulness practices into training and daily routines, the Armed Forces aim to build a more resilient, focused, and emotionally balanced force, better equipped to handle the challenges and demands of their duties [31].

14.3 Conclusion

Addressing common neurological ailments across all ranks in the military is paramount for maintaining mental health in today's stressful environment. Disorders such as headaches and strokes are prevalent in services and veterans often exacerbated by the unique pressures and service requirements of military life. Environmental factors, such as extreme heat, infections, and exposure to foreign-related complications and toxins, further increase the risk of neurological disorders. Lifestyle choices, including poor nutrition, alcohol and smoking habits, obesity, and diabetes, compound these issues, highlighting the need for comprehensive health strategies.

Epilepsy and other seizure disorders present significant challenges, particularly in the high-stress and demanding environments typical of military operations. Occupational neurology addresses the unique neurological impacts associated with specific military roles, emphasizing the importance of tailored preventive and treatment measures. Autoimmune neurological disorders and degenerative diseases, while less common, require specialized care and ongoing management to ensure affected service members can continue to serve effectively.

Mental well-being remains a critical component of overall neurological health. The intense stress and psychological demands of military service necessitate robust mental health support systems. As noted in "On Combat" by Lt. Col. Dave Grossman, "The mind can only endure what the body can tolerate." This illustrates the intrinsic link between physical and mental resilience, underscoring the need for integrated healthcare approaches that address both aspects simultaneously.

In the context of the armed forces, maintaining mental health is not merely about addressing individual disorders but about fostering a supportive environment that promotes the overall well-being. The Indian Army ethos, "Service before self," encapsulates the dedication and sacrifice inherent in military service, emphasizing the critical importance of supporting the health and well-being of each soldier. By adopting a prophylactic approach to both physical and mental health through comprehensive care and preventive measures, the Armed Forces can better support the resilience and effectiveness of their personnel. Therefore, it is imperative to have a holistic approach to mental health in the Armed Forces; this will ensure that those who serve are not only capable of fulfilling their duties but are also able to lead healthy, fulfilling lives during and after their service.

References

1. Jankosky CJ, Hooper TI, Granado NS, Scher A, Gackstetter GD, Boyko EJ, et al. Headache disorders in the millennium cohort: epidemiology and relations with combat deployment. Headache. 2011;51(7):1098–111.
2. Ryali V, Bhat P, Srivastava K. Stress in the Indian armed forces: how true and what to do? Medical Journal Armed Forces India. 2011;67(3):209–11.
3. Jha SK, Anand AC, Sharma V, Kumar N, Adya CM. Stroke at high altitude: Indian experience. High Alt Med Biol. 2002;3(1):21–7.
4. Anadure RK, Wilson V, Sahu S, Singhal A, Kota S. A study of clinical, radiological and etiological profile of cerebral venous sinus thrombosis at a tertiary care center. MJAFI. 2018;74(4):326–32.
5. Mehta S, Muthukrishnan J, Varadarajulu R, Gupta A. Cerebral venous sinus thrombosis : a great masquerader. MJAFI. 2004;60(3):299–301.
6. Kapoor R, Narula A, Anand A. Treatment of acute mountain sickness and high altitude pulmonary Oedema. MJAFI. 2004;60(4):384–7.
7. Alele FO, Malau-Aduli BS, Malau-Aduli AEO, J Crowe M. Epidemiology of exertional heat illness in the military: a systematic review of observational studies. Int J Environ Res Public Health. 2020;17(19):7037.
8. Deshwal R, Tiwari D, Singh R. Clinical and biochemical characteristics of exertional heat stroke among paratroopers in Agra, India. PubMed. 2017;65(2):57–61.
9. Hart BL, Ketai L. Armies of pestilence: CNS infections as potential weapons of mass destruction. Am J Neuroradiol. 2014;36(6):1018–25.
10. Dhiman S, Baruah I, Singh L. Military malaria in northeast region of India. Def Sci J. 2010;60(2):213–8.
11. Gulati S, Maheshwari A. Neurological manifestations of scrub typhus. Ann Indian Acad Neurol. 2013;16(1):131.
12. Tiwari S, Singh RK, Tiwari R, Dhole TN. Japanese encephalitis: a review of the Indian perspective. Brazil J Infect Dis [Internet]. 2012;16(6):564–73. https://www.sciencedirect.com/science/article/pii/S1413867012002073
13. Kudesia S, Santosh V, Pal I, Das S, Shankar S. Neurocysticercosis—a Clinicopathological appraisal. MJAFI. 1998;54(1):13–8.
14. Freedman DO, Kozarsky PE, Weld LH, Cetron MS. GeoSentinel: the global emerging infections sentinel network of the International Society of Travel Medicine. J Travel Med. 1999;6(2):94–8.

15. Bhoi S, Gupta V, Goel A, Singh J. Clinical profile of venomous snake bites in north Indian military hospital. J Emerg Trauma Shock. 2008;1(2):78.
16. Agrawal T, Bhalla S, Dutt V. Study of prevalence of lifestyle diseases and comparison of their risk factors among serving personnel of army. J Mar Med Soc. 2022;24(Suppl 1):S83. https://doi.org/10.4103/jmms.jmms_71_20.
17. Singh SN. Determining nutritional requirements of Indian soldiers: an outcome of translational research. In: Singh S, Prabhakar N, Pentyala S, editors. Translational research in environmental and occupational stress. New Delhi: Springer; 2014. https://doi.org/10.1007/978-81-322-1928-6_10.
18. Saldanha D, Chaudhury S, Pawar A, Ryali V, Srivastava K, Sood M, et al. Changing pattern of alcohol abuse in the army before and after AO 3&11/2001. MJAFI. 2007 Apr;63(2):160–2.
19. Todnem K, Nyland H, Skeidsvoll H, Svihus R, Rinck P, Kambestad BK, et al. Neurological long term consequences of deep diving. Occup Environ Med [Internet]. 1991. Apr 1 [cited 2020 Feb 6];48(4):258–66.
20. Patil G, Taneja N. Retrospective analysis of initial medical examination of aircrew applicants in the Indian air force. Ind J Aerospace Med. 2006;50(1):44–9.
21. Deussing EC, Jankosky CJ, Clark LL, Otto JL. Estimated incidence of multiple sclerosis among United States armed forces personnel using the defense medical surveillance system. Military Med [Internet]. 2012. [cited 2022 Nov 7];177(5):594–600.
22. Scott GD, Neilson LE, Woltjer R, Quinn JF, Lim MM. Lifelong Association of Disorders Related to military trauma with subsequent Parkinson's disease. Mov Disord. 2023;38(8):1483–92.
23. Veitch D, Friedl K, Weiner M. Military risk factors for cognitive decline, dementia and Alzheimer's disease. Curr Alzheimer Res. 2013;10(9):907–30.
24. Bhat PS. Combat psychiatry: Indian perspective. MJAFI. 2017;73(4):404–6.
25. Harrison P, Cowen P, Burns T, Fazel M. Shorter oxford textbook of psychiatry. 7th ed. London: Oxford University Press; 2017.
26. Gupta S, Kohli K, Padmakumari P, Dixit PK, Prasad AS, Chakravarthy BS, et al. Psychological health among armed forces doctors during COVID-19 pandemic in India. Ind J Psychol Med [Internet]. 2020;42(4):374–8. https://www.ncbi.nlm.nih.gov/pmc/articles/PMC7375358/#
27. Suicide and Fratricide Cases in the Armed Cases [Internet]. Gov.in. [cited 2024 Jun 22]. Available from: https://pib.gov.in/Pressreleaseshare.aspx?PRID=1558887
28. Harris, et al. Exercise and mental health: a review of the literature. J Clin Psychol. 2017;73(1):1–14.
29. Telles S, Gupta RK, Verma S, Kala N, Balkrishna A. Changes in vigilance, self rated sleep and state anxiety in military personnel in India following yoga. BMC Res Note. 2018;11(1)
30. McCaslin SE, Herbst ED, Armitage NH, Allen I, Neylan T, Becket-Davenport C, et al. Deployment anxiety reduction training: a pilot study of acceptability and feasibility in current or recent active duty service members. Mil Med. 2018;183(suppl_1):371.
31. Velayudhan A, Benedict NJ. Mindfulness training for Indian armed forces. Ind J Positive Psychol [Internet]. 2013. Mar 1 [cited 2022 Nov 8];4(1):163–8. https://www.i-scholar.in/index.php/ijpp/article/view/49912

Critical Strategies for Human–Machine–Environment Interaction: Anthropometric and Ergonomic Approaches for Optimum Soldier Performance

15

L. Robert Varte, Deepti Majumdar, Shweta Rawat, Tirthankar Chatterjee, and Debojyoti Bhattacharya

Abstract

Ergonomics focuses on studying how people interact with their work environments to minimize stress, prevent injuries, and improve performance by designing tasks, spaces, and equipment that suits users' physical abilities. This improves overall health and well-being. Rapid technological advancements without considering local needs often result in failure and worker alienation. About 77 years after independence, India has seen increased awareness and research in ergonomics, expanding beyond industrial applications to health, consumer products, defense, and agriculture. However, a major challenge remains: the absence of a categorized anthropometric database, which limits the development of products and facilities suited to the diverse needs of India's population. There is a pressing need for comprehensive research into the physical, physiological, cognitive, and psychological characteristics of India's varied ethnic groups.

Ergonomics can help evaluate both indigenous and imported technologies to ensure their effective use in India. This highlights the necessity for a strong Indian knowledge base, including standards, recommendations, and procedures for occupational health and safety. Given India's diverse factors like climate, demographics, and infrastructure, customized research across sectors is essential. DIPAS, as a leader in military ergonomics research, focuses on enhancing soldier performance and health, applying cognitive and physical ergonomics to improve systems and optimize human–machine–environment interactions.

L. Robert Varte (✉) · D. Majumdar · S. Rawat · T. Chatterjee · D. Bhattacharya
Defence Institute of Physiology and Allied Sciences (DIPAS), Defence Research and Development Organization (DRDO), Timarpur, Delhi, India
e-mail: robvarte.dipas@gov.in

S. Srivastava et al. (eds.), *Health and Sustenance in Extreme Environment*,
https://doi.org/10.1007/978-981-96-8642-1_15

Armed forces personnel operate in diverse environments, facing both physical and mental challenges. Air force operations involve G-forces, rapid maneuvers, and communication, while naval operations include long-term sea activities, sensory changes, and seasickness. Both settings contribute to high cognitive stress due to environmental factors and constant mental demands. Physical ergonomics is crucial in designing human–machine interfaces (HMIs) to enhance user comfort, performance, and safety. It involves tailoring designs based on anthropometric data, ensuring that controls, displays, and seating accommodate users' body dimensions. Ergonomic considerations minimize physical strain, reduce repetitive injuries, and optimize task efficiency by streamlining workflows. Well-designed HMIs also improve safety by ensuring users can easily interact with machines. Additionally, accessibility for diverse users, including those with disabilities, is considered. Critical elements should be within optimal visual fields, and designs customized to users' needs increase satisfaction by reducing frustration and enhancing the overall experience.

Human biomechanics applies principles of mechanics to study the human body's response to forces, combining engineering with physiology and anatomy. It includes kinematics (movement) and kinetics (causes of movement). In defense, biomechanics aids in understanding musculoskeletal systems, muscle activation, and optimizing performance using tools like force platforms and electromyography.

Keywords

Ergonomics · Work environments · Performance enhancement · Anthropometric database · Cognitive ergonomics · Occupational health · Military ergonomics · Human–machine–environment interaction · Biomechanics

15.1 Introduction

Since time immemorial, humans have used ergonomic principles to improve comfort and efficiency for centuries, from hunter-gatherers sharpening tools for hunting and protection to early cavemen using flint to ignite the first fire. Ergonomics, the scientific study of people in their work environments, focuses on reducing stress, preventing injuries, and improving overall performance and well-being. This is accomplished by designing tasks, workspaces, equipment, controls, and displays to match users' physical capabilities, supporting their mental and physical well-being over extended periods.

The significance of ergonomics is underscored by its ability to minimize stress-related issues. For instance, an ergonomically designed keyboard can help prevent repetitive stress injuries like carpal tunnel syndrome, which results from repetitive actions such as typing. Ergonomic keyboards often feature a split design and structures that support the palms and wrists, reducing strain. Globally, ergonomics and human factors are increasingly crucial in designing products, procedures, and work environments. The impact of human factors is most noticeable when it is lacking, rather than when systems function seamlessly. Interest in ergonomics often peaks following disasters or accidents, highlighting deficiencies in safety measures and prompting improvements.

India, one of the world's earliest civilizations, has long engaged in practices that align with ergonomic principles to improve daily life, even if these practices were not formally documented as in Western countries. Despite this early engagement, the growth of ergonomics in India has been hindered by challenges such as inadequate education, an unstructured industrial and health sector, a developing economy, and a lack of government regulations. The adoption of advanced technologies without adapting them to the environmental, physical, mental, and sociocultural needs of the Indian workforce has often proven ineffective and, in some cases, even harmful. Efforts to advance technology rapidly, while neglecting the characteristics, limitations, and preferences of the local workforce, have frequently led to failed technology transfers. These attempts have often alienated workers and achieved minimal improvements in their living and working conditions.

About 77 years after independence, India has seen increased awareness and research in ergonomics, extending beyond industrial applications to include health, consumer products, defense, and agriculture. Despite this progress, a major challenge remains: the lack of a categorized anthropometric database hampers the development of products and facilities tailored to the diverse needs of the Indian population. There is a pressing need for comprehensive research into the physical, physiological, cognitive, and psychological aspects of India's varied ethnic groups. Ergonomics can play a crucial role in evaluating both indigenous and imported technologies, ensuring their effective application in the Indian context. This underscores the need for a robust Indian knowledge base, including standards, recommendations, and procedures related to working conditions, occupational health, and safety. Given the diverse factors shaping issues in India—such as climate, demographics, work methods, infrastructure, and finance—customized research is crucial across agricultural, industrial, defense, and domestic sectors. To optimize the overall workload, strategies must balance sensory input processes with decision-making, error reduction, and performance improvement. Employing both subjective and objective ergonomic assessment tools can enhance the interaction between humans, machines, and environments, resulting in more compatible, sustainable, and efficient systems.

In Indian military perspective, as a leader in this area of research, the Defence Institute of Physiology & Allied Sciences (DIPAS), Delhi tries to fulfill its role effectively, with a primary focus on enhancing soldier performance while preserving their occupational health. The Institute, as a leader in the field, is dedicated to advancing knowledge in ergonomics research across the country. Its vision and programs aim to increase awareness of ergonomic concepts among the Indian populace, contributing to the overall development and modernization of Indian society.

To effectively understand complex military systems and the corresponding user needs, including limitations and capabilities, the principles of ergonomics or human factors engineering must be applied at all stages of development, design, and usage, beginning from the conceptual phase. As traditionally practiced, Ergonomics and

Human Factors Engineering research at DIPAS follows the International Ergonomics Association categorization of the subject area, in accordance with the nature of the issues to be studied, and is categorized as cognitive ergonomics, physical ergonomics, and organizational ergonomics. The third category, that is, research under organizational ergonomics is not within the scope and mandate of the DIPAS charter of duties. Therefore, all research activities undertaken by DIPAS pertain to the issues pertaining to cognitive ergonomics and physical ergonomics.

15.2 Complex Cognitive Ergonomics Issues in Military Operations

Cognitive ergonomics is a crucial area of ergonomics focused on optimizing the overall workload of operators. Daily activities—such as household chores, recreational activities, sports, leisure, and professional work—each involve unique cognitive demands. Operators encounter varying levels of cognitive load depending on their interaction with task elements. In contrast to domestic and recreational activities, occupational and sports environments often impose spatial and temporal constraints that elevate cognitive demands. Military operations, in particular, present significant challenges due to extreme environments and high-stress situations, where soldiers must make critical decisions under intense time pressures. This context necessitates a detailed exploration of how ergonomics can effectively manage cognitive loads in defense scenarios. A conceptual framework will be discussed to address these challenges and enhance cognitive ergonomics in military operations.

15.2.1 The Complex Military Scenarios

Military operations commonly pertain to land (armed forces), water (navy), and air (air force), each having unique cognitive characteristics owing to specific mode of environmental exposure and demand of tasks. It is often observed that soldiers' cognitive performance is dominantly influenced by four factors, viz., (1) exposure to *unstructured environments*, (2) *equipment status*, (3) *climatic conditions*, and (4) *individual factors* (Fig. 15.1).

As given in Fig. 15.1, an *unstructured environment* is mostly characterized by hostile terrain conditions, stringent alignment to temporal and spatial constraints, and poor illumination conditions. Diverse terrain conditions—such as undulating rocky surfaces and snow-covered landscapes at high altitudes, shifting desert sand dunes, dense jungles with slippery terrain, and submerged coastal wetlands—are major sources of physiological and biomechanical stress. Poor illumination conditions are generally considered an integral part of many military operations, where the attempted target is identified as a faint/blurred stimulus. The outdoor illumination factor is more challenging than the indoor conditions in the context of exposure

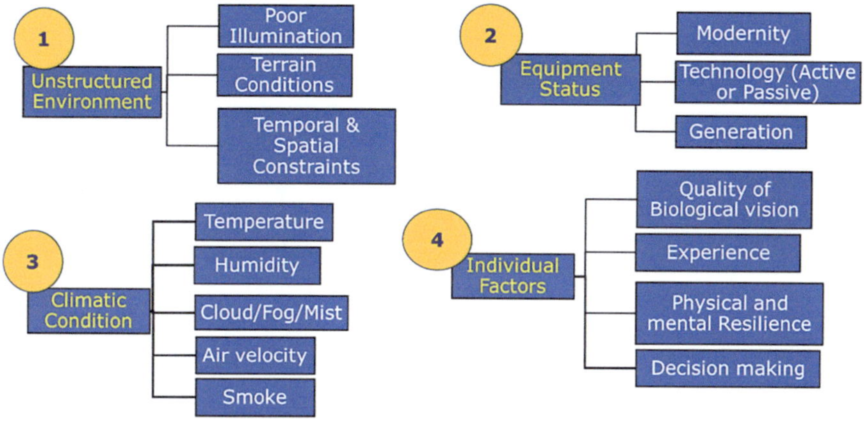

Fig. 15.1 Critical challenges of military operations

to unstructured conditions. On the other hand, profuse illumination of outdoor daylight condition contributes to reflection, and glare causing overstimulation of photoreceptors and may impact the soldiers' own hideout process. On the contrary, tiny illumination of natural nighttime conditions, i.e., Star Light (10^{-3} lx), Half Moon Light (5×10^{-2} lx), and Full Moon Light (10^{-1} lx) creates a degraded visual environment and may interfere with the detection of the hidden enemy, even after engaging the night vision equipment.

Equipment status implies the use of contemporary or state-of-the-art weapon systems, command-control stations, and accessory gadgets. However, cognitive demands for operations of active (powered) and passive (nonpowered) equipment or gadgets are entirely different, depending on the complexity of the system. *Adverse climatic conditions*, such as extreme heat, cold, humidity, hypoxia, fog, mist, smoke, and strong wind chill, create significant challenges for information processing and communication, leading to increased cognitive stress. Finally, *individual factors or intrinsic human factors* are considered which indicate that humans are finite beings having some unique capacities and limitations. These individual factors, e.g., physical and mental capabilities, education status, intelligence, decision-making, etc., result in individual and population variations and are linked to an individual's daily life survival. Therefore, it is perceived that military operations are multitasking in nature imposing multidimensional stresses on physiological, biomechanical, and cognitive architecture.

With a clearer visualization, it is predicted that soldiers' performance resides at the center of a triangular loop of human–machine–environment interaction (Fig. 15.2). The single-direction arrows demonstrate the direct influence of the individual component on soldiers' performance. Bidirectional arrows between "environment–human" and "machine–human" demonstrate the coupled interaction while performing specific tasks, which, in turn, affect soldiers' performance.

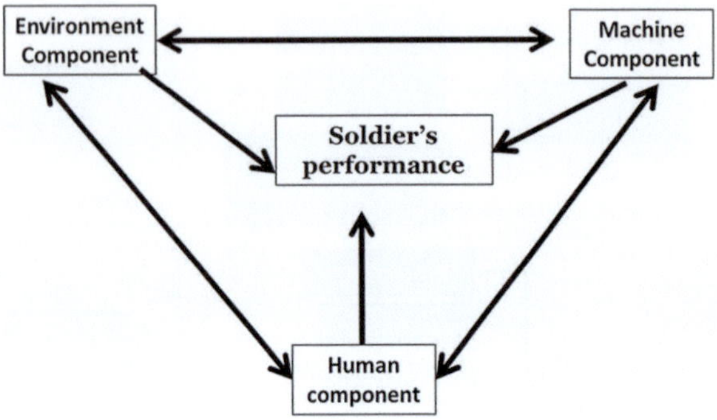

Fig. 15.2 Human–machine–environment interaction loop posing critical workload to the soldiers

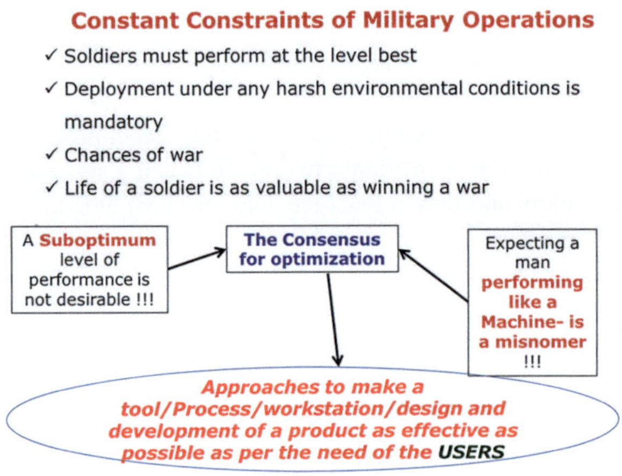

Fig. 15.3 Constraints of military operations and process of optimization

15.2.2 Amalgamating the Ergonomic Principles

Given this background, let's explore various mitigation strategies, keeping in mind the inherent constraints of military operations. Soldiers must perform at their peak despite harsh environmental conditions, the looming threat of conflict, and the fundamental truth that a soldier's life is as valuable as victory itself. Hence, achieving optimal performance is critical, contrasting with the simplistic notion that soldiers should operate like machines. Optimization is key in addressing this challenge. It involves refining tools, processes, workstations, and product designs to maximize effectiveness according to user needs (see Fig. 15.3). This requires a collaborative, interdisciplinary approach involving experts from fields such as physiology, anthropology, biomechanics, cognitive sciences, physiotherapy, design, data science, biomedical engineering, mechanical engineering, electronics, psychology, machine learning, and safety sciences. Each expert plays a vital role in developing solutions that enhance performance and ensure safety in demanding military contexts.

15.3 Cognitive Loading in Tri-forces

Operations of the armed force personnel span multiple terrains, i.e., high altitude, desert, and jungle including both daytime and nighttime. Most of their operations are physically as well as mentally demanding. In the case of air force operations, it is mentally challenging, as the personnel experience effects of G-force during high-speed maneuvers combined with pitch, roll, and yawing of the aircraft, rapid communication to command-control stations through audiovisual interfacing, rapid target identification, and decision-making. In contrast, naval operations are characterized by long-term deep sea (blue water) as well as off-shore activities where personnel have experienced major alterations in sensory cues, i.e., abrupt changes in visual environments (near point and far point (horizon) surrounding is only constituted with water) and lack of on-ground movements. Continuous exposure to pitch, roll, and yaw due to strong interaction between waveforms and the naval vessel, contributing to seasickness. Sharing the same onboard platform for habitability as well as occupational activities contributes to acute cognitive stress.

However, it is imperative to mention that cognitive stress is always felt prominently/dominantly over other forms of stress which is irrespective of operational platform specificity. This is because the soldier must make the right decision under any adverse condition, follow strict adherence to personal and group safety issues, and strive for the overall successful accomplishment of the mission. Under these circumstances, human senses, both general and special, exclusively interact with physical attributes of workstations and work environments, i.e., temperature, humidity, illumination, noise, vibration, work surface, equipment, furniture, and complex command-control system Thermal sensation, touch, and pressure are the most frequently exploited general sensation while performing acts like gripping, pulling, pushing, pressing, pinch-holding, turning, twisting, hammering, rubbing, etc. Conversely, the visual sensation is the most exploited special sense in military scenarios where various components of the vision system interact with the natural world and various indoor environments of command-control stations. Intensive human–computer interaction (HCI) in simulated training and modern warfare, particularly within the Command-Control-Communication-Computer-Intelligence-Surveillance-Reconnaissance (C4ISR) loop—often referred to as "The Military's Nervous System"—creates scenarios where screen-based information sharing becomes essential. Advanced C4ISR system provides desired prediction of situational awareness, knowledge of the adversary and environment, and shorten the time between sensing and response. Onscreen physical attributes like font/symbol characteristics (type, size, color, orientation, etc. placement of information, foreground–background color combination), speed and direction of dynamic information display, display size, etc. are important independent factors that may result in elevated cognitive stress and fatigue.

15.3.1 Complex Multitasking Military Environment Needs Critical Executive Brain Functioning

Intensive multitasking military environments engage almost three major domains of nervous systems, i.e., *peripheral, autonomic, and central nervous* system for every instance that occurs. Memory, attention, and reaction time are considered as dominant cognitive functions in military tasking. Attentional resources span over selective and sustained attention while doing vigilance tasks. The fastest and correct decision-making critically relies on 10–15 s duration specifically utilizing the resources of *working memory (visuospatial).* A rigorous unstructured environment stimulates the peripheral nervous system by propagating the kinesthetic sensation through various mechanoreceptors during various postural and movement maneuvers. Subsequently, cardiorespiratory and vascular systems get activated through the autonomic nervous system due to increased musculoskeletal activities. Finally, this subcortical information reaches various cortical areas where integration of information is done by collecting specific inputs from specific areas, e.g., the visual cortex, auditory cortex, and association cortex. Finally, a critical decision is made to continue with the task by regulating all the sensory routes. Furthermore, decisions are also influenced by an individual's emotional/motivational status through the interference of the limbic system.

15.4 Physical Ergonomics

Human–machine interfaces (HMIs) are very crucial for any workstation design as they directly influence how effectively and comfortably users can interact with a given technology. Physical ergonomics assessment includes the application of measurement tools for anthropometry, physiology, and biomechanics on the target user population for identifying ergonomic interventions and solutions for optimizing HMI issues and maximizing the performance and occupational health of the users, especially Indian soldiers, for whom research at DIPAS are directed [41–44].

Different Aspects of Human–Machine Interfaces Include:

1. *Anthropometry (or Users' Body Dimensions):* Design customization for target users in terms of anthropometric measurements is important for designing workstations compatible with the physical dimensions of the user population. For instance, the placement of controls, screens, and seating must accommodate various body sizes and shapes to ensure comfort and usability for most users. Such customization also helps enhance the comfortable reach and accessibility of all necessary control and display components, e.g., head-up displays and right and left multifunctional display panels, in cockpit design, where precise accessibility can impact operation efficiency and safety.
2. *User Comfort:* To minimize physical strain, common HMI issues that need to be considered for any workstation design are the positioning of controls and dis-

plays/monitors, keyboards, seating and standing console dimensions, space constraints, etc. with respect to limitations and capabilities of target users. These considerations can help reduce the risk of repetitive strain injuries and musculoskeletal disorders.
3. *Task Efficiency and Productivity (in Terms of Time of Task and Error):* Ergonomically designed workstations, processes, and equipment can streamline multiuser interaction interfaces and workflows, leading to more efficient use of the machine. By creating anthropometrically compatible interfaces and omitting unnecessary steps in the process workflow of the given task, users' efficiency, in terms of reduction in time taken and number of errors made, increases manifolds.
4. *Safety:* Ergonomic considerations also enhance safety. Properly designed HMIs can prevent accidents by ensuring that users can easily and safely interact with machines, with clear warnings and safe operational procedures integrated into the design.
5. *Accessibility:* Ergonomics considers the diversity of users, including those with disabilities, though this is not commonly considered for soldiers. However, certain research deliverables that can be used as spin-off benefit for civilian masses necessitate such considerations. Thus, designing HMIs that accommodate a wide range of physical abilities ensures inclusivity and allows more people to use the technology effectively.
6. *Visual Fields and Obscuration:* For enhanced performance and reduction in errors and accidents, it is mandatory that critical display screens/monitors, controls, and other interactive elements of the workstation should be within the optimal visual fields and should not be obscured by any part of the workstation, workspace, or users' body part under any condition.
7. *User Satisfaction:* Designing a product that is customized in terms of anthropometry, physical limitations, and capabilities of user's results in a more personalized experience, increasing user satisfaction and reducing frustrations, and physical and mental fatigue.

15.5 Anthropometry

Incorporating anthropometry in ergonomically designed HMI involves:

1. User-cantered design: Ensuring that the design process considers the physical and cognitive limitations, capabilities, and needs of users from diverse backgrounds.
2. Iterative testing: Testing prototypes with real users to gather feedback on comfort, accessibility, and usability, and incorporation of data obtained for further modifications in design for maximized user fit.
3. Flexibility: Designs need to be adjusted or customized to accommodate different user profiles, such as adjustable seats, screens, and control placements.

15.5.1 Importance of Anthropometric Measurements and Sizing in Design

Understanding anthropometric data are essential for designing products and systems that accommodate the physical dimensions, limitations, and capabilities of the target user population. Such ergonomic workspaces with optimized equipment and product interfaces used under optimal environmental conditions support and ensure efficient and safe use.

15.5.2 Application in the Defense Sector

In the defense sector, anthropometry plays a crucial role in designing military equipment, uniforms, vehicles, and systems. The goal is to ensure that military personnel should be able to operate effectively and comfortably under various operational conditions. One most important and critical distinction between the design needs of civilians and military/paramilitary personnel is the uniform. To date, no standardized indigenous sizing system exists for civilians in India. However, within the defense framework, size standards set by the general staff quality requirement (GSQR) are typically followed, and these standards require regular updates. This is because trends evolve over time, and if sizing standards aren't periodically adjusted, the discrepancies between actual and standard sizes may increase, eventually leading to sizing issues. A well-fitted garment or uniform should be comfortable, properly tailored, and offer sufficient room for ease of movement, free from unnecessary wrinkles or bunching. It should neither be overly tight nor excessively loose and should present a visually appealing appearance. Achieving a good fit relies on selecting a garment that aligns with precise body measurements.

"Fitting" involves adapting clothing designs to the human body, considering key measurements such as waist, hips, shoulder length, and other body dimensions according to design specifications. These measurements collectively determine the appropriate garment size. Clothing that is too tight impedes air circulation around the body, which can adversely affect comfort and mobility. Since individual preferences for "fit" may vary, what constitutes a "good fit" is inherently subjective. Differences in body size and shape in any population are largely influenced by factors such as nutrition, ethnicity, age, and gender. To ensure that clothing continues to meet needs effectively, it is crucial to regularly monitor and adjust to these variations. Secular trends, unsatisfactory uniform fitting, and color mismatch make it necessary to revise the sizing for combat uniform. Periodic updates to anthropometric measurements for creating sizing systems are undertaken by many countries approximately every 15–20 years. These updates are driven by secular trends and temporal variations in anthropometric dimensions. Consequently, it is essential to continually refresh anthropometric databases, especially when they are utilized for product design purposes, to ensure their relevance and accuracy.

Various nations have adopted sophisticated methodologies for the extensive collection of anthropometric data to formulate and refine sizing systems. Prominent among these are the United States, Germany, France, Italy, and several Asian countries. These countries regularly conduct comprehensive surveys to reassess and update their sizing standards, thereby ensuring that they remain aligned with contemporary anthropometric trends. In India, there has been a longstanding gap in the establishment of a standardized garment sizing system applicable across the entire country. In the past, anthropometric measurements have been taken conventionally to improve the fit and use of military equipment. An anthropometric study of US Army personnel conducted by [18] measured 1774 men and 2208 women to aid in the design and sizing of clothing and equipment. In the project like the United States and European (CAESAR project) [40] and the United Kingdom (Size UK survey) [4], 3D scanning was used for the similar purpose. Globally, the adoption of 3D scanning technologies for compiling extensive databases of anthropometric measurements, encompassing both civilian and defense personnel, has revealed significant cross-national variations. These differences are primarily attributed to the distinct ethnic compositions of various countries. Consequently, there is a pressing need for a comprehensive pan-India sizing system formulated from the body dimensions data collected using contemporary anthropometric measurement technologies. Some of the research work carried out by DIPAS in this field is given below:

I. *Sizing of Combat Uniform of Army*

Any sizing system refers to people, not to garments, and is intended to identify the size of an individual so that garments can be selected which fit him/her. The initial step in the sizing process involved gathering anthropometric measurements. The project titled "Survey of Size Rolls of Troops of the Indian Army Using 3D Human Body Scanner" was carried out by DIPAS with the collaboration from Master General of Ordnance and Sena Bhawan for developing the Indian Army Combat Uniform. The previous uniform was bought in certain sizes, and this had led to problems with standard size and fit. The project attempted to rectify these issues by conducting a sizing survey to generate a set of issue uniform that uniformly fitted soldiers. The first objective of this research was to update the sizing system for Indian Army combat uniform by measuring soldiers' body dimensions employing the advanced technology of 3D digital whole-body scanning. The second objective was to establish a sizing system for combat wear; namely shirts and trousers that are worn as a complete set that will improve on fit, comfort, and standardization across the army. To address these issues, this project was conducted by DIPAS on 7956 soldiers (males, age group 18–55 years) who were scanned with the help of a whole human body 3D scanner. Following this, the data underwent a thorough cleaning process to eliminate any outliers. A pivotal part of developing the sizing system was to categorize individuals with similar body measurements. This was achieved through the creation of distinct size classes.

The primary goal of establishing size classes is to group individuals with closely matching measurements, allowing them to be assigned a standardized garment size. This method aims to cover a wide range of body types with a minimal number of sizes while accommodating variations within a defined tolerance range. To determine the most critical variables for garment sizing, the correlations among various measurements were analyzed using ISI standards, and experienced tailors were consulted. This approach ensures that the size classes are both practical and reflective of actual fitting needs. The ethnic diversity of Indian people was the primary concern in this survey, which made the size of the sample large enough to meet the population characteristics.

Critical parameters for sizing were identified through an analysis of anthropometric variables, a review of existing literature, and consultations with tailors. For shirts, the key measurements were chest girth and arm length. For trousers, navel waist girth and navel height were used. For the combined shirt and trouser set, chest girth and navel waist girth were the primary parameters. Applying statistical methods 13 sizes of uniforms were designed intending to cover 97% of users. This makes each size a combination of chest girth and navel waist girth and of course provides a better fit for both garments. Additionally, tolerance in seams and fabric for undergarments and changes in size and figure were also considered. This new system offers a more innovative and accurate method of having a uniform size with great fit and comfort for the army personnel (L. [37]).

The key steps involved were:

I. *Identification of Critical Measurements:*

- *Shirts:* Chest girth and arm length were identified as the key measurements.
- *Trousers:* Navel waist girth and navel height were deemed critical.
- For the combined shirt and trouser set, chest girth and navel waist girth were selected as the primary parameters.

II. *Data Analysis:*

- The data were analyzed using the SPSS software, focusing on bivariate analysis to determine the most frequent combinations of chest and navel waist girths within the population. This analysis led to the creation of size classes.
- Five size intervals were established for chest girth (ranging from 75 to 124.99 cm) and six for navel waist girth (ranging from 65 to 124.99 cm). These intervals were then used to create 12 distinct uniform sizes, labeled 1A, 1B, 2A, 2B, and so on, with the numbers representing chest girth intervals and letters representing navel waist girth intervals.

III. *Uniform Sizing System:*

- Some of the new features implemented in the new sizing system include: a sizing system that effectively covers 97% population, and each one of the sizes signifies a new combination of chest girth and navel waist girth.
- Shirts: for instance, the size 1A shirt is said to measure chest girth of 85–94 cm with detailed measurements of inter-acromion length, arm length, and neck Girth.

- Trousers: in the same manner, the size 1A trouser is associated with a navel waist girth of 65–74.99 cm, with outside leg length, defined body rise with other dimensions as specified.

IV. *Design Adjustments and Allowances:*

- The system also has provisions for having slightly looser fits for undergarments and/or for additional clothing layers, for when the wearer is dressed to operate in different environmental conditions.
- Other modifications that were deemed necessary included restraining mechanisms like buckles, belts, or velcro to cater for people with smaller waist size.
- Expertise in detailing was also taken into consideration and can also be modified further here such as for thigh measurements and shirt sleeves.

V. *Implementation and Impact:*

- About 13 sizes as suggested in this system helped to accommodate most of the army personnel, and hence eliminate problems in fitting in uniforms, something that can hamper mobility, comfort, and overall efficiency in the theatre of war.

The primary objective of this survey was to develop an anthropometric database accurately representing the body dimensions of the Indian male population. The study proposed distinct sizing categories for upper and lower garments and recommended size ranges for protective ensembles based on the 3D-scanned measurements of Indian male individuals [38].

II. *Female Full Body Antiriot Gears*

Globally, full-body protectors are said to be crucial for police and security personnel, especially in riot situations. And, if the soldiers or police personnel are forced to use armors that do not fit properly, it may impede shooting accuracy, movement, and communication. A lacuna in research for Indian paramilitary forces was observed by DIPAS scientists when antiriot gears were considered, especially for female troops, who had to use the same personal protective ensembles (PPE) as their male counterparts, since PPE customized for Indian females was not available.

In the US military, female soldiers using body armor designed for men have faced multiple problems in the past, such as restricted mobility, reduced firearm accuracy, and difficulty entering or exiting vehicles quickly. Additionally, during hand-to-hand combat, poorly fitted armor can create gaps that can be unduly exploited by the enemies. A Military Medicine report published in 2015 highlighted that among the troops deployed in Afghanistan, about 20% of musculoskeletal injuries in female personnels were linked to incompatible body armor, as compared to only 4% of male personnel who experienced similar issues (Reference Military Medicine, 2015). Ill-fitting armor can significantly impact female service members adversely, particularly in shooting scenarios. In Australia, a survey was conducted to evaluate the effectiveness of body protectors for female troops, focusing on their fit and functionality. The findings indicated that 68% of participants experienced musculoskeletal pain and discomfort in areas such as the neck, upper and lower back, thighs, shoulders, abdomen, and hips, highlighting issues with poor fit. As a result, recommendations were made to redesign body armor specifically for female soldiers [15, 16].

In 2016, the National Conference for Women in Police proposed the introduction of gender-specific protective gear to enhance the safety and effectiveness of women in riot control and other demanding situations. This issue was brought to the attention of the Lok Sabha by the concerned Honorable Member of Parliament, who discussed the challenges faced by female personnel in the police and armed forces. In response, the Ministry of Home Affairs quickly acted by forming a committee to develop a plan for the procurement of these specialized assets. Subsequently, the Inspector General requested DIPAS to analyze and address the design shortcomings of existing protective gear for female troops. Besides analyzing the existing ones, DIPAS designed a full-body protector that conforms to the female anatomy, with adjustable and modular parts to ensure a better fit compared to the previously used protectors that were designed for the male counterparts.

Designing body armor specifically for women involves challenges in shaping the armor plates to fit the female body more accurately. This task is more complex as compared to designing armor for men, as they generally have a more uniform torso shape. In India, there is an urgent need for female-specific equipment, especially for female troops. Often, extra-large male body protectors were issued to female personnel to accommodate the bust size. However, these were too large, leading to operational challenges such as front plates pressing against the chin while seated, and difficulty bending at the waist when entering or exiting military vehicles. Equipment for the future has to involve gender-specific design and up-to-date materials in accordance with recently developed security specifications [34–36].

Workflow in Design of Full Body Protector (FBP)

The project began with the collection of anthropometric data from female police troops of various ethnic backgrounds. Issues faced by female troops while using the FBP designed for males were identified and documented, and feedback from female users was incorporated. Utilizing statistical analysis and anthropometric data, a new design was crafted after comprehensive research. The development process involved selecting key variables to identify typical clusters among female volunteers. Data mining and cluster analysis were used to group similar body contours, ensuring that clusters were internally consistent (with similar members) and externally distinct (differentiated from other clusters). To validate these clusters, a decision tree or regression tree was applied, visually profiling the data and identifying additional significant variables that aligned with the primary dimensions selected through principal component analysis (PCA). This approach helped in profiling both the target variable and the predicted variables.

Cluster analysis: nonhierarchical K-means clustering was used for its simplicity and optimal results, assuming that variables are independent and normally distributed within each group.

Classification tree: regression or classification trees were utilized to classify variables, validate cluster sets, and further profile the data. The cluster group served as the dependent variable, while body measurements acted as the independent variables.

III. *Designing Digital Avatar Using Anthropometry*

Anthropometry measurements of Indian Defense personnel were used to design and develop digital human models (DHM) representing the Indian male soldiers for use in all military product designing, e.g., uniform and PPEs and the future design processes involving human body/tissue characterization or visualization of hazardous environment where actual human being cannot be used, like, blast impact testing, using applications like finite element modeling (FEM) in military applications. A digital avatar of Indian defense personnel has been developed at DIPAS, enabling automatic anthropometric parameter modeling for each body facet. The proposed approach utilizes a limited number of anthropometric parameters (3–5 measurements) as input, simplifying the measurement process and enhancing user-friendliness in real-world applications. For the first time, body measurements of Indian male soldiers ($n = 3995$), aged 18–50 years, were recorded using a 3D whole-body scanner between 2015 and 2016. The age range was selected to represent young adulthood (18 years) to middle adulthood (50 years) [39]. This indigenously designed digital avatar holds significant potential for various industrial design applications, including automotive interior design, occupant cabin design, ready-made garment production, footwear and helmet design, and 3D computer games/animation.

15.6 Physiological Parameters as an Important Physical Ergonomics Evaluation Tools

Indian soldiers are very often deployed in extreme environments, and they are required to carry out operations in diverse and difficult terrains like moderate to high altitude areas of western and eastern parts of the Himalayas and lower altitudes of northern India, to the hilly and jungle areas of north-eastern and central parts to the deserts of western India [1, 19]. During operations under such situations, they must adapt to the challenging environmental conditions and execute their tasks efficiently under different personal and occupational constraints. They use machinery and equipment as per missions' requirements and very often need to carry very heavy backpacks containing weapons for patrolling or guarding borders and premises. Or they might need to use armored fighting vehicles (AFVs) to carry howitzer guns for long distances. Every task carries its own importance and requires stamina and focus for precise execution. The soldier needs to be combat fit under any adverse environment. Most of these tasks usually involve a high physical workload (moderate to heavy, [1, 3, 6, 28]). Therefore, it is important to measure soldiers' physiological stress while executing physically demanding tasks. However, real-time missions are risky and subject to permission and logistics issues. Hence, several scientific studies were carried out in simulated conditions inside defense laboratories and on real terrains with volunteers from the armed forces. Such studies attempted to determine the actual physical demand of the current mission and give the scientific community and the military decision-makers valid information to design the tasks appropriately ([22]; 5). Thus, there have been thorough investigations of the physical demands of different military operations within India (at DIPAS) and across the globe.

At DIPAS, the physical demand of any military task is quantified using proven techniques, namely, metabolic measurement or breath by breath gas analysis, accelerometer-enabled physiological status monitoring, thermal assessment, etc. Studies also reported the use of pre and postexercise measurement of power and strength, calculations of sweat rate and tracking volunteers' movement, speed, and elevation through GPS-enabled devices [7, 8]. Apart from this, extensive studies on the working environment with respect to temperature, relative humidity, air pressure, wind flow, illumination, noise, vibration, etc. have been carried out [20]. These attributes are critical because all of these factors affect physical performance and need to be considered while predicting certain outcomes. These studies are planned in such a way that the physiological variables can be correlated to the subjective feelings of the volunteers which are usually recorded after the whole event [13]. Heart rate variability (HRV) is another popular tool that predicts modulations in the autonomic nervous system (ANS) due to the appearance and disappearance of several work and environment-related stressors. This supports the other conventional technique to find out the real factors behind physical stress imposed during such activities [13]. This part of the chapter discusses the physiological studies carried out at DIPAS to quantify physical demands associated with several military activities in extreme environment and terrains in India.

Carrying moderate to heavy equipment for short and long distances is one of the key activities performed by the soldiers. One of the first studies was carried out in the early 1960s by Ramaswami et al. [29] which mathematically predicted the optimum load for various altitudes and speed combinations. The efficiency of the Nepalese porters was evaluated at their natural working conditions (high altitude) by Nag et al. [21]. Attributes of soldiers' load carriage were evaluated by exploring various combinations of slopes, speeds, and mass by Pal et al. [23, 25]. They suggested the optimal load for carriage at different speeds and gradients for Indian Army soldiers at sea level, considering energy cost, oxygen consumption, and the relative workload of each task. Effects of gradients and load on soldiers' performance during continuous uphill (UH) walking were evaluated by Paul et al. [26, 27] using cardiorespiratory, biochemical, and electromyographic (EMG) approach. The findings showed high physiological, biochemical, and EMG response from UH load carriage. The same group of researchers also separately evaluated the effects of continuous downhill (DH) walking with a load. Optimum loads with respect to continuous UH and DH walking tasks simulating mountain climbing were predicted with a similar scientific approach [9, 10, 14]. Distribution and placement of load on the body while walking at various gradients were studied on soldiers with combinations of gradient, speed, and load mass [5, 10). Two studies were conducted for the Indian Army in field locations to assess changes in physical performance and load carriage capacity due to high-altitude residency, considering variations in load, walking speed, and path slopes. The degree of physiological and EMG responses of important lower limb muscles changed with a rise in altitude, speed, load, and gradient. Based on the physiological responses of participants, recommendations on load carriage were made. Load carriage of 32% of body weight (BW) was recommended for long durations, such as 8 h at 3500 m. For shorter durations of 2 h at the same altitude, carrying 45% of BW was suggested. At 4300 m, during level walking, load carriage of 32% and 45% of BW was advised for 2 h at a speed of 3.5 km/h, among other recommendations [6, 11]. Other terrains like desert and jungle were also explored with respect to soldiers' load carriage. In the jungle, volunteers were subjected to the carriage of equipment up to 30 kg through a 1 km jungle track consisting of DH, plain, and UH jungle track. Uphill load carriage was found to be most critical followed by slippery downhill and muddy plain tracks [17]. Similarly, carrying load on sandy and hard surfaces was explored in another study. Here, optimum loads were determined based on physiological costs and were found that the sandy surface was more difficult than that of the hard terrain [2, 17]. The effect of walking on sand dunes with different magnitudes of loads and various distances was explored with the help of physiological measurements [24]. Another study examined long-distance (6 km) load marching at a self-selected speed while carrying up to 28.27% of body weight. It concluded that a speed of 3.95 ± 0.55 km/h should be limited to a maximum of 30 min [1]. The combination of load, speed, and duration across various terrains provides valuable insights into soldiers' routine load carriage tasks. This information may help optimize equipment weight, reducing the risk of cumulative carriage-related injuries and ultimately enhancing soldiers' operational readiness and mobility [2, 24, 30].

Like the soldiers' load carriage, several special operation training events were also studied separately. Training events were targeted for the jungle and counter-insurgency operations, urban-to-semiurban low-intensity conflicts (LIC), and high-altitude warfare techniques. The actual demand of such practices was quantified. Here short duration events were compared to establish terrain-specific stress and typical requirements [7, 8]. Slow versus fast finishers' physiological and biomechanical perspectives were evaluated in another study [8]. The results confirmed terrain-related difficulties in the jungle and better strategies for working safely. The practices targeted for the LICs involve running with or without equipment followed by activities like rope climbing, shuttle runs, trench crossing, etc. [31]. The events targeted for the high-altitude warfare included skiing through snowy terrains and activities like patrolling, navigating, rescue operations, etc. Scientific evaluation of such activities is fruitful in understanding the actual demand and the inner physiological events and necessary adjustments needed [31].

The operations of heavy military equipment like battle tanks and infantry troop's careers are also explored with respect to physiological evaluations. Such studies provided valuable insights into the adaptability of the soldiers to their working environment inside AFVs and their physical and cognitive workload [12, 20, 32]. The physiological quantification of military activities carries special importance as it provides essential information about soldiers' ability to perform difficult tasks. Their stress levels and ability to manage and withstand the difficulties are portrayed through these studies. In a way, physiological evaluation helps to optimize soldiers' performances by understanding the environment and terrain-related impacts [33]. Also, valid recommendations for carrying out physically intense tasks are derived based on the outcomes.

15.7 Human Biomechanics as a Tool of Physical Ergonomics in Indian Defense

Biomechanics is the science that applies principles of the human body with special reference to mechanics, that is, the study of internal and external forces and their effects on a static or dynamic human or animal body. This approach attempts to combine the science of engineering with human physiology and anatomy. Kinematics is a branch of biomechanics that focuses on studying movement from a geometrical perspective. In contrast, kinetics examines the forces and factors that cause a body to move in a particular way.

15.8 Broad Applications of Human Biomechanics

Biomechanics encompasses the study of everything from the inner workings of a cell to limb movement and development, as well as the mechanical properties of soft tissue and bones. A deeper understanding of the physiological behavior of living tissues enables researchers to advance tissue engineering and develop improved treatments for various pathologies. Knowledge of biomechanics, including kinematics and kinetics, is primarily applied to studying the human musculoskeletal system. This research utilizes force platforms to analyze ground reaction forces and infrared videography to track the trajectories of markers attached to the body for studying three-dimensional motion. Additionally, electromyography (EMG) is used to assess muscle activation, enabling the investigation of muscle responses to external forces and perturbations.

15.9 Three-Dimensional Gait and Motion Analysis: An Important Ergonomics Tool

Human motion results from complex data processing controlled by the central nervous system. This process generates a sequence of coordinated neural commands sent to various muscles (actuators) to produce movement. All voluntary movements, including walking, involve intricate interactions between the brain, spinal cord, peripheral nerves, muscles, bones, and joints. So far, research on different systems of the human body, e.g., cardiovascular, respiratory, excretory, etc. has been carried out in detail. But, till today we know very little about the skeletomuscular system and human/animal motion/movement. In a treatise on the movements of animals, Aristotle (384–322 BC) formulated theories on the control of movement based only on observations. He systematically compared and contrasted animal and human patterns of movement, deducing criteria necessary for motion to occur, the most important being vertical displacement that occurs during upright walking. Most mammals are quadrupedal and when such animals move, one limb progresses while three limbs support the body on the ground providing better stability. As humans adopt an upright posture with a bipedal gait, this stability is diminished, and required an increased neural control to enable efficient locomotion and movement. It is interesting to know that simple human walking demands about 25–60% of one's energy expenditure depending upon the age of the individual. During the seventeenth century, Galileo (1564–1642 AD) discovered the principles of motion, using time and distance to calculate the velocity and acceleration of various objects as they fell through space.

The study of human motion to understand the level of stress on the skeletomuscular system is incomplete without quantification of segmental and whole-body acceleration, velocity, speed, displacements, changes in the center of gravity, different angles, moments, powers, etc. Therefore, integrating various technologies available can provide greater and in-depth information on human motion.

Visual motion analysis is the most primitive method used to analyze motion by the unaided human eye in two dimensions. This method, though versatile, suffers from various limitations: the observations are transient and no permanent record can be kept; the human eye cannot accurately observe high-speed events, as it can only perceive movement without providing information on ground reaction forces. Moreover, observation relies entirely on the skill of the individual, making it highly susceptible to human error.

With the advancement of technology, next came the videotape examination method, which is not entirely objective as it does not provide quantitative data in numerical form. This method typically involves a camcorder for recording and a separate video cassette recorder for replaying the tapes, thus providing some permanent recording of the observations.

Various systems are currently available for recording and analyzing motion for different purposes. While there are notable differences in convenience and technology among cine film, videotape, television/computer, and optoelectronic systems, the data processing methods for three-dimensional motion measurement remain similar. Motion capture can be conducted in either two or three dimensions, with three-dimensional measurements requiring two or more cameras. The simplest kinematic measurements use a single, uncalibrated camera, though such measurements are often inaccurate. To achieve reliable motion analysis, a calibrated three-dimensional system—incorporating multiple viewpoints—is essential.

The quantification of forces and related parameters became possible with the development of force platforms. The first crude force plate was designed by Amar in 1924, followed by an improved mechanical version by Elfman in 1938, where applied force resulted in the proportional movement of a pointer. Modern force plates utilize piezoelectric sensors, offering enhanced sensitivity and accuracy. These platforms measure the total force applied by the foot to the ground but do not provide force distribution details (e.g., heel versus forefoot contact). While some force plates measure only one component of force (typically the vertical or z-axis force), most now provide a full three-dimensional analysis of the average ground reaction force vector. The processed electrical signals yield three force components (lateral, fore-aft, and vertical), the two coordinates of the center of pressure, and moments about the vertical axis. By analyzing the center of pressure from both feet, researchers can identify abnormal foot contact patterns.

The style or pattern of an individual's walking motion is known as gait, and gait analysis is a noninvasive diagnostic method for assessing walking or running mechanics. With computerized gait analysis, biomechanical abnormalities can be detected in various populations, including children with developing feet, athletes, and older adults experiencing mobility difficulties. Common biomechanical disorders identified through gait analysis include excessive pronation, over-supination, ankle instability, pelvic tilt, poor shock absorption, leg length discrepancies, and unusual walking or running patterns.

Gait Analysis in Sports Injuries

Gait analysis is a valuable tool for assessing running biomechanics, particularly in sports-related injuries. Using a treadmill, digital video, and computer software, an individual's running or jumping mechanics can be analyzed frame by frame, making it easier to detect biomechanical faults and overuse injuries. Some of the most common sports injuries identified through gait analysis include:

Shin splints (lower leg pain)
Plantar fasciitis (heel pain)
Achilles tendonitis
Ankle sprains
Knee pain
Stress fractures

By identifying movement inefficiencies, gait analysis helps in injury prevention, rehabilitation, and performance optimization for athletes and individuals with mobility challenges.

The methods of gait analysis have immensely aided in advancing the knowledge of the mechanical and physiological mechanisms of human walking. Human motion are measured and studied in Gait laboratories. The body's energy resources, the internal and external forces acting on the body, and the resulting movements can be expressed in mathematical terms for objective assessments. Gait laboratories contain the necessary instruments to visualize and quantify parameters that are used to describe walking. The evolution of gait laboratories has been intimately linked to advances in technology and scientific methods.

15.10 Application of Human Biomechanics Research in Defense

Several studies in the area of physical ergonomics with special emphasis on kinematics, kinetics, and electromyography have been conducted so far in DIPAS, Delhi. A few of them have been briefly discussed below.

15.10.1 Soldiers' Load Carriage

The advent of modern technology has reduced, but not eliminated, the manual handling of loads. Load carriage remains an integral part of a soldier's daily routine, whether stationed in the plains or at high altitudes. With the introduction of advanced equipment, arms, and ammunition designed to enhance lethality and sustainability, the composition and weight of load carriage have become increasingly important. However, battlefield mobility and maneuverability are critical factors that must never be compromised in conflict and combat situations.

Military operations that demand exceptional physiological and psychological capabilities, leading to significant human adaptation for survival and performance, are classified as extreme military activities. These scenarios are primarily encountered in wartime situations. Since direct assessment of biomechanical, physiological, and cognitive demands during battle is not feasible, ergonomic methodologies and principles are employed to study these demands under simulated field conditions. Quantifying data from such simulations enhances our understanding of battlefield conditions and enables the development of evidence-based countermeasures to better equip personnel for extreme military activities.

The Defence Institute of Physiology & Allied Sciences (DIPAS) has conducted comprehensive physical ergonomics assessments of load carriage among Indian infantry soldiers across various environments, including sea level, hot-humid and hot-dry deserts, jungles, and high to extremely high altitudes. These studies have evaluated biomechanical and physiological stresses during load carriage operations with varying load magnitudes and marching speeds, ultimately leading to recommendations for optimal and maximum load capacities for the Indian Army. The findings have contributed to the design and development of ergonomically optimized load carriage ensembles, improving soldiers' efficiency and reducing physical strain.

15.11 Assessment, Design, and Development of Military Workstations, Vehicles, and Arms

With the introduction of proactive ergonomic design techniques, various computer-aided design (CAD) software programs have been developed, enabling digital product design and evaluation using digital human models (DHM). These models provide a significant advantage by allowing designers to assess the potential benefits and limitations of different design configurations without the need for costly physical prototypes, which were traditionally used.

DHM can be created for any population or even an individual. Once it is developed it can be interfaced with the virtual prototype of any product or workstation for ergonomic analysis of the physical fit of the operator to the system concerned. These software tools help reduce product design time, lower development costs, improve quality, increase productivity, enhance safety, and optimize the man–machine interface. A few DHM have been developed at DIPAS based on various ethnic populations of the Indian Armed Forces using human modeling and simulation software in a virtual reality environment. Digital prototypes of a fighter aircraft, a bulletproof military vehicle, a rocket launcher, a futuristic main battle tank, and most recently digital prototype of an indigenously developed weapon system have so far been assessed for optimized human comfort and performance in terms of limitations and capabilities in biomechanics of target population by this technique at DIPAS. Specific design improvement recommendations have been provided to laboratories developing these workstations. This approach allows designers to ensure an optimal human fit and compatibility between the operator, workstation, and working environment. In the long run, it promotes both the mental and physical well-being of users. Additionally, this technique has broad applicability beyond the military, extending to civilian sectors such as industry, health care, sports, and recreation [45].

15.12 Design and Development of Functional Clothing

Clothing is primarily designed to protect the human body from potential injuries. Beyond this basic function, functional clothing enhances the wearer's ability to perform specific tasks more effectively. For instance, soldiers depend on specialized clothing to safeguard against various adversities and threats, including extreme cold, ballistic projectiles, radiation, chemicals, and biological agents. Examples of such protective gear include body armor systems, nuclear-biological-chemical (NBC) war suits, and space suits.

At DIPAS, an indigenous bulletproof jacket (BPJ) was designed with strict adherence to the given surface area and weight of BPJ, to help Indian soldiers during insurgencies or low-intensity conflicts. Using an existing anthropological database of Indian Army personnel, a number of digital human models (DHM) of different sizes were developed using simulation and human modeling software. Once the bulletproof jacket (BPJ) assembly—including soft armor plates, hard armor plates, neck, collar, shoulder, and groin guards—was modeled, it was integrated with the digital human model (DHM). Virtual kinematic maneuvers were then conducted to perform an ergonomic analysis of its physical fit under various operational conditions. This digital design process effectively addressed many of the challenges soldiers would face in real-life scenarios. As a result, the BPJ prototype ensures optimal human comfort and fit, which will ultimately enhance soldiers' performance and sustainability in low-intensity warfare and insurgency situations.

15.13 Conclusions

Overall, the integration of ergonomics and anthropometry in HMI design ensures that technology is user-friendly, efficient, safe, and accessible. By addressing the physical and cognitive needs of users, designers can create interfaces that enhance performance and satisfaction, leading to better outcomes in various applications ranging from consumer electronics to complex industrial systems. Anthropometry the science of measuring human body dimensions is critical in designing clothing and military gear that are safe, comfortable, and effective. Historically, militaries worldwide have used anthropometric data to enhance the design of uniforms and equipment. The research undertaken at DIPAS for optimizing human–machine interfacing issues mandatorily uses the following user-centric approaches:

1. Creation of a Database on Soldiers' Capabilities and Limitations:
    - *Ob*jective: Develop a database documenting soldiers' performance metrics on platform-specific tasks.
    - Considerations: Ensure the data collected encompasses a range of capabilities and limitations, and that it is updated regularly. Include details on both physical and cognitive factors.
    - Action steps: Collaborate with experts in military physiology and psychology. Use simulated tasks to gather data and continually refine the database with real-world performance insights.

2. Ergonomic Case Studies on Existing Platforms:
    - Objective: Analyze current platforms to evaluate ergonomics using time and motion studies.
    - Considerations: Address not only the physical setup but also the interaction between the workstation, environment, and tasks.
    - Action steps: Implement structured work-study methods to quantify task demands and identify areas for improvement.

3. Conducting Human Factors Integration (HFI) Surveys:
    - Objective: Use standardized questionnaires to assess human factors across platforms.
    - Considerations: Develop or adopt validated surveys to ensure comprehensive coverage of human factors.
    - Action steps: Analyze survey results to identify common issues and areas for enhancement in platform design.

4. Creation of Standards for Work, Workstation, and Environments:
    - Objective: Develop standards tailored to indigenous requirements.
    - Considerations: Standards should address all aspects of the work environment, including physical, cognitive, and social factors.
    - Action steps: Involve a diverse team of experts to create and validate these standards.

5. Human Reliability Analysis (HRA):
    - Objective: Analyze potential errors and reliability if platforms adhere to standards.
    - Considerations: Use models that account for human error in different scenarios.
    - Action steps: Perform simulations and real-world testing to evaluate the effectiveness of the standards in minimizing errors.

6. Continuous Collaboration:
    - Objective: Establish a feedback loop involving scientists, users, developers, and policymakers.
    - Considerations: Regular meetings and updates will ensure all stakeholders are informed and can contribute effectively.
    - Action steps: Create formal channels for communication and feedback, such as regular workshops or review sessions.

7. Preparation of Ergonomic Checklists:
    - Objective: Develop checklists for operations and maintenance based on standards.
    - Considerations: Ensure checklists are practical and cover all critical aspects of ergonomic design.
    - Action steps: Regularly review and update checklists based on feedback and new findings.

8. Building State-of-the-Art Training Facilities:
    - Objective: Provide cognitive training through real-time simulations or immersive virtual environments.
    - Considerations: Facilities should replicate military operational scenarios as closely as possible.
    - Action steps: Invest in advanced simulation technology and ensure it integrates with military operational cues.

9. Integration of Neurocognitive Techniques:
    - Objective: Use advanced techniques to collect and analyze brain response data.
    - Considerations: Ensure that the integration of neurocognitive data is compatible with other sensory data for comprehensive analysis.
    - Action steps: Partner with cognitive neuroscientists and engineers to develop and implement these techniques.

10. Conducting Ergonomic Audits:
    - Objective: Regularly review and assess ergonomic practices and designs.
    - Considerations: Audits should be thorough and address both new and existing systems.
    - Action steps: Develop an audit protocol and schedule periodic reviews to ensure ongoing compliance and improvement.

References

1. Arya K, Yadav A, Bhattacharyya D, Chatterjee S, Pal M. Effect of long-distance load marching on physiological responses at desert environment. BLDE Univ J Health Sci. 2022;7:204–10. https://doi.org/10.4103/bjhs.bjhs_73_22.
2. Bhattacharyya D, Pal M, Chatterjee T, Majumdar D. Effect of load carriage and natural terrain conditions on cognitive performance in desert environments. Physiol Behav. 2017:179. https://doi.org/10.1016/j.physbeh.2017.06.01.
3. Bhattacharyya D, Chatterjee T. Military footwear and extreme environment operations: an ergonomics perspective. Chapter 12. In: Tulsawani R, Vohora D, editors. Adaptation under stressful environments through biological adjustments and interventions; 2024. p. 162–84. https://doi.org/10.1007/978-981-99-7652-2_12161.
4. Bougourd, J and Trelaven, P. UK National Sizing Survey - SizeUK. Proceedings of the International Conference on 3D Body Scanning Technologies, Lugano, Switzerland, 2010;19–20.
5. Chatterjee T, Bhattacharya D, Pal M, Majumdar D. Cardiorespiatory Changes with Compact Backpack System and Distributed Mode of Load Carriage. Indian J Physiol Pharmacol 2012; 56(2):130–136.
6. Chatterjee T, Bhattacharyya D, Pramanik A, Pal M, Majumdar D, Majumdar D. Soldiers' load carriage performance in high mountains: a physiological study. Mil Med Res. 2017a;4:6.31.
7. Chatterjee T, Bhattacharyya D, Yadav A, Arya K, Meena R, Pal M. Quantification of mental and physiological workload associated with two specialised military running events on different terrains. Defence Life Sci J. 2022;7(3):169–78. https://doi.org/10.14429/dlsj.7.17841.
8. Chatterjee T, Bhattacharyya D, Yadav A, Pal M. BMJ Mil Health, Epub ahead of print. 2022. https://doi.org/10.1136/military-2022-002154.
9. Chatterjee T, Paul S, Pramanik A, Pal M, Majumdar D. Assessment of muscular fatigue with electromyography on lower back and leg muscles during continuous uphill and downhill load carriage task. Defence Life Sci J. 2017;2(1):21–9. https://doi.org/10.14429/dlsj.2.11018.
10. Chatterjee S, Chatterjee T, Bhattacharyya D, Sen S, Pal M. Effect of heavy load carriage on cardio-respiratory responses with varying gradients and modes of carriage. Mil Med Res. 2018;5:26. https://doi.org/10.1186/s40779-018-0171-8.

11. Chatterjee T, Bhattacharyya D, Pramanik A, Pal M, Majumdar D. Electromyographic and cardiorespiratory responses of load carriage at different high altitudes with varying gradients. J Arch Mil Med. 2017;5:e46044. https://doi.org/10.5812/jamm.46044.
12. Chatterjee T, Bhattacharyya D, Majumdar D, Meena R. Application of heart rate variability to quantify physiological and cognitive stress of several operations of infantry and mechanized forces. Communicated to the proceedings of HWWE, ACED 2023 international conference: Springer Nature (accepted 2024).
13. Chatterjee T, Bhattacharyya D, Yadav A, Pal M. Heart rate variability, task load and perceived exertion associated with a long-distance military ski exercise: a pilot study. Indian J Physiol Pharmacol. 2022;66(3):196–202. https://doi.org/10.25259/IJPP_101_2022.
14. Chatterjee T, Paul S, Pramanik A, Chowdhury B, Pal MS, Majumdar D. Cardio-respiratory and metabolic changes during continuous uphill-downhill load carriage task. 2015. https://www.researchgate.net/publication/315046455.30.
15. Coltman C, Steele JR, Spratford, W, Molloy RH. Are female soldiers satisfied with the fit and function of body armour? Applied Ergonomics 2020;89:103197. https://doi.org/10.1016/j.apergo.2020.103197.
16. Coltman CE, Brisbine BR, Molloy RH, Ball NB, Spratford WA, Steele JR. Identifying Problems that Female Soldiers Experience With Current-issue Body Armour. Applied Ergonomics 2021;94:103384.
17. Ergonomic design of load carriage ensembles and developing load standards for high altitude, desert and jungle environments Report no. DRDO-DIPAS-DLS-TR-06-13, 2015.
18. Gordon CC. (1986). Anthropometric sizing and fit testing of a single battledress uniform U.S. Army men and women. Performance of Protective Clothing, ASTM STP 900, American Society for Testing and Materials.
19. Majumdar D, Pal MS, Majumdar D. Effects of military load carriage on kinematics of gait. Ergonomics. 2010;53(6):782–91.
20. Malhari A, Bhattacharyya D, Arya K, Chatterjee T, Pal M. Assessment of vibration exposure and physiological response of crew members during infantry combat vehicle (ICV) operation: a pilot study. BMJ Mil Health. 2019;165:152–8.
21. Nag PK, Sen RN, Ray US. Optimal rate of work for mountaineers. J Appl Physiol Respir Environ Exerc Physiol. 1978;44(6):952–5.
22. Nindl BC, Billing DC, Drain JR, et al. Perspectives on resilience for military readiness and preparedness: report of an international military physiology roundtable. J Sci Med Sport. 2018;21:1116–24.
23. Pal MS, Majumdar D, Bhattacharyya M, Kumar R, Majumdar D. Optimum load for carriage by soldiers at two walking speeds on level ground. Int J Ind Ergonom. 2009;39:68–72.
24. Pal M, Yadav A, Arya K, Bhattacharayya D, Chatterjee T, Kumar B. Optimization of load carriage at desert environment. Int J Ind Ergon. 2020;79:102986. https://doi.org/10.1016/j.ergon.2020.102986.
25. Pal M, Majumdar D, Pramanik A, Chowdhury B, Majumdar D. Optimum load for carriage by Indian soldiers on different uphill gradients at specified walking speed. Int J Ind Ergon. 2014;44:260–5.
26. Paul S, Bhattacharayya D, Chatterjee T, Majumdar D. Effect of uphill walking with varying gradient and speed during load carriage on muscle activity. Ergonomics. 2016;59(4):514–25.
27. Paul S, Bhattacharayya D, Chatterjee T, Pal M, Majumdar D, Singh S, Majumdar D. Physiological and biochemical responses during incremental uphill load carriage. Int J Ind Ergon. 2015;50:26. https://doi.org/10.1016/j.ergon.2015.08.010.
28. Roy TC, Knapik JJ, Ritland BM, Murphy N, Sharp MA. Risk factors for musculoskeletal injuries for soldiers deployed to Afghanistan. Aviat Space Environ Med. 2012;83:1060–7.
29. Ramaswamy SS, Dua GL, Raizada VK, Dimri GP, Vishwanathan KR, Madhavaiah, Srivastava TN. Study of load carriage at high altitudes. Part-I. Relative effects of the magnitude of the load carried and the steepness of the terrain on the optimum speed of march. Ergonomics. 1963;30(5):567–75.

30. Technical Closure Report of project entitled "High Altitude Rapid Induction and Acclimatization Strategies" (Report No. DRDO-DIPAS-DGLS-TR-03-2022) DIP-261.
31. Technical Closure Report of project entitled "Human Machine Interface issue for performance improvement of Indian soldiers under different occupational environment" (Report No. DRDO-DIPAS-DGLS-TR-02-2022) DIP-266.
32. Malhari AA, Bhattacharyya D, Chatterjee T, Mani KV, Pal M. Effect of combat vehicle operation on cognitive workload and performance. Defence Life Sci J. 2021;6(2):177–86. https://doi.org/10.14429/dlsj.6.16818.
33. Chatterjee T, Majumdar D, Bhattacharyya D, Pal M, Meena R. Heart rate variability and subjective response during a simulated battle tank operation: a pilot study. The proceedings of HWWE 2022 through Springer Nature network (accepted in 2024).
34. Robert Varte L, Kakkar D, Rawat S, Singh I, Chaudhary Y, Kumar B. Data mining anthropometric parameters for the design and sizing of female full body protector. DRDO-Defence Institute of Physiology and Allied Sciences (DIPAS), Delhi – 110 054, India. Defence Life Sci J. 2021;6(4):275–83. https://doi.org/10.14429/dlsj.6.16602. DESIDOC.
35. Robert Varte L, Rawat S, Singh I, Choudhary S, Singh S, Kakkar D, Chaudhary Y, Kapoor N, Kumar B. Personal protective ensemble reference size development for Indian male defence personnel based on 3D whole body anthropometric scan. J Text Inst. 2021;112(4):620–7. https://doi.org/10.1080/00405000.2020.1770517.
36. Robert Varte L, Rawat S, Singh I, Choudhary S, Singh S, Kakkar D, Chaudhary Y, Kapoor N, Kumar B. Sizing Indian female full body protector ensemble. Faridabad: FIPS; 2021.
37. Robert Varte L, Rawat S, Gautam P, Singh I, Chaudhary Y, Kakkar D, Sharma YK. Size India Defence Personnel and Development of Sizing Software. Int J Sci Eng Appl. 2023;12(12):67–75, ISSN:- 2319–7560. https://doi.org/10.7753/IJSEA1212.1017. www.ijsea.com 67.
38. Robert Varte L, Rawat S, Singh I, Choudhary S, Singh S, Panjwani U. Lower and upper clothing ensemble sizing using a 3D whole body scanner in Indian adult males. Defence Institute of Physiology and Allied Sciences (DIPAS) Defence Research and Development Organization (DRDO), Lucknow Road, Timarpur, Delhi- 110054, India. J Hum Ergol. 2017;46:29–38.
39. Robert Varte L, Rawat S, Singh I, Chaudhary Y, Kakkar D. Digital Avatar of Indian defence males generated from 3D anthropometric scan. NDT. net Issue: 2022–04.NDE 2021-Virtual Conference & Exhibition, 09–11 Dec 2021 by Indian Society for NDT (ISNDT) (NDE-India 2021). Special Issue of e-Journal of Nondestructive Testing (eJNDT) ISSN 1435–4934. 27(4).
40. Robinette KM, Daanen HAM, Paquet E. The CAESAR project: A 3D surface anthropometry survey Proceedings of the Digital Imaging and Modeling, Ottawa, Canada. 1999.
41. Sen RN. The society's lecture 1983, The Ergonomics Society. Ergonomics. 1984;27(10):1021–32.
42. Van Cott HP. High technology and human needs. Ergonomics. 1985;28(8):1135–42.
43. Wisner A. Ergonomics in industrially developing countries. Ergonomics. 1985;28(8):1213–24.
44. Wisner A. Understanding problem building: ergonomic work analysis. Ergonomics. 1995;38(3):595–605.
45. Zachariah T, Kishnani S, Parmanik SN, Selvamurthy W. Body measurements-design applications and body composition. Defence Institute of Physiology and Allied Sciences, Delhi, DRDO, MOD, Delhi 110054, India, 2001.

Further Reading

Ganguli A. Ergonomics in India – Focus and Change, Proceedings of International Conference on Humanizing Work and Work Environment, 2009, Kolkata, India.

Haslam RA. Targeting ergonomics interventions-learning from health promotion. Appl Ergon. 2002;33:241–9.

Moray N. Ergonomics and the global problems of the twenty-first century. Ergonomics. 1995;38(8):1691–707.

Role of Precision Medicine in Soldier's Health Management: Application of Artificial Intelligence and Machine Learning

16

Ashutosh Katiyar, Manish Sharma, and Gaurav Kumar

Abstract

Modern warfare exposes soldiers to extreme environmental conditions and mental stress, which can significantly impact combat outcomes if effective health management of soldiers is not implemented. In order to ensure the optimal deployment of soldiers, it is necessary to employ precision medicine techniques that involve the use of extensive multidimensional biological datasets for accurate diagnosis, therapeutic intervention, and prognosis monitoring. This chapter addresses the present health obstacles encountered by soldiers and examines the advantages of implementing artificial intelligence–machine learning (AI–ML) approaches in precision medicine, using artificial intelligence and machine learning in enhancing the health of soldiers. AI has the potential to enhance precision medicine for the management and monitoring of soldiers' health by facilitating the identification of disease subtypes, forecasting disease progression, selecting personalized treatments, and enabling remote monitoring. This chapter also explores several machine learning models and their clinical significance in predicting illness risk using clinical and biological data, which are also applicable to the health of soldiers. Moreover, we also discuss the existing problems and complications that may impede technological progress in this field. In summary, this chapter highlights a new age of medical science using AI–ML in the health and disease management of soldiers, which has seen significant advancement in the prediction and diagnosis of soldiers' health and has far-reaching consequences in deployment and post-deployment health hazard.

A. Katiyar · G. Kumar (✉)
School of Life Sciences and Biotechnology, CSJM University, Kanpur, Uttar Pradesh, India
e-mail: gauravk@csjmu.ac.in

M. Sharma
Defence Institute of Physiology and Allied Sciences (DIPAS), Defence Research and Development Organization (DRDO), Delhi, India

S. Srivastava et al. (eds.), *Health and Sustenance in Extreme Environment*,
https://doi.org/10.1007/978-981-96-8642-1_16

Keywords

Physiological stress · Artificial intelligence · Supervised learning · Unsupervised learning · Deep phenotyping · Machine learning

16.1 Introduction

Contemporary combat involves intense cognitive and physical abilities, combined with long and severe physical and stress situations with little room for mistake [1]. It takes a significant toll on the health of soldiers in adverse environmental conditions like heat, cold, and high altitudes; stresses, combined with rigorous labor, while carrying large loads or wearing body armor and chemical protection equipment, may make them vulnerable to acute and chronic disease [2]. This calls for better health management to gain tactical advantage during deployment of soldiers.

Notably, genetic background and exposure to the environment are important underlying pillars of soldiers' health but are often overlooked; rather, the one-size-fits-all principle is used for disease management in soldiers [3]. Precision medicine assumes that subpopulations with a different biology and other features may be properly recognized and treated differently [4]. The recent development of precision medicine provides new hope for individualized treatment; this approach requires records of multidimensional data; its training uses different AI–ML models like supervised and unsupervised learning, unlike the current treatment protocols. Recently, computer capabilities have developed tremendously, and the worldwide scientific community has made massive multidimensional biological and clinical databases accessible. Cancer and cardiovascular disease prediction systems using artificial intelligence have demonstrated encouraging results, predicting illness risk more accurately [5, 6]. The success achieved in different biomedical domains can also be translated into the field of soldier health management.

Proper implementation of precision medicine in AI–ML requires reliable physiological measurements, environmental component quantification, and genetic variant identification with known penetrance. Moreover, AI can help organizations capitalize on precision medicine and expand our understanding of several chronic illnesses faced by soldiers like high-altitude pulmonary edema (HAPE), high-altitude cerebral edema (HACE), frostbite, mountain sickness, hypoxia, dehydration, heat stress, anemia, PTSD, depressive disorder, and combat stress. The success of precision medicine and AI–ML implemented for soldiers' health monitoring and measurement will require the use of better data-acquitting models of genetic, physiological, and environmental conditions [7]. Next is data integration and analysis, i.e., data collected from various sources like genomics, wearable devices, and medical records needs to be analyzed comprehensively, which can ensure the data is interpreted accurately to form a diagnosis [8] This chapter summarizes soldiers' health challenges and opportunity of AI–ML in the efficient management and prevention of soldiers' health concern.

16.2 Current Health Challenges Faced by Soldiers

Soldiers face a variety of health challenges due to the extreme environments in which they are deployed. These challenges vary from physical to psychological, which affect their overall health and their efficiency to perform operations. Not only are there challenges related to physical, mental, and nutritional conditions [2], but there are also several other challenges related to technology and logistics, like limited access to diagnostic tools, i.e., in extreme and remote environments (deserts, mountains, forests), access to advanced tools is difficult, which affects the process of obtaining accurate diagnosis results [9, 10]. Medics working in the military need to be trained to recognize and diagnose vast conditions with limited resources, ensuring that they have the necessary experience [11]. It is pertinent to collect data from multiple sources like wearable devices, medical records, genetic data in order to ensure accurate diagnosis. The following section discusses various challenges faced by soldiers in different environment conditions and limitation in diagnosing several acute and chronic disorders.

16.2.1 Physiological Health

Soldiers deployed at high altitude face a risk of cold-related injuries like frostbite [12] and other high-altitude related illnesses like mountain sickness [13] and hypoxia [12]. The climate change continuously at high altitude can cause HAPE, which causes coughing, chest tightness, difficulty breathing, and fatigue. It can be diagnosed with the help of lung crackles through a stethoscope and confirmed by doing a chest X-ray. Another altitude-related problem is HACE, which causes headaches, cognitive impairment, and a lack of coordination, and if not treated timely, it can also lead to coma [14]. It can be diagnosed or detected with the help of a CT scan or MRI [14]. When soldiers are deployed in extreme dry environments, like deserts, they face dehydration, heat stroke, and heat stress, which can cause a rise in body temperature, weakness, fatigue, and sometimes lose consciousness [15]. It can also cause vomiting and metabolic acidosis, i.e., excessive accumulation of acid in the body, which can be either because of an increase in acid production or a decrease in bicarbonate levels. This causes an imbalance that disrupts normal cellular function and can lead to various health issues. When in grassy areas like jungles and marshy areas, soldiers are prone to several diseases like malaria and dengue fever [16, 17].

Even though conventional diagnostic methods are extensively used, there are chances of misdiagnosis. In high-altitude regions, diagnosing conditions like HPAE and HACE can be challenging in the earlier stage by conventional methods as the symptoms might overlap with symptoms of other conditions like coughing, chest tightness, difficulty breathing, headaches, and cognitive impairment [18]. These symptoms can be mistaken for less serious conditions. The harsh environment can make tools like chest scans and brain imaging difficult to access, which can delay the diagnosis and treatment. In an extreme heat environment, symptoms like an increase in body temperature and weakness are caused by dehydration, heatstroke, and heat stress, which are common in soldiers under physical exertion [19]. It can be difficult to differentiate these symptoms from other illnesses and can be challenging without immediate access to medical facilities and diagnostic equipment. In areas like jungles and marshy areas, malaria and dengue fever are very frequent, and diagnosing them in the early stages is very crucial, but symptoms like fever, headache, and fatigue are common to many other illnesses, which complicate the diagnosis process, and limited access to diagnostic tools also limits accurate diagnosis [16, 17, 20].

16.2.2 Mental Health

Other than physical stress, a soldier also faces mental stress, which affects the mental health of the soldier. Mental stress can be caused by separation from the family, exposure to any fatal accident, or constant alertness in the combat field [21]. Post-traumatic stress disorder (PTSD) is common in soldiers with combat experiences that cause them to have flashbacks of the combat, nightmares, and unwanted thoughts; this can also lead to combat stress, depressive disorder, and anxiety disorder [22].

There are challenges in diagnosing psychological conditions like PTSD, severe depressive disorder, and anxiety disorder due to stigma and cultural barriers, i.e., the negative attitude toward mental health in military culture [23]. People are ashamed to talk about it, which leads to soldiers not reporting these symptoms and avoiding seeking help, which causes the lack of physical activity and an increase in mental health issues that cause the worsening of the symptoms [24]. It also became difficult for health-care providers to diagnose and identify the psychological condition at an early stage. Symptom overlaps, i.e., symptoms of PTSD, anxiety, and depression, are often mistaken for each other due to being very common, which makes it difficult to differentiate between these conditions, and various physical symptoms like fatigue and irritability can be related to both physical and psychological health, which can further complicate the diagnosis [25].

16.2.3 Nutritional Health

The physical and operational efficiency of the soldiers depends on the amount of nutrition they received. The military lifestyle possesses a challenge for soldiers to maintain a balanced diet. They depend on ration packs, which help them meet the basic dietary needs, but they lack a variety of essential nutrients, which cause a deficiency of various vitamins and minerals that affect the physical and cognitive functions [26]. For example, lack of iron can lead to anemia, which leads to a decrease in energy and strength [27]. Deficiency of calcium and vitamin D can cause weakening of bones [28]. During a combat situation, there are irregular eating patterns and high energy expenditure, which can cause worsening of the tissues and lead to weight fluctuations, exhaustion, and weakening of the immune system [29–31].

There are several challenges in diagnosing the nutritional deficiency and health-related issues, like limited dietary variety, i.e., soldiers do not receive enough vitamins and minerals as they rely on ration packs, which can lead to conditions like anemia and bone health issues. These diseases have nonspecific symptoms like fatigue and weakness, and diagnosis is difficult without a detailed nutritional assessment. Irregular eating patterns, i.e., the demanding and unpredictable lifestyle of soldiers can lead to irregular eating patterns, which can cause weight fluctuations and deplete energy levels. These changes can hide the fact of nutritional deficiencies and hinder the diagnostic process.

16.3 Relation Between Precision Medicine and AI

Precision medicine or personalized medicine is made according to individual genetics and lifestyle, while traditional medicine was made on the principle that one size fits all, which means that the treatment and the medications are created in such a way that they are applicable to the large population, despite individual differences [3]. Precision medicine focuses mainly on the customized treatment plans, which differ from patient to patient. Therapies are developed on the basis of individual genetics, environment, and lifestyle [4]. To understand individual genetics, genomics and bioinformatics techniques were used. For the better management of the disease, genotypic, phenotypic, and behavioral information was used [4]. They can also help in the detection and prevention of the disease by understanding individual risk. The efficiency of precision medicine can be enhanced by the use of artificial intelligence, which helps us in analyzing complex datasets like genetic information, i.e., genetic data can be used in identifying the mutations and markers associated with disease [32]. Artificial intelligence can analyze data and predict sensitivity to conditions, which helps in drug selection and personalizing treatment plans. Electronic health records, for example, contain information such as past medical history and previous or ongoing treatments. Artificial intelligence can filter this data, identify patterns, assist in tracking disease progression, and suggest the most effective medications [33]; lifestyle data include data related to diet, exercise, sleep, and other lifestyle factors that are important for a holistic view of health. Artificial intelligence can be used to integrate this data and provide changes to be made in one's lifestyle that can help in improving health [34, 35]. Environmental factors, i.e., exposure to pollutants, climate, and living conditions, can affect individual health. Artificial intelligence can analyze these factors to identify the risk and suggest some preventive measures. Different types of artificial intelligence techniques are used for extracting the data that help us in getting better insight into the data [36]. As illustrated in Fig. 16.1, AI is based on machine learning algorithms which learn from the data and can be helpful in predicting outcomes; it can also help in classifying the different conditions and identifying the risk factors [37].

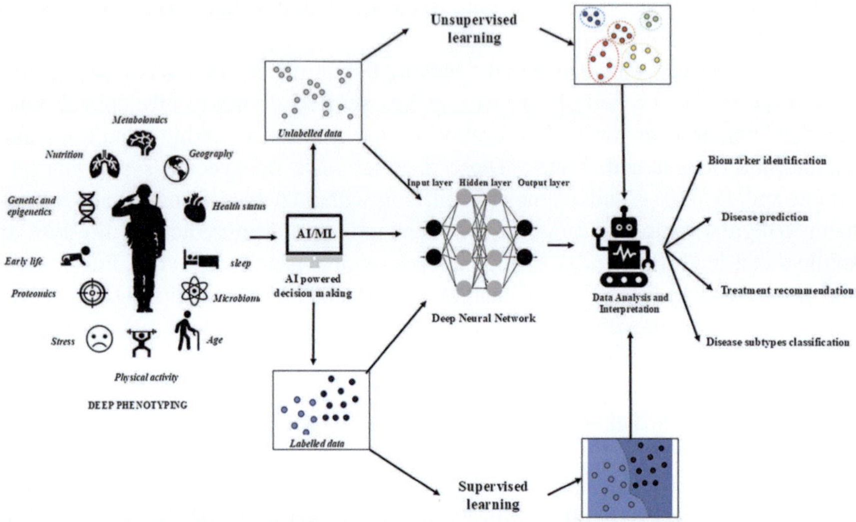

Fig. 16.1 Schematic representation of AI–ML models in analyzing phenotypic data; input data include a variety of phenotypic factors for supervised, unsupervised learning approaches and deep neural networks for AI-powered decision-making

Various machine learning techniques include supervised and unsupervised learning, which use labeled and unlabeled data, respectively [38–40]. Deep learning is a subcategory of machine learning that uses neural networks, which use many layers of hidden neural networks [41]. It is helpful in handling large and heterogeneous data, which is helpful in health care. Various artificial techniques and algorithms are used for extracting patterns and getting insight on the data, the application of supervised and unsupervised learning across different conditions is outlined in Table 16.1 and detailed in the subsequent discussion.

16.3.1 Supervised Learning

This model is trained based on labeled data, which is useful in predictive modeling and classification of soldiers based on their health and performances. One of the various supervised learning algorithms is linear regression, which is used for predicting continuous outcomes [38]. For example, it can help in predicting the time taken by the soldier to recover fully by studying its genetic makeup and the type of injury the soldier endured [72]. Another supervised learning algorithm is logistic regression, which predicts the outcome in binary forms such as yes and no [73]. For example, it can help in predicting whether the soldier will develop PTSD (post-traumatic stress disorder) based on the pre-deployment assessment. Various factors that are evaluated include mental health history, previous exposures to combat, and

Table 16.1 A tabular representation illustrating distinct categories of machine learning together with biomedical applications

AI/ML techniques	Explanation	Biomedical applications	Challenges
Supervised learning			
Decision trees	Tree-like models of decisions with branches representing choices.	Identifying risk factors for PTSD [42]	Need for accurate and timely data collection: ensuring model relevance across different units.
Support vector machines (SVM)	Constructs hyperplanes in high-dimensional space to separate classes.	Decision support systems for medical evacuations [43]	Managing unbalanced and dynamic data while integrating with real-time operational environments is a crucial task.
Random forests	Ensemble learning method using multiple decision trees to improve accuracy.	Forecasting soldier health conditions [44]	Complex parameter tuning: the possibility of overfitting noisy data
Naive Bayes classifier	Probabilistic classifier assuming independence between features.	Disease diagnosis [45] Health monitoring [46] Predictive modeling [46] Medical imaging [47]	The strong feature independence assumption may not hold in all cases.
Neural Networks	Models simulating the human brain's interconnected neuron structure.	Image analysis for injury detection [48] Assessment of psychological conditions [49]	Complex model architecture: requires significant computational resources.
Logistic regression	Predicts binary outcomes using a logistic function to model probability.	Hypoxemia risk analysis [50] Suicide ideation prediction [51]	Sensitivity to outliers: affects model performance and interpretability
Gradient boosting machines	Builds models sequentially to correct errors of prior models.	Predicting PTSD [52] Health monitoring [53] Personalized medicine [53] Disease risk prediction [53]	Training complexity: requires careful tuning of boosting parameters
Linear regression	Predicts continuous outcomes based on linear relationships between variables.	Estimating blood glucose levels: diabetes management [54]	Assumption of linear relationships: may not capture nonlinear patterns effectively.

(continued)

Table 16.1 (continued)

AI/ML techniques	Explanation	Biomedical applications	Challenges
K-nearest neighbors (KNN)	Classifies based on similarity measures between input data points.	Predict the survival time and classification of various stages of oral cancer [55]	Sensitivity to irrelevant features: impacts classification accuracy.
Ensemble methods	Combines multiple models to improve prediction accuracy, such as AdaBoost or Bagging.	Evaluating training effectiveness [56]	Model interpretability: blending multiple models complicates result analysis.
Unsupervised learning			
Hierarchical clustering	Divides data into nested clusters, creating a hierarchy.	Health and oral health-compromising behaviors (HOHCBs) [57] Combat readiness and health promotion [57] Clustering patient characteristics [58]	Determining the optimal number of clusters: interpreting hierarchical relationships.
Association rule mining (a priori)	Identifies relationships between variables in large datasets.	Injury and illness correlation [59] Preventive health strategies [59] Medication effectiveness [60]	Efficient handling of large datasets: interpreting complex associations.
K-means clustering	Divides data into non-overlapping subgroups based on similarities.	Personalized health interventions [61] Operational readiness assessment [62]	Initial cluster centers are sensitive; determining the optimal number of clusters can be difficult.
Principal component analysis (PCA)	Reduces data dimensionality while retaining important information.	Analyzing gene expression data in soldiers [63]	Assumes linear relationships: complex interpretability of reduced dimensions.
Gaussian mixture models (GMM)	Probabilistic models assuming data points are from a mixture of Gaussian distributions.	Genetic variation analysis [64] Disease Association Studies [64]	Determining the optimal number of components: model selection challenges.
t-Distributed stochastic neighbor embedding (t-SNE)	Reduces dimensions for visualization while preserving data structure.	Genetic data analysis [65] Single-cell RNA sequencing [66]	Computational cost: scalability issues with large datasets.

(continued)

Table 16.1 (continued)

AI/ML techniques	Explanation	Biomedical applications	Challenges
Self-organizing maps (SOM)	Produces a low-dimensional representation of input data.	Single-cell data analysis [67] Personalized medicine [67] Health monitoring [68]	Training complexity: requires fine tuning of map structure and learning rate.
DBSCAN (density-based spatial clustering of applications with noise)	Clusters dense regions of data points.	Neuroimaging analysis [69]	Sensitivity to parameters: requires careful tuning of epsilon and minimum points.
Factor analysis	Examines underlying relationships between variables, useful for data reduction.	Symptom clustering [70]	Assumption of linear relationships: may not capture complex interactions effectively.
Isolation forest	Anomaly detection method separating observations by randomly selecting features.	Health monitoring [71] Medical imaging [71] Wearable health devices [71] Genetic data analysis [71]	Interpretability: understanding the rationale behind anomaly scores.

other stress [52, 74, 75]. Next is the support vector machine (SVM), which is helpful in the classification of the problem [76]. Help in identifying soldiers who have a high risk of getting disease based on genetic and health data. It can be helpful in separating soldiers with high and low risk of disease [77]. Another supervised learning algorithm used is random forest [78]. It uses multiple decision trees to improve the accuracy of predictions and helps classify soldiers into different risk categories based on different health issues. These categories are created based on genetic, lifestyle, and medical data. Different types of supervised approaches are discussed in Table 16.1.

16.3.2 Unsupervised Learning

In this approach, the model is trained on unlabeled data and discovers hidden patterns within the dataset. It is useful in exploratory data analysis and clustering of soldiers based on health and performance. The various unsupervised learning algorithms used are K-means clustering, which uses the unlabeled data and clusters them in different categories based on similarity [79]. For example, it can help in clustering soldiers with similar genetic and health profiles, which will help researchers identify health conditions and performance traits by studying different clusters. Other algorithms that can be used for the study of the soldier's health are hierarchical classification, which clusters data in the form of a dendrogram (treelike structure) [80]. It helps in establishing a relation between different sets of data. It can help in finding the relationship between genetic markers and health outcomes in soldiers. For example, it can help in understanding various genetic factors that contribute to healing; next is principal component analysis (PCA). It helps in reducing large datasets while retaining the most significant variables. It simplifies the process of data analysis and highlights the variables that have great impact [81]. For example, it can simplify a vast amount of data related to health and performance into the principal component that has the most valuable and critical information. Examples of unsupervised learning and their use in different conditions are tabulated in Table 16.1.

16.4 Deep Phenotyping of Soldier's Health

Deep phenotyping is the comprehensive analysis and characterization of an individual's phenotypic traits; it includes a wide range of characteristics, including molecular, cellular, physiological, and behavioral aspects [82]. This approach provides a detailed understanding of how genetic and physiological variations are present in observable traits and how these traits are influenced by environmental and lifestyle factors. When a genome-wide association study (GWAS) is integrated with deep phenotyping, it becomes a powerful tool in precision medicine in understanding complex disease and tailoring individual strategies, as it offers detailed insights into biological pathways and mechanisms underlying diseases [83]. It helps in understanding the complex interplay between genes, environment, and lifestyle, which can help medical professionals develop highly personalized treatment plans. Deep phenotyping can help in early detection of disease by identifying minute phenotypic changes even before clinical symptoms appear. AI and ML methods necessitate the wider analysis of the phenotypes by employing advanced tools like genomics, wearable devices, and medical records.

16.4.1 GWAS

GWAS is a way of identifying genetic variations that are related to particular traits or diseases. It examines millions of single-nucleotide polymorphisms (SNPs) on the entire genome in a large group of people to find SNPs that occur more frequently in those with a certain disease compared to those without disease [84]. Its main focus is on SNPs and major human diseases or traits like height and blood pressure. Participants are classified based on their phenotypic, i.e., quantitative trait, and genotypic characteristics, i.e., DNA, using SNP arrays [85]. If an SNP is statistically more common in people with disease, it is associated with disease. Associated SNPs mark regions of the genome that may influence disease risk, but they do not necessarily identify casual genes.

Various machine learning algorithms help in identifying patterns and correlations in phenotypic data, which may be indications of disease starts or risk factors.

It can also be used as a predictive model that can predict disease progression and response to the treatment based on an individual's phenotypic profile. Artificial intelligence can also help in overcoming various limitations of the GWAS, as it enhances the GWAS process by improving the analysis and interpretation of the complex data by ensuring accurate and efficient processing [86]. It can help in identifying patterns and correlations between genetic variations and disease that might have been overlooked by the traditional statistical methods. Artificial intelligence can also help in the prediction of an individual developing a disease based on their genetic makeup, which can lead to proactive health-care measures, not just overcome the limitations. Here are some case studies that show the efficient use of GWAS incorporated with AI.

GWAS has advanced our understanding of the genetic basis of various health conditions, and the integration of artificial intelligence in analyzing GWAS data has further refined the identification of genetic loci and their association with various health conditions like PTSD, depressive disorder, and cardiovascular disease. Morder research has helped in identifying various genetic loci and specific genes that cause the risk of developing PTSD. Genetic variations in PACAP (ADCYAP1R1) have been associated with PTSD, especially in women. Artificial intelligence models have been used to refine our understanding of this gene's role in the stress response [87]. FKBP5 is a gene that is involved in regulating the stress hormone system; variations in this gene have been linked to PTSD; artificial intelligence analyses explain how these variations influence PTSD risk through gene-environment interactions [88]. Another gene associated with PTSD is SLC18A2, which encodes a vesicular monoamine transporter [89]. Its role in neurotransmitter regulation has been identified through AI-driven genome-wide association studies (GWAS). In addition to PTSD, several genes linked to depression have also been discovered using AI-assisted GWAS. One such gene is SIRT1, which plays a role in regulating cellular and stress responses [90]. Variations in this gene have been associated with an increased risk of depression. AI helped in finding its role in mood regulation; CRHR1 encodes the corticotrophin- releasing hormone receptor; it is found to be associated with depression, particularly in reference to stress [91]. Various AI models have helped in redefining the understanding of its role in stress-related depression. PLCLO is a gene involved in synaptic vesicle trafficking and neurotransmitter release has been involved in depression. An AI-driven GWAS has explained its contribution in synaptic function and mood disorder [92]. In a recent GWAS, numerous genetic loci associated with cardiovascular disease were identified, which include LDLR, a low-density lipoprotein receptor that is critical for cholesterol metabolism; mutations in this gene is linked to familial hypercholesterolemia and an increased risk of coronary artery disease [93]. PCSK9, i.e., proprotein convertase subtilisin/Kexin type-9, variants can lead to elevated LDL cholesterol levels which contribute to the higher risk of cardiovascular events [94], SORTI1, i.e., sortilin 1, is associated with lipid levels and has been connected to the risk of coronary artery, particularly concerning triglyceride metabolism [95]. APOE, i.e., Apolipoprotein E, is known for its role in lipid metabolism; variations in this gene are related to atherosclerosis and cardiovascular disease risk [96]. Many cardiovascular risk factors are also present in extreme environments, such as high altitude and high temperatures. Genomic analysis of these factors can be valuable for monitoring soldiers' health in such conditions.

16.4.2 Wearable Devices

Wearable technologies offer comprehensive insights that can be utilized for the purpose of health management in soldiers. Wearable devices have the capacity to monitor many physiological and mental indicators for in-depth analysis, prediction, and comprehension of soldiers' health challenges in diverse, severe environments [97]. Unlike existing health checkups that only cover a limited amount of soldiers' health data, wearables continuously monitor consumer and patient data over extended periods of time. This allows for real-time feedback, leading to the accumulation of comprehensive datasets and the possibility of generating more significant insights. Soldiers' data can be collected by wearable sensors that constantly measure vital indications, including heart rate, body temperature, and hydration levels [98]. Measure muscular activity, muscle exhaustion, speed, acceleration levels, and distance traveled. Also monitor blood flow to the brain and biochemical markers such as glucose, cortisol, lactates, electrolytes, and pH from body fluids like blood, saliva, and tears [99]. Artificial intelligence algorithms can be employed to evaluate the vast quantity of data gathered by wearable devices, allowing health-care personnel to discern patterns, forecast health outcomes, and make informed decisions regarding patient treatment. By analyzing an individual's activity level, sleep patterns, and heart rate, artificial intelligence systems can predict the probability of various diseases and combat stress. Artificial intelligence (AI) and machine learning (ML) can be used to analyze time series data. The data collected from these sensors serves the dual purpose of safeguarding the health of soldiers and providing commanders with valuable information to make well-informed decisions and preemptive actions regarding the well-being of their troops [100]. In addition, wearable devices offer a significant advantage in the training and performance monitoring of soldiers, which is a crucial part of warfare when it comes to selecting the most suitable persons for a mission [101].

16.5 Challenges in Implementation of Precision Medicine Based on AI–ML

The incorporation of AI and ML has great advantages in precision medicine. At the same time, this evolving area faces several limitations. Major concerns are categorized into technical challenges, ethical privacy concerns, quality of data, adequate training dataset, ease of interpretation by health professionals, and data reproducibility. The effectiveness of the AI and ML models depends mainly on the quality of the data used for training in health care. Datasets acquired from soldiers can be incomplete, inconsistent, or biased because of low sample size, collection of data in extreme environment, inconsistencies of wearable device. Substandard data quality can lead to inaccurate and unpredictable outcomes, which ultimately affect the soldier's health. The model may perform poorly if the data used for training does not represent the whole population. AI systems are capable of replicating and even amplifying biases present in the data, which can cause variations in the treatment recommendation and outcome across different groups.

Precision medicine requires the assimilation of various data, including genomic, clinical, and psychological data. It is important to use the same analysis methods used by different researchers or under different conditions to ensure that the findings are reliable and can be verified independently, which is important for validating scientific research. However, variations in methods of data collection, instruments, and protocols can cause inconsistency in data quality, which make it difficult to reproduce the same results. Change in analysis techniques, software versions can also lead to changes in results even with the same data. When the sample size is small or inadequate, controls can also result in irreproducible results. However, there are sources that may use incompatible formats or standards, which can be a challenge to create a dataset for analysis. The development of robust AI models is often hindered by the absence of diverse datasets. This is partly due to isolated healthcare systems that restrict data sharing across platforms, making it difficult to compile the wide-ranging data necessary for effective AI training—ultimately reducing the potential for personalized treatment.

There are several concerns related to ethics and privacy regarding the use of genetic information in AI applications. An ethical concern refers to an issue that challenges moral principles and values, focusing on what's considered right and wrong. This includes the violation of one's privacy, risk of personal data being leaked, unauthorized access and likeliness of data being misused. Genetic information is sensitive and mishandling it can cause violation of privacy. There is a challenge in ensuring patents' confidentially while using their genetic data. It is essential to get patients' consent to use their data for health care, but it can also be complex as patients do not understand how their data will be used by AI, which can cause an ethical dilemma. There is a risk of data leakage. Various security measures must be deployed by AI system to protect unauthorized access and potential misuse of patient data. The ease of interpretation of data by health-care professionals is a crucial factor affecting their adoption in clinical settings. Many AI models, especially deep learning algorithms, operate as black boxes, which means they produce accurate predictions, but the process behind these predictions is not transparent, which may worry health-care professionals and confuse clinical service users. The data used in model preparation is heterogenous, consisting of electronic health records, medical imaging data, genomics data, or wearable devices. Without the proper understanding of algorithms and data processing methods, it is difficult for medical personnel to explain AI-generated data. Understanding AI–ML limitations is crucial for their successful integration into clinical practice. For developing effective AI models, high-quality, representative data is essential. Continuous efforts to improve data collection, standardization, and integration are necessary to enhance the reliability of AI in precision medicine.

16.6 Conclusion

In this review, we explored the primary uses of AI–ML-based precision medicine in improving the health of soldiers. We highlighted significant obstacles in the execution of this approach, which requires careful consideration before these technologies may be utilized for the management of soldier's health. The application of artificial intelligence in data analysis for precision medicine represents a significant change in biomedical science and is demonstrating encouraging outcomes in various health-care domains and has great potential in soldiers' health management. If this technology is implemented, the military health-care profession will be able to account for the soldiers' health in relation to exposure to stress in different environments and their genetic history before providing treatment to soldiers. This will primarily involve the implementation of a strategy for managing physiological data under various environmental situations, the development of large-scale omics datasets, and the utilization of advanced wearable devices for data collecting. Moreover, Precision medicine, combined with AI, has the potential to address the health challenges faced by soldiers by development of targeted policies for managing the health of soldiers, which is currently approached in a more generalized manner. At the same time, the utilization of various AI models, such as supervised and unsupervised strategies, for the interpretation of multidimensional datasets should be critically evaluated for revealing concealed patterns in the phenotype and genetic diversity of soldier's disease. Significant investment is required for the development of new treatments that can be applied to various disease categories and subcategories. Additionally, researchers should move away from the current diagnostic frameworks and adopt a more intermediate, biologically driven level of phenotype identification that can be easily interpreted by the clinicians. This approach has already been shown to increase the overall efficiency of soldiers' understanding of diseases, the detection of biomarkers, and the real-time monitoring of disease, thereby enhancing the health of soldiers during both peace and wartime. The results obtained may lead to an improvement in efficiency of soldiers and significantly impacts combat outcomes, which in turn has far-reaching consequences in managing the deployment of soldiers and managing post-deployment health hazards.

References

1. Zueger R, Niederhauser M, Utzinger C, Annen H, Ehlert U. Effects of resilience training on mental, emotional, and physical stress outcomes in military officer cadets. Mil Psychol. 2023;35(6):566–76.
2. Korzeniewski K, Nitsch-Osuch A, Chciałowski A, Korsak J. Environmental factors, immune changes and respiratory diseases in troops during military activities. Respir Physiol Neurobiol. 2013;187(1):118–22.
3. Katschnig H. Modern medicine and the one-size-fits-all approach: a clinician's comment to Alexandra Pârvan's "Mind Electric" article. J Eval Clin Pract. 2018;24(5):1079–83.
4. Naithani N, Sinha S, Misra P, Vasudevan B, Sahu R. Precision medicine: concept and tools. Med J Armed Forces India. 2021;77(3):249–57.

5. Chiarito M, Luceri L, Oliva A, Stefanini G, Condorelli G. Artificial intelligence and cardiovascular risk prediction: all that glitters is not gold. Eur Cardiol. 2022;17:e29.
6. Kumar Y, Gupta S, Singla R, Hu YC. A systematic review of artificial intelligence techniques in cancer prediction and diagnosis. Arch Comput Methods Eng. 2022;29(4):2043–70.
7. Sahu M, Gupta R, Ambasta RK, Kumar P. Chapter three – Artificial intelligence and machine learning in precision medicine: a paradigm shift in big data analysis. In: Teplow DB, editor. Progress in molecular biology and translational science. 190: Academic Press; 2022. p. 57–100.
8. Dingre S. Data integration: exploring challenges and emerging technologies for automation. Int J Sci Res (IJSR). 2023;12:1395.
9. Anticona Huaynate CF, Pajuelo Travezaño MJ, Correa M, Mayta Malpartida H, Oberhelman R, Murphy LL, et al. Diagnostics barriers and innovations in rural areas: insights from junior medical doctors on the frontlines of rural care in Peru. BMC Health Serv Res. 2015;15(1):454.
10. Committee on Diagnostic Error in Health C, Board on Health Care S, Institute of M, The National Academies of Sciences E, Medicine. In: Balogh EP, Miller BT, Ball JR, editors. Improving Diagnosis in Health Care. Washington (DC): National Academies Press (US). Copyright 2015 by the National Academy of Sciences. All rights reserved.; 2015.
11. Raeeszadeh M, Goodarzi H, Moghaddam JA, Machian MSB, Gooshki HS, Khoshi A. The role of military medicine and the significance of training military medicine. J Family Med Prim Care. 2022;11(8):4184–9.
12. Joshi K, Goyary D, Mazumder B, Chattopadhyay P, Chakraborty R, Bhutia YD, et al. Frostbite: current status and advancements in therapeutics. J Therm Biol. 2020;93:102716.
13. Imray C, Wright A, Subudhi A, Roach R. Acute mountain sickness: pathophysiology, prevention, and treatment. Prog Cardiovasc Dis. 2010;52(6):467–84.
14. Abhishek K, Sachin S, Rajesh K. Frostbite: a conundrum in high altitudes. In: Nikolai VG, editor. Current topics on military medicine. Rijeka: IntechOpen; 2021. p. Ch. 5.
15. Parsons IT, Stacey MJ, Woods DR. Heat adaptation in military personnel: mitigating risk, maximizing performance. Front Physiol. 2019;10:1485.
16. Bulugahapitiya U, Siyambalapitiya S, Seneviratne SL, Fernando DJ. Dengue fever in travellers: a challenge for European physicians. Eur J Intern Med. 2007;18(3):185–92.
17. Hesse EM, Martinez LJ, Jarman RG, Lyons AG, Eckels KH, De La Barrera RA, et al. Dengue virus exposures among deployed U.S. Military personnel. Am J Trop Med Hyg. 2017;96(5):1222–6.
18. Taylor AT. High-altitude illnesses: physiology, risk factors, prevention, and treatment. Rambam Maimonides Med J. 2011;2(1):e0022.
19. Leiva DF, Church B. Heat Illness. StatPearls. Treasure Island (FL) ineligible companies. Disclosure: Ben Church declares no relevant financial relationships with ineligible companies.: StatPearls Publishing. Copyright © 2024, StatPearls Publishing LLC.; 2024.
20. Gulati S, Maheshwari A. Dengue fever-like illnesses: how different are they from each other? Scand J Infect Dis. 2012;44(7):522–30.
21. Lim I, Tam WWS, Chudzicka-Czupała A, McIntyre RS, Teopiz KM, Ho RC, et al. Prevalence of depression, anxiety and post-traumatic stress in war- and conflict-afflicted areas: a meta-analysis. Front Psych. 2022;13:978703.
22. Corvalan JC, Klein D. PTSD: diagnosis, evolution, and treatment of combat-related psychological/psychiatric injury. Mo Med. 2011;108(4):296–303.
23. Office of the Surgeon G, Center for Mental Health S, National Institute of Mental H. Publications and Reports of the Surgeon General. Mental health: culture, race, and ethnicity: a supplement to mental health: a report of the Surgeon General. Rockville (MD): Substance Abuse and Mental Health Services Administration (US); 2001.
24. Moore MJ, Shawler E, Jordan CH, Jackson CA. Veteran and military mental health issues. StatPearls. Treasure Island (FL) ineligible companies.
25. Gros DF, Price M, Magruder KM, Frueh BC. Symptom overlap in posttraumatic stress disorder and major depression. Psychiatry Res. 2012;196(2–3):267–70.

26. Sotelo-Díaz I, Blanco-Lizarazo CM. A systematic review of the nutritional implications of military rations. Nutr Health. 2019;25(2):153–61.
27. Kumar SB, Arnipalli SR, Mehta P, Carrau S, Ziouzenkova O. Iron deficiency anemia: efficacy and limitations of nutritional and comprehensive mitigation strategies. Nutrients. 2022;14(14):2976.
28. Cashman KD, Vitamin D. Deficiency: defining, prevalence, causes, and strategies of addressing. Calcif Tissue Int. 2020;106(1):14–29.
29. McAdam J, McGinnis K, Ory R, Young K, Frugé AD, Roberts M, et al. Estimation of energy balance and training volume during Army Initial Entry Training. J Int Soc Sports Nutr. 2018;15(1):55.
30. Adnan D, Trinh J, Bishehsari F. Inconsistent eating time is associated with obesity: a prospective study. EXCLI J. 2022;21:300–6.
31. Tornero-Aguilera JF, Robles-Pérez JJ, Clemente-Suárez VJ. Effect of combat stress in the psychophysiological response of elite and non-elite soldiers. J Med Syst. 2017;41(6):100.
32. Uddin M, Wang Y, Woodbury-Smith M. Artificial intelligence for precision medicine in neurodevelopmental disorders. NPJ Digit Med. 2019;2(1):112.
33. Lee S, Kim HS. Prospect of artificial intelligence based on electronic medical record. J Lipid Atheroscler. 2021;10(3):282–90.
34. Yadav M. Diet, sleep and exercise: the keystones of healthy lifestyle for medical students. JNMA J Nepal Med Assoc. 2022;60(253):841–3.
35. Alowais SA, Alghamdi SS, Alsuhebany N, Alqahtani T, Alshaya AI, Almohareb SN, et al. Revolutionizing healthcare: the role of artificial intelligence in clinical practice. BMC Med Educ. 2023;23(1):689.
36. Rojas-Rueda D, Morales-Zamora E, Alsufyani WA, Herbst CH, AlBalawi SM, Alsukait R, et al. Environmental risk factors and health: an umbrella review of meta-analyses. Int J Environ Res Public Health. 2021;18(2):704.
37. An Q, Rahman S, Zhou J, Kang JJ. A comprehensive review on machine learning in healthcare industry: classification, restrictions, opportunities and challenges. Sensors (Basel, Switzerland). 2023;23(9):4178.
38. Jiang T, Gradus JL, Rosellini AJ. Supervised machine learning: a brief primer. Behav Ther. 2020;51(5):675–87.
39. Eckhardt CM, Madjarova SJ, Williams RJ, Ollivier M, Karlsson J, Pareek A, et al. Unsupervised machine learning methods and emerging applications in healthcare. Knee Surg Sports Traumatol Arthrosc. 2023;31(2):376–81.
40. Duarte JM, Berton L. A review of semi-supervised learning for text classification. Artif Intell Rev. 2023:1–69.
41. Sarker IH. Deep learning: a comprehensive overview on techniques, taxonomy, applications and research directions. SN Comput Sci. 2021;2(6):420.
42. Walker P, Mehalick M, Glueck A, Tschiffely A, Cunningham C, Norris J, et al. A decision tree framework for understanding blast-induced mild traumatic brain injury in a military medical database. J Defense Model Simul. 2017:14.
43. Krygier J, Lubkowski P, Maslanka K, Dobrowolski AP, Mrozek T, Znaniecki W, et al. Smart medical evacuation support system for the military. Sensors (Basel, Switzerland). 2024;24(14):4581.
44. Vinodhini S, Pallavi J, Sandeep J, Smera C. Forecasting and decision-making system for military resource allotment. 2021.
45. Wickramasinghe I, Kalutarage H. Naive Bayes: applications, variations and vulnerabilities: a review of literature with code snippets for implementation. Soft Comput. 2021;25(3):2277–93.
46. Gohari K, Kazemnejad A, Mohammadi M, Eskandari F, Saberi S, Esmaieli M, et al. A Bayesian latent class extension of naive Bayesian classifier and its application to the classification of gastric cancer patients. BMC Med Res Methodol. 2023;23(1):190.
47. Demichelis F, Magni P, Piergiorgi P, Rubin MA, Bellazzi R. A hierarchical Naïve Bayes Model for handling sample heterogeneity in classification problems: an application to tissue microarrays. BMC Bioinformatics. 2006;7(1):514.

48. Hudson D, Cohen M. Neural networks and artificial intelligence for biomedical engineering. 2000.
49. Galán JJ, Carrasco RA, LaTorre A. Military applications of machine learning: a bibliometric perspective. Mathematics. 2022;10(9):1397.
50. Schober P, Vetter TR. Logistic regression in medical research. Anesth Analg. 2021;132(2):365–6.
51. Lin GM, Nagamine M, Yang SN, Tai YM, Lin C, Sato H. Machine learning based suicide ideation prediction for military personnel. IEEE J Biomed Health Inform. 2020;24(7):1907–16.
52. Papini S, Norman SB, Campbell-Sills L, Sun X, He F, Kessler RC, et al. Development and validation of a machine learning prediction model of posttraumatic stress disorder after military deployment. JAMA Netw Open. 2023;6(6):e2321273.
53. Speller J, Staerk C, Gude F, Mayr A. Robust gradient boosting for generalized additive models for location, scale and shape. ADAC. 2023.
54. Aishah A, Jamaludin U, Zainuriah M, Norhilda A. Multiple linear regression model analysis in predicting fasting blood glucose level in healthy subjects. IOP Conf Ser: Mater Sci Eng. 2019;469:012050.
55. Siddalingappa R, Kanagaraj S. K-nearest-neighbor algorithm to predict the survival time and classification of various stages of oral cancer: a machine learning approach. F1000Research. 2022;11:70.
56. Zhang W, Lu M, Liu Y, Mei Y, Cai X, Zhao J. Application of ensemble learning in evaluating the effectiveness of military training. 2024. 1560–4 p.
57. Azil AA, Yusof ZYM, Marhazlinda J. Clustering of health and oral health-compromising behaviours in Army Personnel in Central Peninsular Malaysia. Healthcare (Basel, Switzerland). 2023;11(5):640.
58. Liu X, Chen Y, Cheng AS, Zeng Y, Ullah S, Feuerstein M. Conceptualizing problems with symptoms, function, health behavior, health-seeking skills, and financial strain in breast cancer survivors using hierarchical clustering. J Cancer Surviv. 2022;16(4):751–9.
59. Ma H, Ding J, Liu M, Liu Y. Connections between various disorders: combination pattern mining using apriori algorithm based on diagnosis information from electronic medical records. Biomed Res Int. 2022;2022:2199317.
60. Miswan NH, Sulaiman IM, Chan CS, Ng CG. Association rules mining for hospital readmission: a case study. Mathematics. 2021;9(21):2706.
61. Dinov I. k-means clustering: biomedical and health applications using R. 2018. p. 443–73.
62. Saikia M. K-means clustering machine learning approach reveals groups of homogeneous individuals with unique brain activation, task, and performance dynamics using fNIRS. IEEE transactions on neural systems and rehabilitation engineering : a publication of the IEEE Engineering in Medicine and Biology Society. 2023.
63. Kuan P-F, Yang X, Clouston S, Ren X, Kotov R, Waszczuk M, et al. Cell type-specific gene expression patterns associated with posttraumatic stress disorder in World Trade Center responders. Transl Psychiatry. 2019;9:1.
64. Ko Y, Zhai C, Rodriguez-Zas SL, editors. Inference of gene pathways using Gaussian Mixture Models. 2007 IEEE international conference on bioinformatics and biomedicine (BIBM 2007); 2–4 Nov 2007.
65. Li W, Cerise JE, Yang Y, Han H. Application of t-SNE to human genetic data. 2017:114884.
66. Han H, Zhang T, Benton M, Li C, Wang J, Li J. Explainable t-SNE for single-cell RNA-seq data analysis. 2022.
67. Lim HS, Qiu P. JSOM: jointly-evolving self-organizing maps for alignment of biological datasets and identification of related clusters. PLoS Comput Biol. 2021;17(3):e1008804.
68. Augustijn E-W, Zurita-Milla R. Self-organizing maps as an approach to exploring spatiotemporal diffusion patterns. Int J Health Geogr. 2013;12(1):60.
69. Rangaprakash D, Odemuyiwa T, Narayana Dutt D, Deshpande G, Alzheimer's Disease Neuroimaging I. Density-based clustering of static and dynamic functional MRI connectivity features obtained from subjects with cognitive impairment. Brain Inform. 2020;7(1):19.

70. Kelton ML, LeardMann CA, Smith B, Boyko EJ, Hooper TI, Gackstetter GD, et al. Exploratory factor analysis of self-reported symptoms in a large, population-based military cohort. BMC Med Res Methodol. 2010;10(1):94.
71. Rassam MA. Isolation forest based anomaly detection approach for wireless body area networks. In: Hybrid intelligent systems. Cham: Springer Nature Switzerland; 2023.
72. Liu S, Lu M, Li H, Zuo Y. Prediction of gene expression patterns with generalized linear regression model. Front Genet. 2019;10:120.
73. Shipe ME, Deppen SA, Farjah F, Grogan EL. Developing prediction models for clinical use using logistic regression: an overview. J Thorac Dis. 2019;11(Suppl 4):S574–S84.
74. Schultebraucks K, Qian M, Abu-Amara D, Dean K, Laska E, Siegel C, et al. Pre-deployment risk factors for PTSD in active-duty personnel deployed to Afghanistan: a machine-learning approach for analyzing multivariate predictors. Mol Psychiatry. 2021;26(9):5011–22.
75. Pyne JM, Constans JI, Wiederhold MD, Gibson DP, Kimbrell T, Kramer TL, et al. Heart rate variability: pre-deployment predictor of post-deployment PTSD symptoms. Biol Psychol. 2016;121(Pt A):91–8.
76. Cervantes J, Garcia-Lamont F, Rodríguez-Mazahua L, Lopez A. A comprehensive survey on support vector machine classification: applications, challenges and trends. Neurocomputing. 2020;408:189–215.
77. Mittag F, Büchel F, Saad M, Jahn A, Schulte C, Bochdanovits Z, et al. Use of support vector machines for disease risk prediction in genome-wide association studies: concerns and opportunities. Hum Mutat. 2012;33(12):1708–18.
78. Rigatti SJ. Random Forest. J of Insur Med (New York, NY). 2017;47(1):31–9.
79. Ikotun AM, Ezugwu AE, Abualigah L, Abuhaija B, Heming J. K-means clustering algorithms: a comprehensive review, variants analysis, and advances in the era of big data. Inf Sci. 2023;622:178–210.
80. An G, Akiba M, Omodaka K, Nakazawa T, Yokota H. Hierarchical deep learning models using transfer learning for disease detection and classification based on small number of medical images. Sci Rep. 2021;11(1):4250.
81. Salem N, Hussein S. Data dimensional reduction and principal components analysis. Procedia Comput Sci. 2019;163:292–9.
82. Robinson PN. Deep phenotyping for precision medicine. Hum Mutat. 2012;33(5):777–80.
83. Wright JT, Herzberg MC. Science for the next century: deep phenotyping. J Dent Res. 2021;100(8):785–9.
84. Uffelmann E, Huang QQ, Munung NS, de Vries J, Okada Y, Martin AR, et al. Genome-wide association studies. Nat Rev Methods Primers. 2021;1(1):59.
85. Marees AT, de Kluiver H, Stringer S, Vorspan F, Curis E, Marie-Claire C, et al. A tutorial on conducting genome-wide association studies: quality control and statistical analysis. Int J Methods Psychiatr Res. 2018;27(2):e1608.
86. Enoma DO, Bishung J, Abiodun T, Ogunlana O, Osamor VC. Machine learning approaches to genome-wide association studies. J King Saud Univ – Sci. 2022;34(4):101847.
87. Lind MJ, Marraccini ME, Sheerin CM, Bountress K, Bacanu SA, Amstadter AB, et al. Association of posttraumatic stress disorder with rs2267735 in the ADCYAP1R1 gene: a meta-analysis. J Trauma Stress. 2017;30(4):389–98.
88. Hawn SE, Sheerin CM, Lind MJ, Hicks TA, Marraccini ME, Bountress K, et al. GxE effects of FKBP5 and traumatic life events on PTSD: a meta-analysis. J Affect Disord. 2019;243:455–62.
89. Branco RC, Burkett JP, Black CA, Winokur E, Ellsworth W, Dhamsania RK, et al. Vesicular monoamine transporter 2 mediates fear behavior in mice. Genes Brain Behav. 2020;19(5):e12634.
90. Raynes R, Brunquell J, Westerheide SD. Stress inducibility of SIRT1 and its role in cytoprotection and cancer. Genes Cancer. 2013;4(3–4):172–82.
91. Wasserman D, Sokolowski M, Rozanov V, Wasserman J. The CRHR1 gene: a marker for suicidality in depressed males exposed to low stress. Genes Brain Behav. 2008;7(1):14–9.

92. Mbarek H, Milaneschi Y, Hottenga JJ, Ligthart L, de Geus EJC, Ehli EA, et al. Genome-wide significance for PCLO as a gene for major depressive disorder. Twin Res Hum Genet. 2017;20(4):267–70.
93. Jiang L, Sun L-Y, Dai Y-F, Yang S-W, Zhang F, Wang L-Y. The distribution and characteristics of LDL receptor mutations in China: a systematic review. Sci Rep. 2015;5(1):17272.
94. Hachem A, Hariri E, Saoud P, Lteif C, Lteif L, Welty F. The role of proprotein convertase subtilisin/kexin type 9 (PCSK9) in cardiovascular homeostasis: a non-systematic literature review. Curr Cardiol Rev. 2017;13(4):274–82.
95. Goettsch C, Kjolby M, Aikawa E. Sortilin and its multiple roles in cardiovascular and metabolic diseases. Arterioscler Thromb Vasc Biol. 2018;38(1):19–25.
96. Mahley RW. Apolipoprotein E: from cardiovascular disease to neurodegenerative disorders. J Mol Med (Berl). 2016;94(7):739–46.
97. Canali S, Schiaffonati V, Aliverti A. Challenges and recommendations for wearable devices in digital health: data quality, interoperability, health equity, fairness. PLOS Digit Health. 2022;1(10):e0000104.
98. Prieto-Avalos G, Cruz-Ramos NA, Alor-Hernández G, Sánchez-Cervantes JL, Rodríguez-Mazahua L, Guarneros-Nolasco LR. Wearable devices for physical monitoring of heart: a review. Biosensors. 2022;12(5):292.
99. Friedl KE. Military applications of soldier physiological monitoring. J Sci Med Sport. 2018;21(11):1147–53.
100. Kumar P, Lee HJ. Security issues in healthcare applications using wireless medical sensor networks: a survey. Sensors (Basel, Switzerland). 2012;12(1):55–91.
101. Kodam S, Bharathgoud N, Ramachandran B. A review on smart wearable devices for soldier safety during battlefield using WSN technology. Mater Today: Proc. 2020;33:4578–85.

MIX
Papier aus verantwortungsvollen Quellen
Paper from responsible sources
FSC® C105338

If you have any concerns about our products,
you can contact us on
ProductSafety@springernature.com

In case Publisher is established outside the EU,
the EU authorized representative is:
Springer Nature Customer Service Center GmbH
Europaplatz 3, 69115 Heidelberg, Germany

Printed by Libri Plureos GmbH
in Hamburg, Germany